THE BRAIN, IN THEORY

The Brain, In Theory

ROMAIN BRETTE

PRINCETON UNIVERSITY PRESS
PRINCETON & OXFORD

Published by Princeton University Press
41 William Street, Princeton, New Jersey 08540
99 Banbury Road, Oxford OX2 6JX
press.princeton.edu

GPSR Authorized Representative: Easy Access System Europe - Mustamäe tee 50, 10621 Tallinn, Estonia, gpsr.requests@easproject.com

ISBN 9780691281377
ISBN (pbk.) 9780691281384
ISBN (e-book) 9780691281391

Library of Congress Control Number 2025946691

British Library Cataloging-in-Publication Data is available

Editorial: Hallie Stebbins and Chloe Coy
Production Editorial: Ali Parrington
Jacket/Cover Design: Karl Spurzem
Production: Erin Suydam
Publicity: William Pagdatoon
Copyeditor: Jennifer Harris

Jacket/cover images: Circuit board © Raimond Spekking / CC BY-SA 4.0 (https://commons.wikimedia.org/wiki/File:Doro_PhoneEasy_312cs_-_main_board-0067.jpg), colorized and cropped by Karl Spurzem. Neuron web by ALol88. Both courtesy of Wikimedia Commons.

This book has been composed in Arno Pro.

10 9 8 7 6 5 4 3 2 1

CONTENTS

ACKNOWLEDGMENTS

IT ALL STARTED with a blog, about a dozen years ago. As a theoretician in an overwhelmingly experimental field, I experienced what I can now diagnose as an epistemological clash. Trained in a formal field (theoretical computer science and mathematics), I thought about science in an apparently very different way from my experimental collaborators. A mathematician or a computer scientist produces knowledge by manipulating symbols and making logical inferences. The experimental biologist observes and manipulates natural phenomena, and tends to see theory as speculations, largely a waste of time. For the young theoretician, this dismissive attitude toward theory is very confusing and frustrating. If logical implications drawn from what biologists present in textbooks are just speculations, then what does that say about that textbook knowledge? At some point, I felt the urge of taking some time to reflect on very basic questions. What is theory, and why does it matter? What is a model, and what is a good model? It seemed that my own scientific activity would become meaningless if I could not answer those questions.

Thus, I decided to set aside one day per week to think and write about broad epistemological questions. I would post my writing on my blog, with no further goal than clarifying my own thoughts. This was an amazing experience of intellectual freedom, free from the usual constraints of academic work. I am grateful that the French institutions made this possible by offering permanent faculty positions at a relatively young age, not conditioned to the continued operation of a paper mill. Unfortunately, this basic academic freedom has become scarce in France, just like everywhere else.

My questioning started with the philosophy of science, but as I wondered about what makes a good model and what theory is supposed to explain, I quickly came to be critical of core concepts of computational neuroscience, which I previously adopted without question—codes, information, computation, and so on. I then developed an appetite for scientific books with a philosophical edge. I certainly owe a great intellectual debt to a number of authors I discovered at that time. My first epiphany was probably with James Gibson's book (*The Ecological Approach to Visual Perception*), who pinpointed what was wrong with the concept of information—it should be understood as empirical laws, not codes. The second epiphany was with Francisco Varela's immensely creative work, which developed a truly biological perspective on the theory of mind, in great contrast with computationalism. It is Lakoff and Johnson's classic book on language (*Metaphors We Live By*) that made me realize that the words we use in science are so important, and shape the way we think about phenomena. It is indeed hard to articulate criticisms and alternative views when common words (such as "information") have been preempted by one particular school of thought. Much later, reading Mark Bickhard's charge against cognitivism's commitment to substance metaphysics was a revelation: many conceptual difficulties come from our tendency to think of mental phenomena as "things," sorts of objects residing in the brain, when we should think in terms of processes.

As it turned out, many other scientists shared my interest in these questions, which are barely discussed in the scientific literature. Many thanks to all the readers of my blog. Many of these subjects were discussed with lab members and colleagues: Victor Bénichoux, Bernhard Englitz, Bertrand Fontaine, Sarah Goethals, Dan Goodman, Ali Hosseini, Jonathan Laudanski, Olivier Marre, Charlotte Le Mouel, Kevin O'Regan, Matilde Perrino, Jonathan Platkiewicz, Marc Rébillat, Cyrille Rossant, Marcel Stimberg, Pierre Yger, and many others. Special thanks to Farid Zahnoun for his critical reading of a first version of this book, and for our many philosophical discussions.

I had been wanting to write this book for a long time, and a conversation with György Buzsáki gave me the final encouragement to "just do it." I am grateful to my editor, Hallie Stebbins, who was enthusiastic about the project from the start.

On a personal level, I certainly owe my love of books and critical thinking to my parents, who I am sure would have been proud. Finally, I am especially grateful for the support and love from Marion, Paul, and Clara.

1

Lifeless Brains

The Brain, in Theory

In modern mainstream culture, both popular and scientific, the brain is a sort of computer, a machine that processes information. It acquires data in the form of sensory signals, encodes them into some electrical format, then processes the data with neural algorithms. It broadcasts the information to specialized processing modules: the visual cortex for visual processing, the hippocampus for memory storage and retrieval, the prefrontal cortex for decision making and planning. Eventually, it outputs motor commands to the muscles. Obviously, the brain is not a conventional computer with transistors, hard drives, and USB ports, but a "biological computer" optimized by evolution. The goal of neuroscience, then, is to "reverse engineer" the brain, to understand its functional organization and biological implementation.

All these concepts are borrowed from the engineering domain. This source of inspiration predates the era of computers. In the seventeenth century, brains were likened to hydraulic mechanisms; in the nineteenth century, the nervous system was a telegraph (Cobb, 2020, 2021). In much of the twentieth century, the brain was a computer applying formal rules to mental symbols. Nowadays, the brain might be a neural network, but the kind that engineers run on massive computers with graphics cards: a vector of values updated by series of matrix multiplications, with parameters tuned to minimize a formally defined error. In fact, the modern neuroscience literature simultaneously embraces all of those engineering concepts: neurons are mechanisms (like hydraulic machines) that communicate with codes (like telegraphs); they compute (like computers) with parameters tuned by learning algorithms (like formal neural network models).

Theoretical neuroscience, the activity of building mathematical models of the nervous system, heavily borrows from engineering theories: computer science, signal processing, data analysis, optimization, information theory, control theory. In fact, the main subfield of theoretical neuroscience is called *computational neuroscience*, which aims at understanding how (not whether) neurons compute.

Engineering concepts have indeed been very fruitful in understanding the logic of living beings, and of nervous systems in particular. For example, telegraph theory has been used to develop the biophysics of action potential propagation in the 1950s by Hodgkin, Huxley, Katz, and colleagues (Hodgkin, 1964), as axons share similarities with electrical wires. In fact, the theory of electrical propagation in neurons is traditionally called "cable theory" (Rall, 2011). Optimization principles have been shown to be relevant to understand the structure of living organisms (Rosen, 1967) and of nervous systems in particular (Sterling and Laughlin, 2017). Indeed, the structure of living organisms appears to be particularly efficient at various functions that are especially important for the survival of the organism, such as harvesting and saving energy. This is why biology has in turn been an inspiration for engineering.

But it is one thing to borrow relevant concepts from engineering to understand brains, and another entirely to claim that brains actually *are* engineered. Many general views on mind and consciousness are indeed based on a strict identification between brains and engineered devices (mostly computers). For example, since we are computers and computers are not conscious, then consciousness must be an illusion (eliminativism). Or conversely, since we are computers and we are conscious, computers must be conscious after all, so consciousness may actually be everywhere to different degrees (panpsychism). If intelligence is just an input–output mapping fitted on large amounts of data, then surely with more data and computing power, "artificial intelligence" will soon outrun human intelligence, leading the human species to extinction or slavery (an event called the "technological singularity"). If minds are algorithms, then we should be able to upload minds in a computer simulation, indefinitely extending our lives (transhumanism). In fact, we might already be living in a simulation right now, without knowing it. If not, since mind simulation would allow us to create an astonishing number of new happy human lives, we should make all possible efforts to ensure it happens (longtermism).

Yet, if we were to explicitly ask a modern neuroscientist whether the brain is actually an engineered device, she would certainly strongly object. Brains are not the result of intelligent design. This is a religious view of life that has been discredited by Darwinism. Why then are we to "reverse engineer" brains, if brains were not engineered in the first place?

This terminology is typically excused by adding that brains are engineered *by evolution*, not by God. But Darwin's insight is precisely that evolution is *not* a case of engineering. Engineering is the use of knowledge to solve technical problems. It presupposes an external mind that plans and assembles machines according to a preexisting goal. But evolution has no goals, plans, or knowledge; in other words, it is not an engineer.

Thus, living organisms are not *really* engineered. Therefore, they are not really machines, which are engineered objects, and brains are not really computers, which are kinds of machines. Of course, there are features of machines and computers that are shared by living organisms and brains, which is why engineering concepts can be relevant in biology. But if the idea that we are the result of intelligent design is to be scandalous to a modern scientist, then surely this should at least make *some* difference to the way we conceive brains?

It is the main aim of this book to explore these differences, in particular in the context of making models of the brain. It appears indeed that, in mainstream neuroscience and cognitive science, the idea that we are not engineered is simultaneously an extremely important opinion to hold publicly as well as a theoretically insignificant fact. Hillary Putnam, a major philosophical figure of cognitivism, explicitly claimed that our biological nature is insignificant: "we could be made of Swiss cheese and it wouldn't matter" (Putnam, 1975).

To set the stage, I will briefly outline the main modern theoretical frameworks to think about brains and cognition, starting with *computationalism.*

Computationalism

Computationalism holds that cognition is a form of computation, seen as the manipulation of formal symbols with rules. Brains are said to *implement* such computation, where symbols are represented by the state of some neurons, while brain processes change neural states in such a way that the corresponding symbols are changed according to the formal rules of the computation. Usually, the relevant states are believed to be the firing activity of neurons (how many action potentials they fire per second). As we will see in chapter 8, this is problematic because activity is not a state, let alone a computational state. Unorthodox computational accounts propose instead that symbols are represented by stable molecules such as polynucleotides (Gallistel, 2017). Regardless of the physical basis of computational symbols, it is the computation that matters for cognition, not its implementation. This doctrine is known as *functionalism.* (See Zahnoun [2023] for a critique.) Brains merely support computations; how they do so is largely irrelevant to understand cognition.

This functionalist perspective comes from the fact that a computer is a machine, and what matters for the behavior of a machine is the functional specification of the components, not so much their material basis. An electric car is still a car, because the electric motor produces a rotating motion transferred to the wheels, even though it works differently from a combustion engine. Accordingly, computationalism relies on a distinction between hardware (the brain) and software (the mind). Cognition is defined at the level of algorithms, while neurons only implement those algorithms. Thus, biological implementation is secondary for the understanding of cognition: the mind can run on any material support, as long as the functional organization of computational states, identified to mental states, is preserved. Thus, with some imagination, the brain could be made of Swiss cheese.

Computationalism developed in reaction to *behaviorism*, which was the dominant conceptual framework about brains in the first half of the twentieth century. Behaviorism saw behavior as nested reflexes adjusted by experience, strengthening or weakening associations. But as early cognitivists pointed out, behavior is highly structured and goal-directed, and appears to depend on abstractions rather on the details of proximal stimuli, just like computations. This is obviously so in human reasoning, but it is also a well-documented feature of animal behavior. For example, bees can recognize whether two objects are the same or different (Giurfa et al., 2001) and can count up to four (Dacke and Srinivasan, 2008). Many species such as ants can return to their nest in a straight path after foraging (Wehner, 2020), meaning that they implicitly integrate their own displacement—an ability called dead reckoning. This does not seem to be possible by the mere association of physical cues.

While the cognitivist critique of behaviorism is relevant, it was hardly new. Merleau-Ponty, a phenomenologist philosopher, already pointed out in *The Structure of Behavior* (1942) that behavior is made of actions, not reactions. An action is performed by an agent with certain goals, and therefore it depends both on the organism's internal state and on some abstract features of the situation—for example, whether the given pattern of light is identified as a source of food. Organisms do not respond automatically to proximal stimuli. Rather, behavior is anticipatory: actions are taken as a function of their expected consequences. Computation is indeed also directed toward a goal, which is its result (the thing that we compute), but that is hardly surprising, given that computation is a kind of behavior—the kind we try to emulate in computers. However, the converse assertion, that all behavior and cognition are computational, does not follow, as we will discuss in more detail in chapter 4. In the same way, it seems that we can store and retrieve memories just like a computer, but it is the computer that was built to mimic some features of human memory—indeed, the word *memory* originates from the mental domain, not the engineering domain. It does not follow that the computer literally remembers what you wrote when you open a text file.

Computationalism led to the development of symbolic artificial intelligence, also known as "good old-fashioned artificial intelligence" (GOFAI), in particular expert systems, which implemented logical inference on a base of rules gathered from experts. Those systems made spectacular progress in the

1960s to 1970s, raising high hopes, as recounted by Mitchell (2021). For example, in 1960, Herbert Simon predicted that "machines will be capable, within twenty years, of doing any work that a man can do." Skeptics, such as the philosopher Hubert Dreyfus (1978), explained that experts do not actually rely on rules: it is beginners who use rules to guide their learning process. This unpleasant rebuttal was dismissed, but expert systems were eventually abandoned in the 1980s.

Despite the failure of these approaches, the perspective introduced by computationalism has remained dominant: cognition is a form of computation, and neurons encode symbols used by the brain to compute.

One of the difficulties encountered by symbolic artificial intelligence was with perceptual tasks, such as identifying an object. To address this difficulty, a very different approach was introduced, which did not use symbolic rules: *connectionism*.

Connectionism

The precursor of all artificial neural network models is the binary neuron model of McCulloch and Pitts (1943). In that model, the neuron is seen as either active or inactive, symbolized by 0 or 1, a feature inspired by the all-or-none law of neural excitation. It receives inputs from other neurons, and its output activity is calculated as follows (figure 1.1): take the weighted sum of the activity of input neurons (weights are called *synaptic weights*), and output 1 if the sum exceeds a threshold (otherwise 0). This makes the neuron implement a logical function with n inputs and 1 output. One can then build more complicated logical functions by connecting neurons together. In fact, McCulloch and Pitts demonstrated that any logical function from n inputs to m outputs can be implemented with an appropriately wired neural network. Thus, the article was titled "A Logical Calculus of the Ideas Immanent in Nervous Activity."

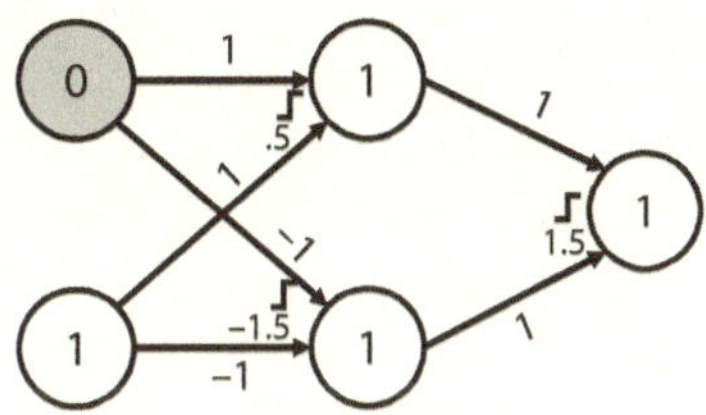

FIGURE 1.1. A network of binary neurons implementing the XOR operation. Binary inputs are multiplied by weights (on edges), and the output is 1 when the result is greater than a threshold.

Philosophically, the model of McCulloch and Pitts stands with classical computationalism (Dupuy, 2013): the state of each neuron represents a symbol with true or false value, and the model implements propositional calculus. Mental states are made of logical propositions. But in the 1950s and 1960s, Frank

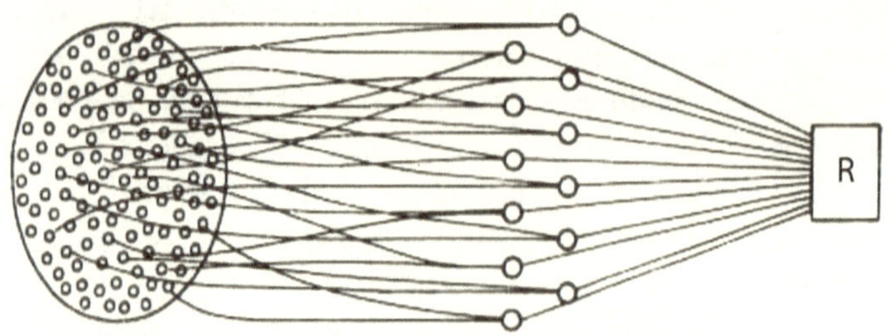

FIGURE 1.2. Rosenblatt's perceptron (from Rosenblatt, 1962).

Rosenblatt started to apply it to visual tasks, under the name "perceptron" (Rosenblatt, 1962; figure 1.2). There, the input variables represented light intensity at photoreceptors, the output represented the recognition of an object, and crucially, the synaptic weights were learned by association. The model did not implement logical inference anymore. Instead, Rosenblatt interpreted the model "in terms of probability theory rather than symbolic logic" and called his approach "connectionist" (Rosenblatt, 1958).

Despite initial interest in connectionism, it was abandoned a few years later in favor of symbolic approaches, when Minsky and Papert (1969) demonstrated the fundamental limitations of the perceptron. When expert systems were abandoned in the 1980s, there was a renewed interest in connectionism, triggered by the design of efficient learning algorithms for multilayer perceptrons, such as backpropagation (Rumelhart et al., 1986), still in use in modern artificial neural networks. Connectionism fell out of fashion again in the artificial intelligence community in the 1990s, in favor of more efficient statistical learning algorithms, such as support vector machines (Cortes and Vapnik, 1995). It was revived in the 2010s, when improvements in model design, software engineering techniques (such as automatic differentiation), as well as computing power and data availability led to impressive results in different areas, such as image processing (LeCun et al., 2015).

According to connectionism, cognition arises from the interaction of many neurons, seen as simple stereotypical input–output devices. Learning consists in modifications of the association strength between pairs of neurons, summarized by a single parameter. Thus, connectionism is explicitly associationist, and therefore conceptually closer to behaviorism than to computationalism. Cognition is not logical calculus anymore, but a form of calculation more akin to linear algebra. Furthermore, the activity of neurons in inner layers is not associated to mental symbols anymore, but rather to intermediate computational variables.

These differences remain a major source of mutual criticism between the two approaches. On one hand, (symbolic) computational models are essentially incapable of dealing with real sensory inputs, such as images. On the other hand, connectionist models have great difficulties dealing with relational tasks, such as deciding whether an image contains two identical objects (Kim et al., 2018), or with compositional tasks (Dziri et al., 2023), or generally tasks that rely on abstraction (Lewis and Mitchell, 2024).

Despite these differences, computationalism and connectionism are conceptually related in many ways. Indeed, the model of McCulloch and Pitts was an explicit inspiration of John von Neumann's work on the electronic computer in 1945 (von Neumann, 1993), as well as of Rosenblatt's first connectionist model. Both computationalism and connectionism see cognition as a form of computation, consisting in applying a series of elementary operations to an input. This means in particular that cognition is an input–output process, which takes data and maps it to a response. As Hendriks-Jansen pointed out already in 1996, "most of the connectionist systems that have been built to date are models in which the inputs and outputs are assigned by the programmer following analysis of a particular task domain" (Hendriks-Jansen, 1996)—this is still the case at the time of writing.

This view preserves the behaviorist concept of the stimulus: behavior is made of responses to stimuli, except that there is now "cognition" between perception and action—the "classical sandwich" model of cognition, as Susan Hurley put it (Hurley, 2001). Indeed, in standard experimental sensory neuroscience, neural activity is almost invariably reported as a response to stimuli, and activity unrelated to the stimulus is called "noise"—as opposed to the autonomous activity of the organism. This is obviously the experimenter's perspective.

The way computations are performed differs greatly between classical computationalism and connectionism. Indeed, in a deep neural network model that identifies faces, neurons of the hidden layers do not represent anything in particular. This is why a common complaint about modern artificial networks (deep learning in particular) is that they are not *explainable*: we cannot easily explain what they do because the results of intermediate calculations are not meant to be interpretable as symbols. However, neurons of the output layer do *represent*: in a face recognition model, their activity represents the occurrence of a particular face. Therefore, the output remains symbolic, just like in classical computationalism. Furthermore, since these output symbols must be the inputs to some other computational networks—for example, those responsible for uttering the name of the face—connectionism still generally commits to a symbolic view of cognition, both for inputs and outputs. There are still "neural representations" or "neural codes" of mental content, in the form of the activity of specific neurons or groups of neurons, but not all neurons encode; only those at the input and output of designated cognitive functions. Thus, classical connectionism has a somewhat confused view of the symbolic nature of cognition.

Connectionism also preserves the hardware/software distinction at the heart of computationalism. In this case, software is the set of synaptic weights. Neurons are input–output devices with a few knobs, but are otherwise rigidly specified. This is a key requirement of modern connectionist models, where the tuning of synaptic weights relies on formal differentiation of neural input–output functions, as we will see in chapter 4.

Thus, although connectionism describes cognition in terms of the operation of "neurons," the biological nature of neurons, or of the organism, plays exactly no role, just like in classical computationalism. The facts that the organism lives and that brains develop (as opposed to being assembled) are peculiarities of "implementation" with no theoretical significance.

Because biology is just implementation, both computationalism and connectionism start from the cognitive problem being solved and then try to figure out how the brain might solve it. This approach is typically called "top-down"—the top being the mind. This is of course in line with the engineering mindset: first, we describe what the machine should do; second, we design its functional organization; third, we implement the functional description by assembling components with the right specifications. This is essentially what David Marr, a pioneer of computational neuroscience, proposed as the methodology for modeling brains (Marr, 1982): start with the "computational level," the task that the model is supposed to achieve; then describe the "algorithmic/representational level," the algorithm that solves the task, at an abstract level; and finally worry about the "implementation level," how the algorithm is realized in the brain.

Of course, this methodology makes perfect sense for artificial intelligence, since in that case, we are indeed engineering the models. In neuroscience, an alternative kind of methodology, which brands itself as more empirical ("data-driven"), consists in measuring the different components of the brain as well as the way they are assembled. This kind of approach is often called "bottom-up." It is in fact also inspired from engineering, because parts of a living organism are conceived as parts of a machine.

Bottom-Up Neuroscience

An example of a bottom-up approach in neuroscience is the Human Brain Project, which aimed at simulating an entire brain based on systematic large-scale measurements of the properties of neurons and synapses. In this case, the neuron models are not classical connectionist models with abstract variables such as the "activity" of a neuron, but biophysical models taking the form of dynamical systems with measurable variables, such as the membrane potential. Those models were obtained from electrophysiological measurements in animals. In the Human Brain Project, the measurements were statistical: models of typical neurons, and average connectivity between brain areas.

Other bottom-up projects rely on more systematic measurements. For example, a technical approach known as *connectomics* aims at systematically measuring the detailed synaptic connections between neurons in an entire region or in the whole brain. The graph of connections is called the *connectome*. According to its strongest supporters, connectomics should bring a decisive contribution to the understanding of brains and cognition. For example, Morgan and Lichtman (2013) assert that "it might not be so unrealistic to hope that in staring into such a map we might get a glimpse of the human mind," and Seung (2012) claims that you literally *are* your connectome. Of course, this is simply the expression of connectionism in its most radical form: cognition is essentially specified by the connections between neurons.

Thus, bottom-up approaches also often embrace some variation of connectionism, as well as the general framework of computationalism—in particular, its terminology. That is, brains are described as implementing computations, processing information, and so on. But in contrast with top-down approaches, models of the brain are established by measurement, independently of what the brain is supposed to achieve. Function is assumed to follow from those measured properties. The implicit assumption is that, like in a machine, the properties of parts are independent of the system in which it is embedded (the "top"), and of what that system does. First come the parts with their specified properties, and then they are assembled according to a plan.

But of course, this analogy with machines is fragile, because in a living organism, parts always grow within a functional system, and so the relation between "bottom" and "top" is circular, not unidirectional. As we will see in chapter 3, this explains why the hopes of bottom-up approaches have not been realized so far.

Brains Beyond Engineering

The Neurocomputational Patchwork

In practice, models of neuroscience (as opposed to artificial intelligence) do not strictly adhere to either computationalism or connectionism in their classical form but instead borrow concepts from both approaches. In the same way, modeling is rarely purely top-down or bottom-up. For example, bottom-up approaches generally use properties of the "top" as constraints, although this is rarely acknowledged (as we will see in chapter 3). Conversely, connectionist approaches often take inspiration from structural peculiarities, such as the modular organization of the brain or the presence of dendrites. Some parts of brain and mind studies are dominated by connectionism, such as systems neuroscience, while others are dominated by classical computationalism, such as cognitive science. The most empirically driven models of neuroscience are in fact dynamical systems that are neither symbolic nor connectionist (such as the Hodgkin-Huxley model).

Thus, brain theory consists of a heterogeneous patchwork of approaches and models. Nonetheless, they share a common terminology borrowed from engineering, in particular computer science: brains and neurons compute, implement algorithms, encode objects and properties, represent and process information, and so on. For example, Sydney Brenner, who pioneered the neurogenetic study of *C. elegans*, a microscopic worm with 302 neurons, describes his approach as follows:

> Behaviour is the result of a complex set of computations performed by nervous systems and it seems necessary to decompose the problem into two: one is concerned with how the genes specify the structure of the nervous system, the other with questions of how nervous systems work to produce their outputs. (Brenner, 1973)

Unlike bees, *C. elegans* cannot count, and so far, no one has found hints of symbolic representations in its neurons. Thus, Brenner meant "computation" in a much broader sense than classical computationalists do. Apparently, it is not just that the animal *can* compute, but *all* behavior results from a kind of computation implemented by the nervous system. This is typical of modern neuroscience literature, a view that I shall refer to as *neurocomputationalism*: neurons are conceived as formally specified input–output devices that compute and implement the algorithms of cognition. But what is meant exactly by "compute," "implement," and "algorithms" is often rather vague, and indeed may differ substantially between approaches.

This terminology is not decorative: it is a theoretical commitment that forms the scaffold of reasoning about brains as well as of model building. Because the precise meaning of those words is often left unspecified, this scaffold is fragile and often incoherent. (Do neurons compute in the sense of connectionism, in the sense of computationalism, or do they just do something useful?) And because brains are not actually engineered, this scaffold is often poorly fitted to the subject, as we will see. In this book, we will explore the meaning of those words, and the extent to which they make sense when talking about brains, including computation (chapter 4), representations and codes (chapter 5), information (chapter 6), prediction (chapter 7), and implementation (chapter 8).

First, I want to make it clear why this choice of words is indeed a theoretical commitment about how brains work.

Of Words and Theories

Many words we use to talk about brains come from our ordinary human experience. For example, we say that neurons communicate, or send messages. These words originate from our social experience as a speaking species. We use them for neurons because we recognize some features of communication in the biological phenomenon: neurons of the retina produce electrical spikes (*action potentials*) that are specific to the image being presented, and this sequence of spikes then travels along the axon, unchanged, up to the axonal terminals, as if they were Morse code messages being delivered through the nerves, from one neuron to the next. On the other hand, we know very well that the receiving neuron does not literally "read the message," neither does it imagine the image that the message is supposed to stand for. There are features of messages that seem relevant to describe the electrical activity of neurons, and others that are not. When we say that spikes are messages, we focus on those features that we find relevant. This point about language was made eloquently by Lakoff and Johnson in their classic book *Metaphors We Live By* (Lakoff and Johnson, 1980): "What metaphor does is limit what we notice, highlight what we do see, and provide part of the inferential structure that we reason with." For this reason, choosing a particular word from another domain is a theoretical commitment. We will discuss communication metaphors in more detail in chapter 5, in the context of neural codes.

The most important engineering metaphor in biology is the machine metaphor. In the modern view, living beings are machines, and brains are computers, which are kinds of machines. We know this is a theoretical commitment because the idea that living beings are machines is supposed to be an *insight*. We know quite well what machines are in real life. If we were to point out a rabbit to a ten-year-old and say, "look at this machine," she would certainly object that it is not a machine but an animal. A machine is something made by humans to do something useful for them, it is not autonomous, it does not grow, it does not feed, and it does not feel. A ten-year-old, as well as most adults, would certainly put machines and animals in different categories. Thus, when the biologist Jacques Monod insists in *Chance and Necessity* (Monod, 1970) that *actually* a living organism is a molecular machine, he wants to convey something important and not obvious about life. It is not just a decorative term but a theoretical claim.

What was so important to Monod? Mainly, he wanted to oppose *vitalism*, according to which organisms live thanks to a nonphysical vital fluid. By claiming that living organisms are machines, he meant that biological matter follows the same ordinary laws of physics and chemistry as inert matter, and, like machines, it is by virtue of its organization that the organism does what it is supposed to do (living and reproducing), not thanks to a special substance. A machine is made of components interacting together in certain ways so as to support the function of the machine. In the same way, a biological organism consists of a functional arrangement of organs—the digestive system, the circulatory system, and so on—in the service of the maintenance and reproduction of the organism. Thus, Monod's theoretical claim is that living organisms are goal-directed functional organizations of ordinary matter.

This is not a trivial claim at all. Surely, we can recognize parts such as organs in animals, but those are very unlike the parts of machines. Organs grow, for example. At the microscopic level, the molecular content of a cell changes in composition, number, and localization, at timescales of milliseconds to years. It is not so obvious how this molecular maelstrom can be conceptualized as components to which we can assign functions, like the functional diagram of a machine.

In fact, Monod also meant that living organisms are machines in the sense that their processes are essentially mechanical—that is, that their parts follow deterministic local interactions between discrete elements, mostly based on shape, like the solid macroscopic objects of our ordinary experience. Monod used the word "clockwork." This is the idea of the standard "key-and-lock" concept of molecular biology, according to which the shape of a protein determines its function. However, the claim that living processes are essentially mechanical in this narrow sense is demonstrably false, as Daniel Nicholson has clearly argued (Nicholson, 2019), and as we will see in the next chapter. A common example in the brain is the action potential, which is produced by spatially separated ionic channels that interact nonspecifically at a distance.

Thus, by claiming that living organisms are machines, Monod makes three assertions, corresponding to three features of machines. The first is that, like machines, living organisms are made of ordinary matter, following the same laws of physics and chemistry as inert matter. This is fairly consensual. The second is that living organisms are organized like machines, with parts arranged so as to ensure the function of the whole system. This is questionable or at least ambiguous (what are "parts"? what is "function"?). The third is that living processes are mostly mechanical, essentially deterministic local interactions between discrete objects (he had proteins and nucleic acids in mind). This is demonstrably false.

This illustrates several important points about words and theories. First, the choice of engineering words is a theoretical commitment. When we use the word *machine* to designate living beings, we refer to some features of machines that we think are shared by living beings. This is a convenient way to make theoretical claims about how living beings work. These claims may or may not be correct or may need to be substantiated. Second, strict identification as in "organisms are machines" or "neurons compute" is a great source of confusion. In what sense are living organisms machines? Are they made of parts? Assembled? Are they mechanical? Are they lawful? Are they engineered? These are very different claims. If one needs to carefully explain in what exact sense organisms are machines, and if different people pick different features, then organisms are not actually machines. They are somewhat similar, and somewhat different. This acknowledgment is crucial for conceptual clarification.

Biological Cognition

Cognition is a property of (at least some) living organisms. Perception, cognition, agency, free will, and consciousness are all biological phenomena. Even though we might try to replicate those phenomena in artifacts, the primary empirical source remains biology. Yet, strikingly, the study of cognition appears to be a branch of computer science rather than of biology. This dismissal of biology is even explicitly embraced by classical cognitive scientists, a view known as *functionalism*—biology is just "implementation." In neuroscience, the standard terminology of brain theory largely refers to a nonbiological world, the world of machines made by humans—computation, implementation, algorithms, codes, optimization.... Ironically, scientists have abandoned the idea that living organisms have been designed by God, only to adopt a model of the living based on artifacts made by an engineer. Thus, Monod ridicules vitalism as some sort of magical belief, but then identifies living organisms with machines, those artifacts made by humans for a purpose using knowledge and planning. Is this a scientific view on life, or monotheism rejecting paganism?

The idea that animals result from intelligent design is scandalous to a scientist. Yet, it appears to make very little difference to the way we think about brain and mind. On the contrary, I assert that a proper understanding of life, beyond engineering preconceptions, is crucial to an understanding of its cognitive properties.

Why are living organisms compared to machines in the first place, rather than to any complex physical system like the climate? The reason is that machines are goal-directed, like living organisms. But the goals of machines are just the goals of their engineers, and therefore the machine view does not actually address the issue of goals, which means that the choice of the machine metaphor has no ground. As we will see in the next chapter, the reason why living organisms have goals is not because they are machines, but because they are precarious entities that must exchange matter and energy with their environment in order to maintain themselves. Cognitive properties are rooted in these facts of life, not in their presumed mechanistic nature.

Living organisms must feed. They have no material persistence. They develop by division. They evolve with no plan or direction. They are autonomous. This book explores the consequences of these facts of life for the understanding of brains, cognition, and behavior. I will start by presenting a modern view of life in the next chapter. In the rest of the book, we will use these lessons of life to revisit the standard concepts of brain theory. In chapter 3, I will question the reductionist preconceptions of "bottom-up" (*reverse-engineering*) approaches. In chapter 4, I will argue that brains are not *biological computers* in any useful sense. In chapter 5, I will explain that *neural codes* (or *neural representations*) are a misleading engineering concept, which does not stand empirical scrutiny, and which is theoretically incoherent when applied to brains. In chapter 6, I will show that the neuroscientific concept of *information* is problematic in a biological setting, because it is framed as what the engineer can recover from a signal, and the engineer always uses preexisting knowledge in addition to the signal. In chapter 7, I will argue that anticipation is the core property that theories of cognition try to explain, but that its common identification with prediction is mistaken. Instead, I will develop an account of anticipation as the exploitation of regularities, rooted in the precarious nature of life. In chapter 8, I will show that the concept of *implementation* introduces a biased view of the organization of brain processes, mirroring the way *we* make devices rather than accounting for the autonomy of life. I will end the book on an alternative view of organisms and brains as colonies of living entities, and outline what it implies for the development of brain theory.

2

Life as We Know It

UNLIKE BLACK HOLES, life is familiar. We intuitively recognize as living those precarious entities that change in an apparently purposeful way: a living being seems to have goals (plants seek light), grows, and dies. This prescientific notion of life readily distinguishes life from machines. But the modern biologist knows much that is hidden to the naked eye, such as cells, molecules, and the history of life stretching over billions of years. The scientific view of life has changed considerably since Darwin, and it has also changed considerably in the last decades. (See Ball [2023] for a selection of modern developments.)

In this chapter, I will not delve into the molecular details of modern biology but rather try to give a sense of the general logic of life, as we understand it today. I will start by addressing the fundamental question "What is life?" and how it relates to our prescientific understanding—in particular, the goal-directedness of life. I will then explain in what ways living organisms work differently from machines. Last, I will turn to multicellular life, how it develops and evolves, and how these processes differ from assembly and optimization.

What Is Life?

Living Organizations

What is life? This is not a simple question at all, and we can easily give wrong answers. For example, one may define life by some of its usual capacities, such as reproduction. But a sterile individual is still alive. Or one may focus on structural properties of living organisms, such as the fact that all living cells have DNA. But so do dead people.

In the modern understanding, life is due not to the presence of a special vital substance but to a particular organization of matter. It is in this sense that Jacques Monod identifies living organisms with molecular machines (Monod, 1970). But machines are not alive. What is special, then, about living organizations?

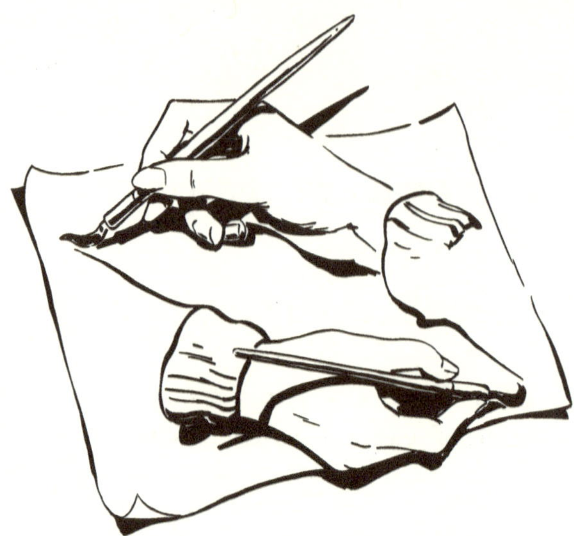

FIGURE 2.1. The living organism as both cause and effect of itself. Inspired by *Drawing Hands*, M. C. Escher (1948).

The prescientific notion of life insists on two fundamental features of life: change (organisms are born, develop, and die) and goal-directedness. Modern developments in theoretical biology elaborate on this prescientific notion, by conceiving life as a normative organization of processes, on different timescales (Nicholson and Dupre, 2018): metabolism, development, evolution. This conception of life was popularized by Varela and Maturana (1973) under the name *autopoiesis*—literally, "self-creation."

Darwin's theory of evolution is probably the most successful theory in biology, but it cannot define life, since evolution presupposes living organisms. Thus, Varela and colleagues motivate the autopoietic model of life as follows (Varela et al., 1974): "instead of asking what makes a living system reproduce, we ask what is the organization reproduced when a living system gives origin to another living unity?" An autopoietic system is a network of processes of production of components with a circular organization, with each component participating in the network that produces it. In addition, the autopoietic network produces a boundary (e.g., a membrane) that defines it as a spatial unity.

This concept has roots in Kant's view of organism as "natural purpose" (Thompson, 2010): something that is both cause and effect of itself (figure 2.1). There are a number of closely related theories of life: autocatalytic sets (Kauffman, 1986); metabolism-repair systems (Rosen, 2005); the chemoton

(Gànti, 2003); the hypercycle (Eigen and Schuster, 1978). Most of them take their inspiration in the molecular networks of metabolism, as those described by Monod. Enzymes are proteins that catalyze reactions between substrates: these reactions would not occur spontaneously without the enzymes. But these enzymes are themselves produced by the metabolism. Enzymes act as self-produced constraints on the physical processes that can occur, and this idea can be generalized beyond metabolism—for example, cell membranes or the vascular system (Montévil and Mossio, 2015). The living system produces its own constraints, which in turn specify the production network. The common theme is *operational closure*: each constituent of the organization is a product of the operation of that organization.

These ideas are rooted in thermodynamics: in a closed system, entropy must increase, which means that the organization degrades. Living systems are thermodynamically open systems, which actively maintain their organization by importing free energy from the outside (food) and releasing heat (waste). Thermodynamically, living systems exist out of equilibrium. This is in fact a key property that might allow us to detect extraterrestrial life with a telescope: at the planetary scale, the molecular constituents of Earth's atmosphere (which can be detected by spectroscopy) are not in equilibrium (Kaltenegger, 2017; Lovelock, 1965). Namely, O_2 and CH_4 spontaneously react to produce CO_2 and water, yet they are present in large quantities, because they are produced by life on Earth.

This implies that a living system consists of active processes that continuously renew its components. This is known in molecular biology as *turnover*. What appears as stable biological objects, such as membranes, organelles, or the brain, are in fact in dynamic equilibrium. For example, the half-life of proteins in the hippocampus of mice is about a week (Dörrbaum et al., 2018), which means that in a week, half of the proteins have been replaced. In the human brain, the half-life of proteins is about 18 days (Smeets et al., 2018). In other words, the brain gets reconstructed every month or so. This is obviously very different from the stability of the components of an ordinary machine. The repair process is also very different. Repair does not happen by someone taking out a protein and putting a fresh one in its place. Instead, there are ongoing parallel degradation and production processes, and somehow this results in a stable molecular organization. In the language of dynamical systems, living structures are attractors of the underlying organization of processes.

It is worth insisting on this point: in a living system, what is stable is not the molecules, not even the processes, but the organization of processes (or more generally constraints)—more precisely, there is a continuation rather than a stability (Nave, 2025). For example, a cell may die because its membrane is ruptured. What causes death in this case is that the membrane does not act as a constraint anymore, so that the extracellular medium and the intracellular medium equalize, by the second law of thermodynamics. Thus, it is a global topological property of the cell membrane—namely, the fact that it is a closed surface with no hole—that allows it to act as a necessary constraint for the network of processes that produces the membrane.

It is in this topological sense that there is a whole beyond the parts. The whole cannot be meaningfully reduced to the set of its parts, not even to the interactions between parts: it is rather a global property of the arrangement of parts and processes. This observation is the source of considerable difficulties with purely reductionist ("bottom-up") approaches, as we will see in chapter 3.

In summary, a living organism is a precarious organization of processes that exchanges matter and energy with its environment in such a way that it maintains itself out of thermodynamic equilibrium. It is this mandatory interaction with the environment that grounds the concepts of goals, functions, and norms in living systems (Jonas, 1953, 1966).

The Normative Nature of Living Systems

An organization is not just some kind of regular structure: it is a particular arrangement of components and processes that achieves something. A machine is organized: its components are arranged in such a way that the machine works. A living being "works" in the sense that it lives, and this is made possible by its particular organization of processes coupled to the environment.

This peculiar constraint is what gives normative properties to living systems. Canguilhem insightfully remarked that life is a normative activity (Canguilhem, 1966). Canguilhem was a physician turned philosopher, and he noted that living beings can have such a thing as a pathological state, which they try to avoid. A living system may then be said to be dysfunctional or pathological when the continued operation of its autopoietic organization is challenged. In this respect, living beings can be likened to machines: machines can also be said to work or to be broken, except that in machines, the norm is extrinsic. Like in machines, it is possible to attribute function to a part of a living system, such as an organ. Function is then the role that the part plays in the continued operation of the living organization. This is why Monod compared living beings to machines (Monod, 1970): like machines, they follow the same fundamental laws as inert matter, and like machines, they are organized according to normative principles—which he called "teleonomy," following Pittendrigh (1958) and Mayr (1961). Monod focused on the "teleonomic project" of reproduction. But given that by the time a human is able to reproduce, her brain would have been materially renewed more than a hundred times (for its protein content) while simultaneously growing—the reproductive project is secondary to the mere maintenance of the living organization.

Crucially, the normativity of a living organization is not an intrinsic property of the organization, but of its coupling with the environment. Living beings are very particular kinds of entities that exist out of thermodynamic equilibrium, and this is possible only because they exchange energy and matter with the environment. Living systems need to feed because their ongoing processes constantly dissipate energy. The normative properties of living beings derive from this thermodynamic openness.

Thus, the way we characterize the environment depends on the considered organism. For example, something qualifies as food if the organism can metabolize it. Environment and organism do not exist separately. This point leads to the famous concept of *Umwelt*, introduced by Jakob von Uexküll (1909): the environment seen as opportunities for interaction and their value for the organism. In psychology of perception, this is related to Gibson's concept of *affordances* (Gibson, 1979), which we will discuss in chapter 6. Thus, the same natural environment may correspond to very different Umwelts for a human, an ant, a plant, and for bacteria. The body of a living organism might correspond to the Umwelt of another one—for example, the human gut for *E. coli* (see Yong (2022) for a tour of this diversity). In the same way, what we call "internal states," such as hunger, make no sense unless we explicitly consider this coupling between organism and environment. Thus, a living organization cannot be described as an isolated object, independently of its environment: because of its thermodynamic openness, a living organism is necessarily a relational concept.

This mandatory interaction with the environment is what we call "behavior." Behavior is made of actions: the organism engages in activities when those activities are beneficial to the organism, and this depends both on the organism's internal state and on some abstract features of the situation—for example, whether the given pattern of light is identified as a source of food. For this reason, behavior is best explained at a global level in terms of goals. Behavior is anticipatory: actions appear to be taken as a function of their expected consequences—anticipation is a difficult notion that we will examine in chapter 7.

We recognize in these remarks some of the key motivations of computationalism, in opposition to behaviorism: behavior is made of actions, which depend on abstract features relative to the organism's goal, and not just reactions to proximal stimuli. However, none of this implies that behavior is computational, as we will discuss further in chapter 4. Computation is indeed also directed toward a goal, which is its result (the thing that we compute), but that is hardly surprising, given that computation is a kind of behavior—the kind we try to emulate in computers. There is no reason at this point to conclude the converse, that behavior is a kind of computation. Indeed, as we will see in chapter 7, what computationalism tries to address is the anticipatory properties of living organisms, but computation is a poor model of anticipation.

Thus, a living organization is defined relative to an environment. But this relation is not symmetrical, because a living system is also autonomous.

Biological Autonomy

The living organism is autonomous in two related senses: it is at the origin of actions, rather than reacting to inputs, and it follows its own laws.

The first point follows from the organism's precariousness: the living organization must be constantly active so as to resist degradation. This holds both for metabolism and behavior. An important feature of the concept of Umwelt is indeed that it is the organism that is generally at the origin of interactions with its environment. The organism has an autonomous activity, even in the absence of any "stimulus," and interacts with the environment when it is useful to maintain its own activity. In neuroscience, stimulus-independent activity is often qualified as "neural noise." This comes directly from framing neural activity as the result of a computation, which transforms an input to an output. Indeed, if this is truly the case, then any activity in the absence of an input is necessarily noise. But this forbids any possibility of an autonomous activity, and in this respect computationalism is still in line with the behaviorist notion of stimulus-driven responses. We will see in chapter 3 and 8 that this view is at odds with empirical observations.

Etymologically, autonomy (from Greek *autos*: self and *nomos*: law) means that the living organism is subjected to its own particular laws, the laws that allow it to live—for example, a particular way of obtaining and digesting food, a particular way of developing itself. This is what Varela (1979) meant by "biological autonomy." (See also Moreno and Mossio [2015].) In chapter 1, I pointed out that biological matter follows the same laws as inert matter. But this remark can be misleading. Microscopically, it seems correct—the laws of chemistry and physics. But macroscopically, it is clearly false: a living being follows macroscopic laws that are specific to living beings and are precisely the subject of biological science. The reason why there can be laws beyond the fundamental laws of physics (Kauffman, 2019) is that no general law ever specifies a particular phenomenon: it must be complemented by particular constraints. For example, corkscrews, because of their particular arrangement of parts, are governed by particular laws: when the handle is turned, the helix descends into the cork, proportionally to the number of turns. There is one theory of mechanics, but many mechanical objects with their own logic. When we recognize constraints that apply to many situations, this logic can in turn form general laws. Microfluidics, for example, has its own laws, not because they are against the fundamental laws of physics, but because constraints on water flows translate into new laws that are specific to the situation.

The same holds for biology. The biologist can distinguish many laws: the way chromosomes duplicate, the way food is digested and then transformed into ATP (e.g., the Krebs cycle), the way muscles contract, the way the heart circulates blood by alternating contractions of atria and ventricles. All of these phenomena are lawful. Many of these laws are species-specific or have notable exceptions, but they derive from a common constraint: the organism lives. The laws that govern a biological organism have this peculiar property that they ensure their own continued operation—biological theoreticians Montévil and Mossio (2015) call this defining biological property "closure of constraints." It is in this sense that a living organism is autonomous.

Machines or tools also follow special laws (like a corkscrew), but those laws are specified by someone else. A living organism operates according to its own laws, which stem from both evolution and individual history: particular ways in which the organism keeps its processes running. This is why there are no universal laws of biology. Surely, we can find principles and rules that apply to groups of individuals, such as species. These shared principles exist for historical reasons (common ancestors) or because of shared constraints—a phenomenon called *convergent evolution*. But these are groupings that the observer makes, not rigid laws that apply to all biological phenomena. It is indeed notorious that all biological laws have exceptions. For example, the "central dogma of molecular biology," as Francis Crick called it, states that genetic information flows from DNA to RNA to proteins—that is, DNA is used as a template to make RNA (transcription), which is then used as a template to make proteins (translation), but the reverse processes do not occur. This "dogma" is as lawful as it gets in biology, and yet is has exceptions, such as retroviruses (e.g., HIV) that can transcribe RNA into DNA. Another example is the *neuron doctrine*, which states that the nervous system is composed of discrete separated units, neurons, which interact through gaps between neighboring neurons, called synapses. Although it is a "doctrine," it again has exceptions. Ctenophores, or "comb jellies," are marine invertebrates with a continuous nervous system, without synapses (Burkhardt et al., 2023). Instead, neurons form a syncytium, a multinucleate cell resulting from the fusion of cells, like muscle cells. These are reasons to be skeptical of claims of universal laws of biology or cognition—there are recent attempts such as the free energy principle (Friston, 2010), which we will briefly discuss in chapter 7. The principle of biological evolution rather seems to be "whatever works." Of course, laws are what sciences in general try to produce, but it must be kept in mind that biological laws necessarily have a (more or less) limited perimeter.

Individuality

A simple consequence of the fact of evolution is that an individual can have a different organization than its parents, sometimes spectacularly so. This is the lesson that we can draw from "monsters," as exposed by developmental biologist Mark Blumberg (2010). Sometimes, a goat is born with only two legs, and it develops into a bipedal goat. In such cases, the entire organization of the animal (skeleton, muscles, nervous system) is reshaped in a way that ensures its continued living. For evolution to be possible at all, there must indeed be a possibility that a structural change is not an error but a different logic, a new law of living. This is an obvious difference with engineered machines, where the plan is the source of rationality, and a divergence from the plan is then necessarily an error. The living organization can even change during individual life, as is the case in development, most spectacularly with metamorphosis, where a crawling caterpillar turns into a flying butterfly.

Of course, interactions with the environment can also induce long-lasting structural changes. In the cognitive domain, these changes underlie memory and learning. This implies that each brain has its own organization. None of this should come as a surprise: we know that people have their own coherent ways of thinking and behaving, which is why there are psychologists and novels. Thus, biological variation is individuality, autonomous changes in organization, not just noise around some ideal template. This is also one of the fundamental principles of behavior, which makes averaging across individuals problematic.

A more subtle consequence of biological autonomy, in multicellular organisms, is that each cell is an individual, with an organization that depends not only on filiation but also on its own history of interactions, with other cells and with the environment. This phenomenon is known in neurobiology under the term *plasticity*. Biological individuality causes considerable difficulties with statistical approaches to brain function, as we will see in chapter 3, as well as with statistical analyses of behavior (Gomez-Marin and Ghazanfar, 2019).

In this respect, living organisms differ from machines: machines are organized, living beings are self-organized. They also diverge in other important ways.

Life Beyond Machines

Materiality

Oddly enough, Varela and Maturana described the cell as an autopoietic "machine." Clearly, no machine continuously renews its components, and machines have extrinsic norms, they are not autonomous. Why then call an autopoietic system a machine? They explained it as follows (Varela and Maturana, 1972): "To envision living systems as machines is to point to their structural condition independent of materiality." By this, they mean that it is only the interrelations between components that matters, and not the material basis of the components themselves, in the same sense that an electric car is still a car, because its electric motor interacts in the same way with the wheels as a combustion engine, despite the different source of energy. Just like in Rosen's relational biology and Monod's view of the cell, we can see in this view the historical influence of cybernetics, which describes regulatory systems based on feedback loops between components. In cybernetics, systems are described in terms of control, information, and so on, independently of physical laws. This is similar to the functionalist view of cognition, which underlies computationalism ("we could be made of Swiss cheese").

This is unfortunate, however, as later theoreticians have recognized (Bickhard, 2016b; Ruiz-Mirazo and Moreno, 2004)—in fact, Varela himself appears to have changed his view toward the end of his life (Thompson, 2004). Seeing living systems as independent of materiality neglects a crucial aspect of living systems: thermodynamics—in this respect, the same critique applies to Rosen's theory of life, although he disagreed that living systems are machines (Rosen, 2005). It is not correct that physical laws are essentially irrelevant to the organization of living systems. As we have seen, thermodynamics implies that a living system must consist of an organization of constantly active processes, and perhaps more importantly, thermodynamics is what explains the relation of living organisms to their environment. Living systems need to feed because their ongoing processes constantly dissipate energy: this is not evidently contained in a relational description of the interaction between components, and yet it is key to explain the normative properties of living systems.

In fact, the original toy model of autopoiesis (McMullin and Varela, 1997; Varela et al., 1974; figure 2.2) suffers from precisely that issue. This model is a cellular automaton—that is, an algorithmic description of the evolution of a two-dimensional grid with local update rules. An enzyme (star) can catalyze a reaction that transforms a substrate (circle) into a piece of membrane (square), and that piece of membrane may spontaneously degrade into

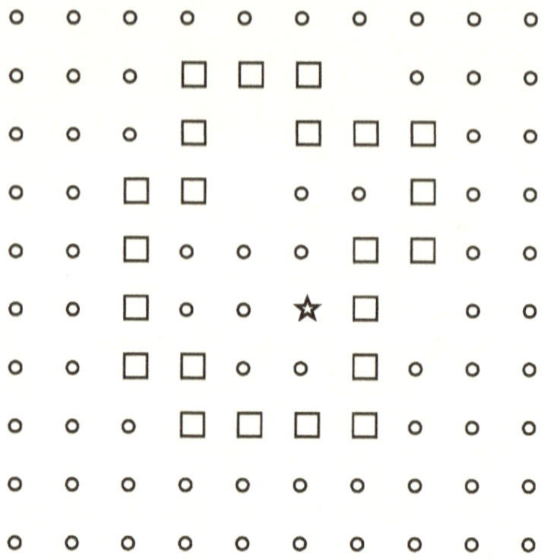

FIGURE 2.2. A minimal autopoietic model (adapted from Varela et al., 1974).

the original substrate. These rules produce a stable autopoietic organization with a membrane enclosing the enzyme. But this organization cannot be realized in the physical world because it is thermodynamically impossible. Indeed, the model organism takes in some substance in the environment and releases the exact same substance later. Thermodynamically, what the system releases (waste) must have lower free energy than what it takes in (food), it cannot be the same thing. It is this material feature that explains why a living organism needs to feed, to act in the environment, and why it has goals, as we will see further in chapter 7. It is therefore crucial to the understanding of life.

Thus, living systems are not just autopoietic organizations, but thermodynamically possible autopoietic organizations. This puts strong constraints on living organizations, beyond their mere relational properties. For example, including these constraints in the model of Varela et al. would imply that the organism would deplete the environment from its food resources and eventually die, unless it moves and seeks food. In other words, some of the fundamental features of behavior that we want to explain with theories of brain and mind come precisely from the fact that, unlike machines, living organisms are thermodynamically open autopoietic organizations.

Thus, the phrase "autopoietic machine" is clumsy. What is meant is that autopoiesis is an abstract characterization of the living organization (namely, its closure). But so is the particular structure that characterizes solids, liquids, and gases, and we do not use the term "machine" to qualify the states of matter.

On closer inspection, we have found that the asserted independence of living organisms from their material realization is false.

Biological Processes Are Integrated, Not Local

The term "machine" can have different meanings. As I mentioned in chapter 1, the phrase "molecular machine" also points to the idea that living processes are essentially mechanical, in the sense that they follow laws that are similar to those of the solid macroscopic objects of our ordinary experience: essentially deterministic local interactions based on shape. This was Monod's view (Monod, 1970), which was motivated by his work on the way enzymes recognize their substrate, in a "lock-and-key" fashion. In this context, "mechanical" means that a molecule can only interact with neighboring molecules (locality) and the result of this interaction is determined by the structure of the molecules (especially their shape) and their relative positions. This is why the recent successes of deep learning tools in predicting the structure of a protein from its amino-acid sequence have been acclaimed as a major advance for molecular biology.

This narrowly mechanical view appears to have been influenced by Schrödinger's reflections on the nature of the molecular processes underlying heredity. Schrödinger, one of the greatest physicists, and known in popular culture for his "Schrödinger's cat" thought experiment, argued in a short influential book called *What Is Life?* that the molecular basis of heredity must be a sort of "aperiodic crystal," something with the properties of a solid, because it appeared necessary for its long-term stability (Schrödinger, 1944). This idea influenced the later discovery of the structure of DNA.

But it is false that biological processes are essentially local interactions. In many important cases, microscopic elements are directly influenced by macroscopic properties—in particular, properties defined at the cellular or organismic level. This phenomenon is often called "downward causation" (Campbell, 1974; Noble, 2012), and it is neither mysterious nor restricted to the living world. An important example is the case of cellular electricity. All living organisms are made of cells, and all cells have an electrically polarized membrane. When a protein opens a pore in the cell membrane, electrically charged molecules (ions) move across the membrane. This local flow then changes the electrical potential all over the membrane, almost instantaneously, and without molecular transport, because the cytosol is an electrical conductor. In turn, changes in membrane potential can activate transmembrane proteins. By these means, two transmembrane proteins can interact at a distance, and not only with their neighbors: interactions are not local. This fact is sometimes underappreciated by modern neurophysiologists, who tend to describe electrical phenomena as some kind of object motion (Telenczuk et al., 2017). For example, when an action potential "travels" along an axon, there is actually very little material motion in the axonal direction. Ions, which carry electrical charges, move mostly across the membrane, and the "traveling action potential" is a propagating wave of charge exchange across the membrane, where the exchange is triggered by distant electrical interactions.

This is why electricity is so important for coordinating the activity of cells, including their molecular processes. To give an example, consider *Paramecium*, a protist (unicellular eukaryote) that swims in fresh water by beating the thousands of cilia covering its body (Brette, 2021). *Paramecium* is large, 100–300 µm depending on species. When it hits an obstacle, its cilia immediately reverse so that *Paramecium* swims backward for a little while, then turns and swims forward again in a new direction. This is called the "avoiding reaction" (figure 2.3). How does it manage to simultaneously reverse all its cilia when it hits the obstacle? Physiologically, what happens is that mechanoreceptors on its front end open and let calcium enter the cell (figure 2.4). This creates a calcium signal localized at the front end. It takes on average about 1 second for a calcium ion to travel across 100 µm by diffusion. So, it would take a couple of seconds for this molecular signal to spread to different sites on the membrane, and it would arrive at different times. Instead, what happens is that the calcium influx raises the membrane potential, simultaneously on the entire membrane (which is isopotential), including the ciliary membrane. As a result, other (voltage-gated) calcium channels located in the cilia open: this is an action potential. Calcium channels let calcium ions enter the cilia, where they bind to motor proteins and make the cilia revert. By these means, a local event on the membrane triggers a global electrical response (an action potential) that coordinates local molecular activity across the entire cell, allowing *Paramecium* to behave coherently as a single unit, as an organism. This

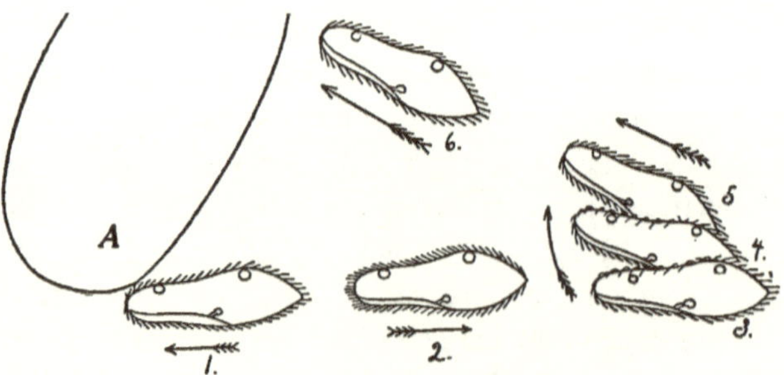

FIGURE 2.3. Avoiding reaction of *Paramecium* (Jennings, 1906).

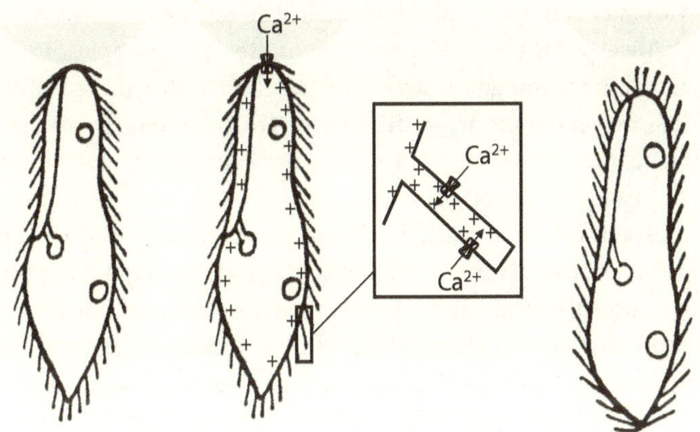

FIGURE 2.4. Electrochemistry of the avoiding reaction of *Paramecium*. The cell hits an obstacle (left). Membrane stress opens a mechanoreceptor, which lets calcium ions in, instantaneously raising the membrane potential (middle). This opens voltage-gated calcium channels in the cilia (inset), triggering ciliary reversal (right).

two-way interaction between the molecular level and the cellular level exists in all cells of all living organisms. In multicellular organisms, the nervous system bridges the cellular and the organismic level. Indeed, unlike most artificial neural network models used in machine learning, nervous systems typically have a *small-world* topology, meaning that any neuron is connected with any other neuron through just a few synaptic connections (Liao et al., 2017). This contributes to the global coordination of the organism.

Beyond electricity, the body's mechanics can also have integrative properties. As Varela and Frenk (1987) pointed out, the parts of an organism that are distinguished in classical anatomy (the legs, the arms, etc.) are those that can be cleanly separated with a knife. In reality, the extracellular matrix produced by cells all over the body is a continuum that determines the shape and mechanical properties of the animal. This implies that local mechanical events can impact distant sites. These events may be external: for example, one can detect a mechanical contact on the arm at positions where there are no nerve endings. This is because the contact triggers a mechanical wave on the skin that propagates to nerve endings at distant sites on the skin. The events may also be internal: for example, when a motoneuron contracts a muscle, the antagonistic muscle is mechanically stretched, which is sensed by the corresponding proprioceptive neuron. Therefore, the motoneuron and the antagonistic proprioceptive neuron interact at a distance, through the body.

More generally, any solid object like the body moves as a whole, so a local mechanical perturbation impacts distant sites. For example, the body's motion results from the integrated action of muscle cells distributed all over the body. In the inner ear, there are vestibular cells that are sensitive to the acceleration of the head. It follows that a motoneuron that contracts a toe impacts neurons in the head with very short latency.

Even at the cellular scale, mechanics plays an integrating role. The cytoskeleton is a network of protein filaments that gives shape and mechanical properties to the cell. Forces can propagate along this network, all across the cell. For example, forces exerted on the membrane can propagate almost instantaneously to the nucleus and influence gene expression (Wang et al., 2009). Obviously, the cytoskeleton is not organized linearly, like a chain of dominoes, and so the effect of a mechanical perturbation on the shape of the cell or on forces exerted at distant sites cannot be adequately described as a sequence of operations. A better analogy is a tent: the different nodes are coupled mechanically, so that their position is co-determined. (See in particular the tensegrity perspective on cell and body mechanics in Turvey and Fonseca [2014].)

In summary, the interaction between the constituents is not local, contrary to what metabolic diagrams might suggest. On the contrary, global properties of the cell or of the organism feed back on their local processes. This bidirectional causality between local and global levels contributes to the coherent behavior of the organism. Thus, biological processes are not local but integrated at different scales. As we will see now, they are also not deterministic.

Harnessing Randomness

The title of Monod's book, *Chance and Necessity* (Monod, 1970), conveys the conventional doctrine of molecular biology, which sees the organism as a set of essentially deterministic molecular mechanisms, while creative randomness is restricted to the domain of evolution, through random mutations of the genetic material introduced at reproduction time: "necessity" for the way the organism works, "chance" for the way it evolves. This is the idea of the "genetic program," a sequence of events that unrolls as written in advance in the "genetic code" ("pro-," before; "-gram," write).

Of course, no one denies that at the molecular level, thermal noise is very substantial, which makes the analogy between molecules and ordinary macroscopic objects rather fragile. But the very word "noise" denies the possibility that it plays any role in the function of the organism or of the cell, except negatively: noise is an unwanted deviation from the ideal process, which hopefully gets averaged out at macroscopic scales, or else corrected. This is false: living processes do not simply tolerate noise, they harness randomness.

Oddly enough, although the original model of autopoiesis was described with a small set of deterministic transformation rules, its implementation (described in the appendix of the paper by Varela et al. [1974]) is in fact a stochastic cellular automaton, and it turns out that randomness is not dispensable for the successful operation of the model (see figure 2.2). The main rule is that two neighboring substrate elements next to the catalyst are transformed into one membrane element. But since the catalyst has only eight neighbors, it can produce at most four membrane elements, which can never constitute a boundary enclosing the enzyme. To make more than four membrane elements, the elements must be able to move, and the source of motion is thermal noise: the Brownian motion of molecules. Without thermal noise, no autopoietic unit can be formed. Thus, what we call "noise" is a crucial component of the model, not a source of inaccuracy of the model.

Again, the materiality of the living organization, the fact that it is made of molecules subject to thermal noise, is a key aspect of the organization. This fact is increasingly appreciated (see in particular Nicholson [2019]). Clearly, at the molecular level, thermal noise is a source of motion and therefore a primary causal factor of various processes. For example, cellular electricity is based on movements of ions across the membrane, by a principle called *electrodiffusion*. For example, when sodium channels open in the membrane of a neuron, they let sodium ions flow into the cell, creating a current that will lead to the action potential and ultimately act on other neurons. These ions flow into the cell because there is a higher concentration of sodium outside than inside, and molecules flow from higher to lower concentrations: this is called "Fick's law of diffusion." But diffusion is not at all a force. It is the effect of thermal noise—that is, molecules randomly moving with no particular direction. This results in a net average flow from higher to lower concentrations, simply because there are more molecules moving from the highly concentrated compartment than from the other one. Osmotic pressure is also due to thermal noise, which results in pressure exerted on the membrane (in the same way as gases exert a pressure).

Of course, at a coarser level, the law of large numbers applies, so that we can often apply the deterministic laws of electrodiffusion to the total ionic current passing through the whole membrane, and forget about the key role of thermal noise at the molecular level—note that this is the case precisely because the effect of ionic movements is integrated at the cellular scale rather than acting locally, as we discussed previously. But the law of large numbers does not apply any more when we go down to the level of genetic expression—that is, the transcription of genes into RNA (the first step to making proteins). Indeed, each version of a gene is available as a single piece of DNA in a cell (with its complementary, which is not used for transcription). It is precisely this extraordinary feature that motivated Schrödinger to postulate that genes are encoded in "aperiodic crystals" with solid-like properties: these molecules have to be very stable since they exist as unique instances (diploid cells have two instances of each gene, but often different ones). It follows that the initiation and termination of transcription of a particular gene are necessarily stochastic events, because they involve single molecules subjected to thermal noise. This is an important case where thermal noise at the molecular level translates to stochastic behavior at the cellular level.

One key consequence is that genetically identical cells can differ, even in the exact same environment. Genetic regulation networks are complex systems of feedback loops, and therefore they display stable states and attractors—in particular, those that define cell types (Alon, 2019). Therefore, noise does not simply result in fluctuations around a mean state, but rather in stochastic hopping between stable states (Eldar and Elowitz, 2010). This phenomenon plays a role in development (Ohnishi et al., 2014). In the early stages of mouse development, cells of the embryo first differentiate randomly into several cell types, forming a heterogeneous mass of cells, and then by a combination of spatial sorting, signal reinforcement, and conditional apoptosis (cell death) that depends on the local environment of cells, a spatially organized pattern of cell types emerges (Nissen et al., 2017; more details later).

In the development of the neuromuscular junction, muscles are initially contacted by many motor axons, but after a competitive process, each muscle receives synapses from a single motoneuron of the matching type (fast or slow) (Sanes and Lichtman, 1999). Thus, the sequence of events is not programmed at the cellular level, but the result appears largely deterministic: the developmental process is largely reproducible, provided that environmental conditions are constant. The development of the neuromuscular junction has also inspired Edelman's theory of brain function in which a variety of neural circuits are formed during development and then selected by synaptic reinforcement (Edelman, 1993), a theory related to Changeux's synaptic selection theory (Changeux et al., 1984).

The combination of cellular variation with conditional proliferation and death evokes a familiar theme: Darwinism. (Note that the theories of Edelman and Changeux are not exactly Darwinian since there is no reproduction involved; see Crick [1989].) Darwin's insight was that a combination of individual variation with differential reproduction may result in an apparently goal-directed process at the population level, where the population appears to adapt to new environments. In the theory of natural selection, individuals do not adapt, and yet populations do, without any planning. This has led Jean-Jacques Kupiec to propose that development is a Darwinian process, in which cells change randomly until they are adapted to their local environment, which modulates their proliferation (Kupiec, 1997). There is indeed a surge in gene expression variability early in the cell differentiation process (Richard et al., 2016).

Darwin's theoretical insight indeed applies well beyond the theory of evolution. For example, in the immune system, a large set of lymphocytes are produced with random antibodies, and when a pathogen's antigen binds to an antibody, the lymphocyte is activated and proliferates—this is clonal selection theory (Burnet, 1957). As a whole, the immune system appears to respond reliably to a new pathogen, but it does so without a deterministic mechanism.

Darwinism works with differential reproduction at the population level, but a similar trial-and-error mechanism can work at the individual level with *differential persistence*. This can be observed for example in the thermoregulatory behavior of *Paramecium*, which we have encountered earlier (Brette, 2021). When temperature increases above 25°C, it changes direction with an avoiding reaction triggered by an action potential, and the same occurs if temperature decreases below 25°C. (This preferred temperature can shift with adaptation (Nakaoka et al., 1982)). Thus, the organism changes direction until it swims toward waters at 25°C, where it remains. *Paramecium* does not use spatial gradients of temperature along its body, and it does not turn toward its preferred temperature: it just tries a new direction when the situation degrades. This is sufficient to navigate toward appropriate temperatures.

This is somewhat similar to bacterial chemotaxis, where the cell modulates its rate of "tumbling" (random directional change) as a function of the temporal gradient of chemical concentration (Berg, 2008). Bacterial chemotaxis is a well-known phenomenon that has been formalized mathematically in terms of stochastic optimization (Keller and Segel, 1971). There have also been a few theoretical propositions about brain function based on the differential stabilization of synaptic fluctuations (Harris, 2008; H. S. Seung, 2003), although these remain speculative.

It is important to grasp the fundamental difference between the mechanisms of a machine, as Monod conceptualized it, which are based on the application of formal rules (like the formalism of chemical reactions), and the stochastic processes of Darwinism and trial-and-error. Consider a phenomenon where the system proceeds from state A to state B. This could be, for example, the organism developing from embryo to adult, or the immune response, going from presentation of a pathogen to its destruction. A mechanistic view would describe the phenomenon as a series of steps from A to X to Y to Z to B, each transition corresponding to a materially identified mechanism operating deterministically, possibly triggered by some input. Formally, this is a *finite-state machine*, which is a classical model of computation. In such a model, it is not only the mapping from A to B that is deterministic, but also the path from A to B (the "computation"). In contrast, in a trial-and-error process, the mapping from A to B is deterministic but the path is random. It does not rely on a reproducible series of predefined operations. In a Darwinian process, the effect of a mutation is unknown: it gets known only once the mutant develops and lives or dies. In contrast, the result of a formal operation, like a computation or a mechanism in Monod's sense, is determined by the operands and the formal rule. In other words, a machine incorporates knowledge about all the detailed operations leading from input to output—namely, in its mechanisms—but life does not always work in this way. We will explore this issue further when we discuss computation (chapter 4) and anticipation (chapter 7).

In summary, living phenomena can be described in terms of physicochemical laws, but not in terms of local deterministic interactions. There are crucial interactions across levels: integration from lower to higher levels (molecules to cells to body) and modulation of lower levels by higher levels. At all levels, various processes harness randomness rather than resist noise.

What remains of the concept of a machine is rather thin: the way a living being works is not by magic, but it is also quite different from a washing machine. We will now turn to multicellular life, in particular their development and evolution.

Multicellular Life

From Bacteria to Animals

Life is incredibly diverse, and animals represent a very tiny fraction. Most of the biomass is made of plants (2/3), then about 10 percent is bacteria, and animals represent 0.4 percent, 3 percent of which are humans (Bar-On et al., 2018). There are more protists (motile unicellular eukaryotes) than animals, even in mass.

What kind of living organism is an animal? To answer this question, it is helpful to examine the evolutionary history of animals. Animals are *eukaryotes,* which means that their cells have a nucleus and other organelles such as mitochondria. For about 2 billion years, which is half of the entire history of life, there were only *prokaryotes* on Earth (such as bacteria), unicellular organisms with no nucleus that reproduce asexually, by division. As animals, we tend to think of reproduction as producing an additional copy of the organism—possibly distorted or mixed. But at the cellular level, this is not what happens. When a cell reproduces, it does not produce a child distinct from itself: it splits in two. Each half is literally a part of the original cell, which then grows. All cells reproduce by splitting, including the cells of animals. Thus, every cell on Earth originates from a previous living cell, recursively for more than 4 billion years.

Eukaryotic cells are very different from prokaryotes. For example, they have organelles called *mitochondria*, which produce ATP, used as an energy source in many metabolic processes. How did mitochondria evolve? In fact, they did not. Around 1.5 billion years ago or more, a cell (possibly an ancestral eukaryote or a prokaryote) engulfed bacteria probably related to *Rickettsia*. These bacteria then lived, reproduced, and evolved inside the host cell. Mitochondria still have their own DNA. This phenomenon is called *endosymbiosis*. This theory was proposed in 1967 by Lynn Margulis (Sagan, 1967), but it was not accepted until a decade later—the article was rejected by fifteen journals, as it directly contradicted the modern synthesis of evolution, which sees evolution as a gradual accumulation of mutations. Thus, what textbooks often call the "power plant" of the cell is more an ecosystem than a factory.

Margulis also correctly proposed that chloroplasts, the photosynthetic organelles of plants, originate from the endosymbiosis of cyanobacteria, which are photosynthetic bacteria. Thus, it is not exactly that prokaryotes turned into eukaryotes, but rather that eukaryotes are made of prokaryotes, a higher-order living organization. This new mode of organization then evolved into various forms.

Eukaryotes diverged about 1.5 billion years ago from a common ancestor, called the *last eukaryotic common ancestor* (LECA), to give rise to all eukaryotic species, including animals and plants. Comparing the genomes of distantly related species enables some kind of "molecular paleontology," inferring the likely molecular toolkit of that distant ancestor (Koumandou et al., 2013). It is thought that the LECA was a marine protist with surprisingly modern cellular structures: a sophisticated cytoskeleton with kinesin and dynein motors, endoplasmic reticulum, Golgi complex, a nucleus, a flagellum, the ability to do both mitosis and meiosis, all known families of ionic channels, mechanoreceptors, and so on. Of course, modern protists are not the same ones as those that lived hundreds of million years ago, but some have very similar ancestors. For example, a fossilized *Paramecium* has been discovered in a 200-million-year-old piece of amber, with all the morphological features of its modern relatives (Schönborn et al., 1999).

How did multicellular life forms appear? There are different kinds of multicellular organizations, including bacterial multicellularity (Lyons and Kolter, 2015). Regarding animals, a plausible scenario is that colonies of protists, probably related to choanoflagellates, developed a transient multicellular stage, which later evolved to permanent multicellularity. This optional multicellularity is observed in modern choanoflagellates (Ros-Rocher and Brunet, 2023; figure 2.5).

Thus, as we go from protist to animals, it is not that the flagellum evolved into legs and the protist grew a brain. Rather, an animal is more like a colony

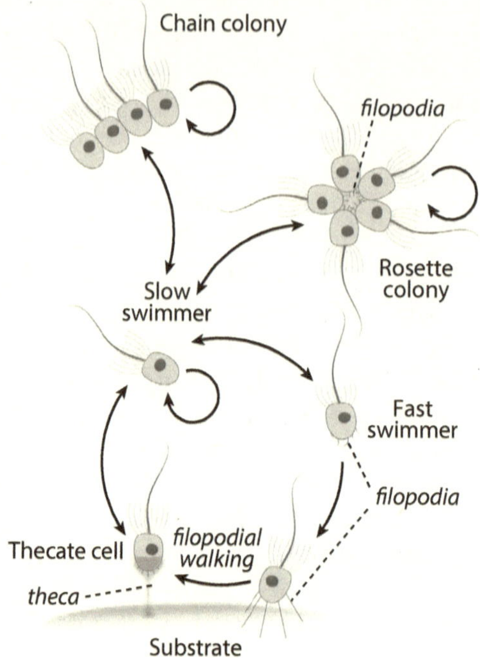

FIGURE 2.5. Optional multicellularity in choanoflagellates. Original figure by Ros-Rocher N, Brunet T. 2023. https://doi.org/10.1007/s10071-023-01776-z / CC BY https://creativecommons.org/licenses/by/4.0/.

of protists that has adopted a cohesive lifestyle. It is a higher-order living organization. An animal is a clonal community: it develops from a single cell, which then divides. All cells in an animal are genetically identical, except for occasional mutations. A neuron has the same genome as a kidney cell, but it is in a different state. These cells produce the body in which they live (their "house")—namely, the extracellular matrix that holds the cells together (collagen fibers, bones, etc.) and forms a continuum (Varela and Frenk, 1987).

Generally, this "house" also hosts a large number of symbiotic unicellular organisms, mostly bacteria. In humans, this microbiome represents about ten times more cells than human cells, and it plays key physiological roles, in particular in the digestive system and on the skin (Turnbaugh et al., 2007). These cells are accepted as part of the organism by the immune system (Pradeu, 2010). As we have seen, animal cells also have symbiotic bacteria inside themselves, namely mitochondria. Thus, an animal is better described as an ecosystem rather than a machine.

Development

When building a machine, we start by gathering its parts, according to specifications, then we assemble them according to a global plan. No cell is ever assembled and neither is a multicellular organism: they develop.

There are two major differences between assembly and development. The first one is that a machine is not functional until all its components are assembled and the machine is switched on, while an organism is a living unit from embryo to adult, which feeds and maintains itself. Even the *zygote*, the first cell of an animal, is itself continuous with a previous cell, and so on over several billion years. Each cell of a developing organism is itself a living unit, which requires continuous material exchanges with the environment to maintain itself.

The second major difference is that development is a self-organizing process. No external agency places the cells at the right position in the adult animal. There is no plan, even in the genome, of the adult organism. When biologists talk about the "body plan" (or bauplan) of an animal, it should be clear that the plan in question is something that the biologist has made as a descriptive tool, not something that is written in the genome and then read. When we "map the brain"—for example, by measuring the connectome—what we measure is not an assembly plan but a snapshot of a process of development, where cells divide, differentiate, migrate, and die as a function of their local environment. What is specified by the genome is not the position and type of the cells in the adult organism, but constraints that lead cells to engage in these different processes depending on their state and the local environment.

A simple example is given by Nissen et al. (2017) for the early development of the mammalian embryo; they show that a few simple local rules can give rise to the blastocyst (a ~five-day-old embryo) (figure 2.6). Over the first few days, the zygote divides with constant volume. At this stage (E3.0), compacted cells on the outer layer start expressing a transcription factor (Cdx2). A *transcription factor* is a protein that regulates the expression of other genes, and this results in moving the genetic system to another state: the cells differentiate (gray versus white). Then cells on the inner side start differentiating into one of two competing states (light gray/dark gray)—that is, the two states produce transcription factors that inhibit each other (E3.5). In addition, each cell in a given state produces an extracellular protein that induces its neighbors to shift to the other state. This local competition produces a "salt-and-pepper" pattern. The red cells are less adhesive than the others, which tends to expel them from the mass (E4.0). Finally, dark-gray cells surrounded by light-gray cells die (apoptosis) (E4.5). The result is a particular pattern of cells of different types, which forms a recognizable five-day-old embryo.

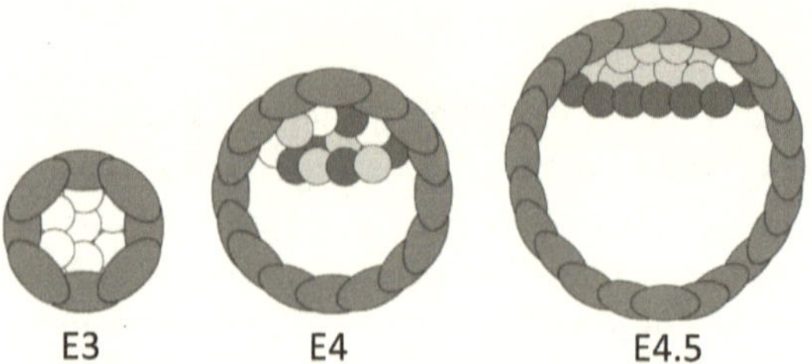

FIGURE 2.6. Early development of the mammalian embryo. Original figure by Nissen SB, Perera M, Gonzalez JM, Morgani SM, Jensen MH, Sneppen K, Brickman JM, Trusina A. 2017. https://doi.org/10.1371/journal.pbio.2000737 / CC BY https://creativecommons.org/licenses/by/4.0/.

We can see in this example that development results from a combination of physical processes (adhesion, mechanical stress, molecular diffusion) and genetic processes. The final position of cells is not specified individually, and indeed as we have seen, the evolution of any particular cell is stochastic. This confers some robustness to the system: if a cell dies, another can replace it; if there are too many cells of a given type, negative feedback will induce differentiation to the underrepresented type.

Thus, the adult morphology of the organism is not explicitly specified by the genome. In fact, it is reproducible only to the extent that the environment is unchanged. If some accident occurs, the final structure can change. A striking example is the case of a ten-year-old patient born with no left-brain hemisphere due to an early dysfunction of embryonic development (Muckli et al., 2009). In normal adults, neurons of the right side of each retina (ganglion cells) project axons to the left hemisphere (thalamus, then cortex). But this patient had no right cortical hemisphere and yet nearly normal vision in both hemifields. Anatomically, retinal ganglion cells of the right hemisphere actually projected to the right hemisphere instead of crossing the optic chiasm, and the person could see.

In *Basic Instinct* (Blumberg, 2006), developmental biologist Mark Blumberg shows that what we call "instincts" are neither written in the genome nor instructed. Rather, they develop from an interaction between the developing organism and its environment. This leads to reproducible outcomes when the environment is constant. For example, ten-day-old rat pups start preferring rat odors to those of other species, not because rat odors are somehow genetically stored in the genome, but through an association between odors and warmth, which usually occurs during development. Similar processes underlie the development of sexual behavior and even thirst behavior. Contrary to popular belief, instincts are not "hard-wired." They are not taught either: they develop.

We can see that even the term "self-organization" is slightly misleading since it involves processes that extend beyond the organism—for rats, it involves social interactions.

These remarks show that the phenomenon of development is not well-captured by the concepts of program and code, with biological structures somehow encoded in the genome—they are not. Instead, Conrad Waddington, a pioneer of evolutionary developmental biology, introduced dynamical concepts to describe development, such as the "epigenetic landscape" (Waddington, 1957), which essentially frames reliable outcomes of development as attractors.

These concepts also apply to the organization of single cells. For example, when the content of a *Xenopus* egg is thoroughly mixed, it spontaneously reorganizes in ~30 minutes into cell-like compartments (Cheng and Ferrell, 2019). This self-organizing process is based on microtubule polymerization and requires ATP, but not nuclei—that is, it does not directly involve genes. Thus, the concepts of program and maps have little relevance to understand the organization of the cell. Some other relevant concepts from physics seem more applicable. For example, when temperature is smoothly varied, the structure and properties of water can change abruptly, from gas to liquid to ice. This is known in physics as a phase transition, and is described in mathematics by bifurcation theory.

Thus, it is clear that the program metaphor that is so widely used in biology is highly misleading (Noble, 2008). Surely, both the result and the time course of development appear to be very reproducible and depend crucially on genes. But development certainly does change substantially when the context is changed. In fact, when the nucleus (and thus the DNA) from a carp cell is transferred to the egg of a goldfish (where the nucleus has been removed), the resulting animal is a hybrid fish, with some aspects of development closer to that of the goldfish, which did not share its DNA (Sun et al., 2005). Thus, genes are not building instructions, but constraints on development.

Clearly, the development of an organism is as distant as could be from an engineering process. Parts do not preexist to the whole and are not explicitly specified in advance. The organism grows and changes through self-organizing processes, it is not assembled from parts. And for sure, evolution is not an engineer.

Evolution

Everyone is familiar with Darwinism: heritable variations combined with differential reproduction of individuals lead to the adaptation of populations. The population evolves in such a way that it survives and grows in changing environments. Thus, there is an appearance of a goal-directed evolutionary process (toward adaptation) even though no single individual evolves.

Evolution is often depicted as an optimization process. It is clearly the case that certain biological structures appear to be optimal or at least very efficient according to important physiological criteria. The detailed geometry of the vascular system, for example, can be accounted for by a few optimality principles, such as maximizing the surface of contact and minimizing the resistance to flow (Rosen, 1967). In the same way, it seems clear that mutations that improve the efficiency of energy harvesting should be selected, provided that they do not degrade other key processes. It follows that extant species should have efficient ways of harvesting energy from their environment, and indeed they do.

But we quickly encounter serious difficulties when we try to make the optimization analogy more precise. To frame evolution as optimization, we need to identify an optimization criterion. In the case of the vascular system, I mentioned the surface of contact and the resistance to flow. It seems reasonable, but of course, organisms are not actually selected based on vascular surface of contact. There is no optimization criterion in evolution: whatever survives and reproduces is, by definition, "selected," and we simply assume that organisms with a more efficient vascular system are more likely to survive and reproduce. But of course, that makes sense only if mutations that are good for the vascular system do not degrade other key processes. Unfortunately, since all cells of an organism have the same DNA, mutations are in general unlikely to affect a single function of interest. Thus, while there is some sense in the idea that some key processes are being optimized by evolution, it must come with the qualification that there are constraints and trade-offs—in other words, organisms are not really optimized for any particular function.

More importantly, this conceptual framework cannot account for the creation of new functions. For example, it is sometimes claimed that the human brain has evolved *for* language (Marcus, 2012), but this is misleading. Language does not preexist to speakers, so it cannot be that language was an evolutionary ("optimization") criterion for the brain unless humans already spoke. Rather, evolutionary change can create opportunities, which might be beneficial for the survival of the species, but there is no predetermined optimization criterion. The engineer starts from the desired functionality, then designs and tunes a structure that fulfills this functionality. In contrast, evolution is a blind historical process: a structure gets created by chance, and might then be conserved and possibly refined if it serves a useful functionality; or some existing structure might change and take on a new functionality. A central argument of Darwin in *The Origin of Species* (Darwin, 1859) is precisely that we find striking anatomical similarities across different species despite the diversity of lifestyles, which only makes sense if we assume a common ancestor:

> What can be more curious than the hand of a man, formed for grasping, that of a mole for digging, the leg of the horse, the paddle of the porpoise, and the wing of the bat, should all be constructed on the same pattern, and should include the same bones, in the same relative positions?

We might want to suggest that at least evolution optimizes survival and reproduction rate. Unfortunately, this proposition has limited scope. Consider cyanobacteria, for example. Those are common photosynthetic bacteria (a sort of microscopic algae) that have been on Earth for several billion years, living today in fresh and salt water. They are responsible for the "great oxidation event" that occurred around 2 billion years ago, when a rise in atmospheric oxygen that they produced caused a mass extinction. This extinction event is an endogenous result of evolution. It clearly did not tend to optimize global survival rate.

Then at some point around that time, some cyanobacteria were engulfed in other cells and started providing energy to those cells. These endosymbiotic cyanobacteria became the chloroplasts of plants, which still resemble modern cyanobacteria. Was this event a case of optimization? If so, which ones are the results of optimization, the chloroplasts or the freely living cyanobacteria that still exist today? Chloroplasts are adapted to their cellular environment, just as freely living cyanobacteria are adapted to seas and lakes, even though they have different reproduction rates.

The same unicellular eukaryote gave rise to trees and humans. Therefore, this evolution cannot possibly be framed as an optimization process. Indeed, evolution is not necessarily competitive, contrary to popular belief. Cyanobacteria do not compete with chloroplasts, for example, and chloroplasts do not compete with their host plant either.

There is a creative component in evolution that the optimization analogy fails to capture. In fact, even the general idea that evolution is a process by which organisms adapt to their environment is problematic, since evolutionary events create changes in environments as well as changes in the structure of organisms. For example, the endosymbiosis of cyanobacteria is an evolutionary event that corresponds to a structural change from the perspective of the host organism, but to a change in environment from the perspective of cyanobacteria. In addition, as we have seen, what we consider as an organism's environment, or more precisely its Umwelt, depends on the organism itself. For example, prey and predators are adapted to the same environment from the perspective of the observer, but to a different Umwelt: the same environment can lead to completely different kinds of adaptation.

Thus, if evolution is an optimization process, then it is one that produces multiple results that are co-dependent, where the criterion is not stationary and depends on what is being optimized. In other words, it is not really optimization. Again, the engineering analogy can have value in some contexts, but it remains an analogy with its limited scope and shortcomings.

Next, we turn to the source of variation in evolution.

How Living Organisms Change

Darwin did not know about the source of variation: simply, he observed that the offspring differ from their parents, while being similar. In fact, his main source of inspiration was the artificial selection of domesticated animals, which represents 0.01 percent of today's biomass (presumably less at his time). In the modern synthesis of evolution, which combines Darwin's theory of natural selection with Mendel's observations of peas, the source of variation is the genetic mutations and recombinations introduced at reproduction time. Each gene corresponds to a sequence of nucleotides, and this sequence can be altered (substitutions, insertions, deletions) or exchanged with the sequence of the other parent.

In the engineering domain, this led to the invention of a bio-inspired optimization technique, "genetic algorithms," where a function of several parameters is being maximized (or minimized). The algorithm starts from a population of "individuals," each with a different set of parameter values. Each individual has a "fitness" value, which is the value of the function for the individual's parameter values. In each step of the algorithm, the individuals with the highest fitness are selected and copied, and variations are introduced by varying or exchanging parameters. These parameters are the analogs of the nucleotide sequences of the genes.

However, the modern synthesis has been substantially enriched and altered since it was formulated. We all know that our ancestors were small mammals. What tends to be less appreciated is the fact that our more distant ancestors, more than a billion years ago, were protists, marine unicellular eukaryotes. Before that, the concept of ancestors becomes blurry. For about 2 billion years, which is half of the entire history of life, there were only prokaryotes on Earth, which reproduced mostly by division—hence the traditional concept of species as a population of sexually compatible individuals does not apply. In prokaryotes, a common form of genetic variation is *horizontal gene transfer*, where an entire gene is transferred to another individual and incorporated in its genome (Smets and Barkay, 2005). This means that the "tree of life" is not exactly a tree, especially near its roots (figure 2.7).

This represents a direct challenge to the view of evolution as parameter tuning: in the case of horizontal gene transfer, a new "parameter" would be

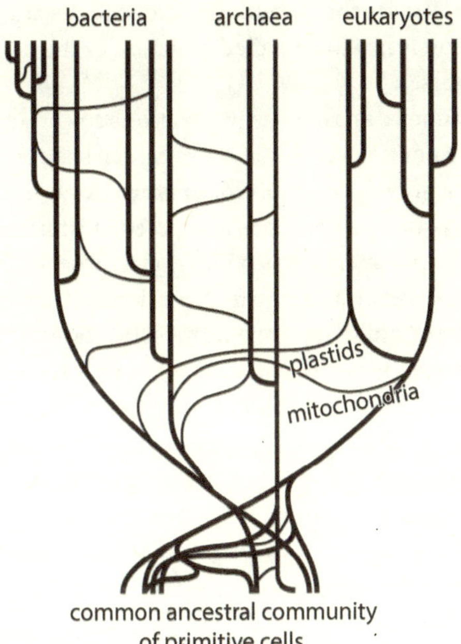

FIGURE 2.7. The updated "tree" of life, with horizontal gene transfer. Modified from figure by Colvin AZ. After Smets, Barkay. 2005. / CC BY https://creativecommons.org/licenses/by/4.0/.

created. Genes can also be duplicated, and occasionally, even entire genomes (Aury et al., 2006). This is what allows proteins of the same family—for example, all voltage-gated ionic channels—to coexist within the same organism. Genes are not parameters: they specify proteins, and proteins are material entities, not variables.

As we have seen, a spectacular example of inheritable variation that does not fit the parameter optimization view is *endosymbiosis*. Mitochondria, often referred to as the "power plants" of eukaryotic cells, did not arise by mutations but by engulfing living bacteria, which then co-evolved with their host. There are other known cases of endosymbiosis, apart from chloroplasts. For example, *Paramecium bursaria* is a species of *Paramecium* that includes a symbiotic alga name *Chlorella*, itself a single-celled green algae including a chloroplast and mitochondria. Endosymbiosis is not a case of tuning.

What is the relation between DNA and the organism's structure? When we say for example that the geometry of the vascular system is optimized, what do we mean exactly? Clearly, there is no map of the vascular system in the genome. This is fortunate, since the vascular system must adapt to the changing morphology of the growing organism in order to remain functional.

Rather, DNA constrains the development of the organism. Thus, what is shaped by evolution is not the detailed structure of the organism but the developmental processes.

DNA is sometimes described as instructions for building the organism. First, over the past 4 billion years, no functional cell has ever been "built." Cells have split, merged, grown, and evolved. In other words, modern cells are the result of 4 billion years of continuous autopoietic history. Thus, the stunning fact is that each cell on Earth has been functional for several billion years, with changes and material renewal but no interruption of its self-sustaining nature: no cell has ever been "built," let alone from instructions.

When a cell divides, it carries with it DNA as well as everything else that is in the cell—in particular, the membrane, cytoplasm, and cytoskeleton (Fields and Levin, 2018). New proteins are produced using both DNA and the cellular machinery, including preexisting proteins (e.g., RNA polymerases). DNA is particularly important for heredity because of its great chemical stability, as Schrödinger had predicted (Schrödinger, 1944). But DNA does not reproduce by itself, and it does not specify organisms either. For example, as we have seen, transferring the nucleus of a carp cell to a goldfish egg leads to a hybrid fish, not a carp cell (Sun et al., 2005).

What DNA does specify, to some extent, is the primary structure of proteins (the sequence of amino acids)—even this must be qualified in eukaryotes, as there is considerable edition after transcription. One of the discoveries enabled by modern genomics is that most families of proteins in modern eukaryotic species have a very ancient origin. The last eukaryotic common ancestor (LECA) was probably a marine protist living about 1.5 billion years ago, with modern cellular structures (Koumandou et al., 2013). For example, the action potential of *Paramecium* is produced by the opening of L-type calcium channels of the Cav1.2 family (Lodh et al., 2016), whose homologs are found in the hearts and neurons of humans. Yet, the common ancestor of *Paramecium* and humans is the LECA, 1.5 billion years ago. All known families of ionic channels expressed in neurons are found in *Paramecium* except sodium voltage-gated channels, which can be found in other protists (Febvre-Chevalier et al., 1986).

The fact that distantly related species have very similar molecular contents led François Jacob to propose that evolution is better depicted as a "tinkerer" than as an engineer (Jacob, 1977). The engineer makes a plan, then collects the appropriate materials with the right properties and puts them at the appropriate places, according to the design. The tinkerer, on the other hand, takes whatever objects are available and uses them in ways that differ from their original use, possibly with alterations, and adjusts the plan along the way. For example, *Paramecium* has NMDA-like receptors (Ramoino et al., 2014). These are receptors also found in the synapses of animals, which bind the main excitatory neurotransmitter of nervous systems, glutamate. In nervous systems, glutamate mediates fast excitatory interactions between neurons. In *Paramecium*, glutamate is an attractant that indicates the presence of bacteria, on which it feeds. Given that the common ancestor of animals and *Paramecium* was a marine protist, it might be that the ancestors of NMDA receptors of neurons were originally used for hunting bacteria.

What Jacob wanted to emphasize with the tinkering metaphor was the strong historical contingency of the structure of living organisms. Tinkering remains an anthropomorphic metaphor, however. Both tinkering and engineering require an external agency, which manipulates things. But the molecules of a living organism are not "used" by that organism: they are the organism. There is nothing else than the molecules and their own activity, and nothing else to arrange them in a different way. If NMDA receptors are "used" for different things by *Paramecium* and humans, it simply means that they take on a different activity when they are embedded in a different network of interaction.

What explains the diversity of animals? A large part of that diversity is due not to mutations in genes encoding ionic channels, metabolic enzymes and other "housekeeping" proteins, but to mutations affecting development (Arthur, 2021). For example, as we have seen, some genes encode transcription factors that can switch other genes on and off. Mutations of such genes amount to changing constraints of development, that lead to different arrangements of the same elements in different animals or plants.

It must be stressed again that the genome does not encode the structure of the animal in the sense of a plan, specifying where each "brick" should be placed. Instead, it specifies proteins and constrains developmental interactions at the cellular level (not the organism level), which also depend on physical interactions. A mutation in a developmental gene can have a global, coherent effect, such as transforming antennae into legs (Schneuwly et al., 1987).

Thus, even in nervous systems that appear identical between individuals, like the 302 neurons of *C. elegans*, evolutionary processes do not tune connections one by one. Instead, evolution shapes the global processes by which neurons grow and identify targets—which are the subject of intense ongoing research (Kovács et al., 2020). In the same way, the density of the different ionic channels in a neuron is not written in the genome, and indeed, they vary considerably between neurons, even of the same type. Thus, the reason why neurons with diverse morphologies can produce functional action potentials cannot be that evolution has tuned the density of each type of channel. It has to be found in the processes by which the cell self-tunes the densities and properties of its channels.

In summary, evolution is very far from an engineering process. Framing evolution as optimization is not entirely useless, but we must recognize its limitations and shortcomings. Evolution alters living organizations by modifying proteins, acquiring new ones, changing developmental processes and occasionally merging different living forms. It does not directly tune biological structures.

Breathing Life into Brain Theory

Theory of brain and mind traditionally downplays biology, and instead draws heavily from engineering as a source of inspiration. But as we have seen in this chapter, living organisms are very different from engineered devices. Like machines, living organisms are made of ordinary matter that owe their properties to their particular organization. But the analogy essentially stops there. Unlike a machine, a living organism is a precarious organization of processes that exchanges matter and energy with its environment so as to maintain itself out of thermodynamic equilibrium. This is what grounds the concepts of goals and functions in living systems, which machines only import from their makers. Living organisms are autonomous in two ways: they have an endogenous activity, and they have their own individual ways of harnessing energy in the environment. Living processes are integrated at different scales and they harness randomness rather than simply tolerate noise. An animal is a clonal community of cellular organisms that build their "house" as they grow, while hosting symbiotic living beings both within the house and within the cells themselves. It develops from a single cell by self-organizing processes defined at the cellular level, combining genetic constraints, physical constraints, and stochastic processes. Evolutionary changes in the organism's structure can be interpreted as adaptation to the environment, or as changes in what counts as the environment. These changes are diverse (mutations, gene transfer, endosymbiosis) and do not independently tune the structural parts of an organism.

All these facts of life challenge the classical concepts of brain theory, which are based on how engineered machines, especially computers, work. If we are to believe the mainstream dogma, then we must reason as if brains were the result of intelligent design, while paying lip service to the idea that they are not. This position is not tenable if we pay attention to biology.

In the next chapters, I will try to breathe life into brain theory while examining classical neuro-theoretical concepts: computation (chapter 4), codes (chapter 5), information (chapter 6), prediction (chapter 7), and implementation (chapter 8). But first, I will start by questioning the preconceptions of so-called bottom-up approaches to "reverse engineering" the brain.

3

Brains from the Bottom Up

THERE ARE two great families of approaches to modeling the brain. One kind of approach starts from behavior and cognition, tries to understand its logic, and then speculates on how neural processes might realize that logic, often under the model of computers (implementation of algorithms). These "top-down" or normative approaches will be the subject of the next chapters. Another major trend of brain research describes itself as "bottom-up" or "data-driven." It takes a systematic empirical approach, by measuring in detail the properties of neurons, axons, and synapses and their arrangement, and then tries to derive the cognitive properties of that arrangement by analysis or simulation.

For example, one might make models of different types of neurons in the hippocampus by measuring their average electrical properties, then measure the number and strength of synaptic connections between neurons, and finally examine the properties of the resulting neural network. The Human Brain Project aimed at systematizing this approach by making large-scale statistical measurements of the properties of neurons and synapses, in order to simulate an entire brain. Henry Markram defended this project by predicting that "scientists could be running the first simulations of the human brain by the end of this decade" (Markram, 2012).

If statistical measurements of connectivity turned out to be insufficient, then one could increase the level of detail—for example, cut the brain in very thin slices so as to precisely measure all the individual connections between neurons, the *connectome*. If connectionism is right, then behavior and mind are determined by the connectome, so that some of its supporters have claimed that you literally *are* your connectome (Seung, 2012). For example, one should be able to read your memories from your connectome.

The key characteristic of bottom-up approaches, as opposed to top-down approaches, is that the modeling process is independent of what the organism achieves. The model is based on direct measurements of the structural elements of the model. Once everything has been properly measured, it is expected that the model reproduces the behavioral and cognitive properties of the organism.

But at the time of writing, scientists did not manage to run simulations of the human brain—the ambition of the Human Brain Project shifted toward the more modest goal of building numerical infrastructure. In fact, the complete connectome of *C. elegans*, a microscopic worm with 302 neurons, was described in 1986 in a (massive) paper with the running head "The Mind of a Worm" (White et al., 1986; figure 3.1). But although this connectome has been known for more than three decades, we still do not know how this relatively simple nervous system works.

In this chapter, I argue that the key limitation of bottom-up approaches is not a lack of data, but a neglect of a fundamental property of life: life is organized, and more precisely self-organized. Living beings are *organisms*, not just structured arrangements of molecules, like crystals. An organization is a particular arrangement of components and processes that achieves something. For example, a work organization might ensure the production of some goods. A living organization ensures that the organism lives. This is a fundamental constraint on the arrangement of its processes, which has been preserved for more than 4 billion years of evolution.

I will start by pointing out that there is no purely "data-driven" science, because all measurements have a theoretical background (*The Myth of Data-Driven Science*). In particular, if the modeling approach is to measure the mean properties of neurons, then it is implicitly assumed that the variability observed among neurons is noise, rather than organization. But organisms are organized (*Living Beings Are Organized*): properties of elements at the "bottom" are not independent, but mutually constrained by the logic of the organism, which lies at the "top." Furthermore, living beings are *self*-organized: their organization develops, rather than being designed and assembled. The fact of development implies that parts do not preexist to the whole: "top" and "bottom" are co-determined. Crucially, the "top" is not defined in terms of structure, but in terms of interactions with the environment (*Brain, Body, and the Environment*). Indeed, a living organization is fundamentally coupled to the environment: living beings exist out of thermodynamic equilibrium, and therefore must exchange energy and matter with the environment in order to maintain themselves. All these facts imply that the brain must be approached as a system—namely, an organization of processes coupled to the body and environment. I will close the chapter on some general notes about scientific explanation (*Science beyond Billiard-Ball Causality*). Contrary to widespread belief, science is not fundamentally reductionist. Physical phenomena are rarely explained in terms of sequences of microscopic events (like billiard balls), but more often in terms of constraints—this is what equations

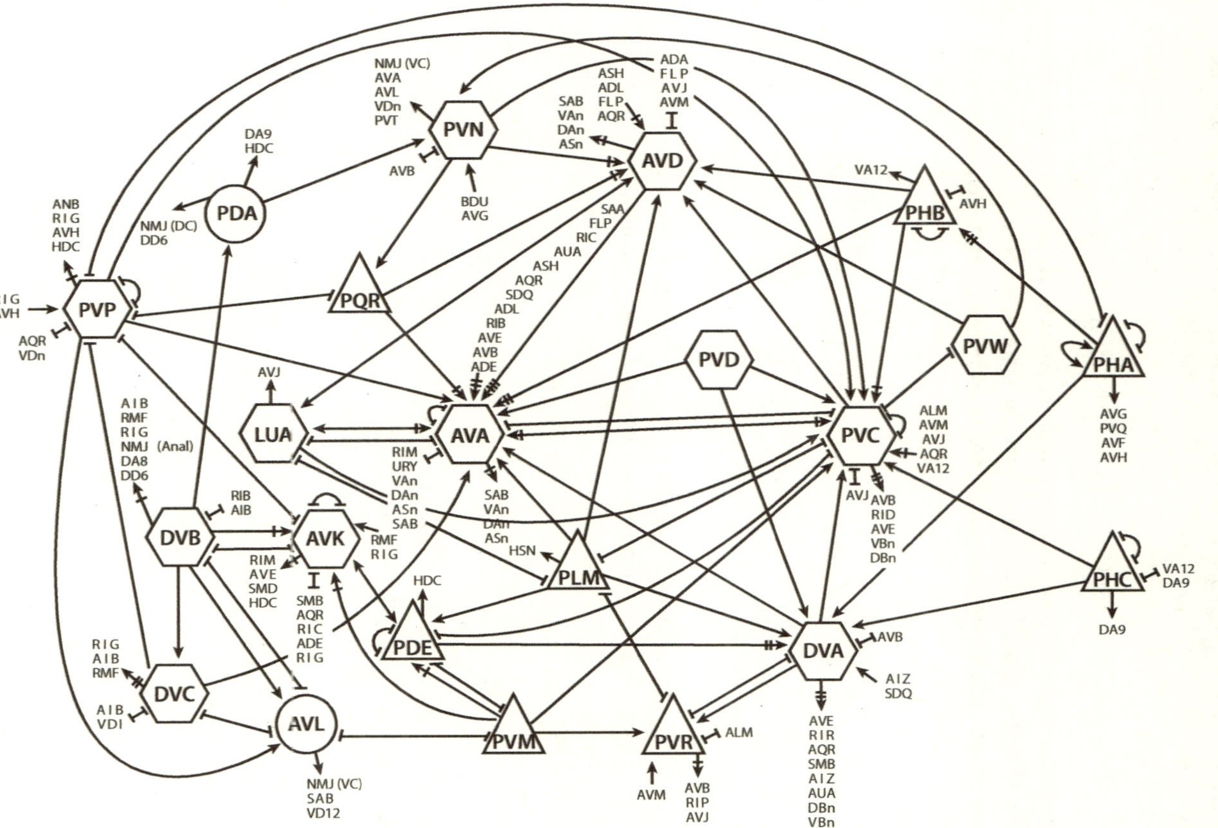

FIGURE 3.1. A small subset of the connectome of *C. elegans*. Used with permission of the Royal Society (UK) from White JG, Southgate E, Thomson JN, Brenner S. 1986. The structure of the nervous system of the nematode Caenorhabditis elegans. *Philos Trans R Soc Lond B Biol Sci* **314**(1165). Permission conveyed through Copyright Clearance Center, Inc.

represent. There are many kinds of constraints, some are local and others are more global. In biology, some important constraints are normative: biology is the science of living organisms, not just of regular arrangements of organic molecules.

The Myth of Data-Driven Science

Since living matter follows the same physical laws as inert matter, it seems that a valid strategy would be to characterize the structure of the brain at a sufficient level of detail, perhaps extract some regularities, and then derive the cognitive properties of this system by analysis or simulation. In a way, there is some humility in this empirical approach: the brain is very complex, we do not know how it works, so we propose to start by looking carefully and systematically at the data, without preconception about how the brain *should* work—in contrast with computationalism, as we will see in more detail in the next chapters. This is the promise of "data-driven science": that it is not biased by prior theory. Indeed, isn't the primary goal of any scientific theory to account for the data?

However, difficulties arise when we try to define more precisely what counts as "data." In bottom-up approaches, this amounts to defining what the "bottom" is. In practice, this level is defined by technical constraints: a structural level where systematic measurement seems feasible.

What Does an Average Brain Think About?

For example, in the Human Brain Project, the idea was to measure statistics of synaptic connections between neurons at a coarse level, rather than the detailed "wiring diagram," as well as typical electrophysiological properties of various types of neurons. This approach is motivated by technical constraints, but also justified by theoretical arguments. One argument was that, empirically, connections appear random at the microscopic scale. But of course, this only means that we failed to find systematic laws of connectivity with the kind of measurements and analyses that we used. Therefore, this justification is circular: we make statistical measurements because we could not see any structure, but to find structure we might need detailed measurements.

Highly structured patterns can easily appear random. For example, the digits of π also appear random (figure 3.2), and it is even conjectured that all sequences of digits are equally distributed (the frequency of each digit is 1/10, the frequency of each pair of digits is 1/100, and so on). Suppose we picked various consecutive sequences of ten digits of π. Even if we examined them very carefully, we would never find any regularity: all possible patterns are

3.14159265358979323846264338327950288419716939937510
5820974944592307816406286208998628034825342117067982
1480865132823066470938446095505822317253594081284811
1745028410270193852110555964462294895493038196442881
0975665933446128475648233786783165271201909145648566
9234603486104543266482133936072602491412737245870066
0631558817488152092096282925409171536436789259036001
1330530548820466521384146951941511609433057270365 7...

FIGURE 3.2. The digits of π appear random.

equiprobable. Yet π is the ratio of perimeter to diameter of a circle, and its sequence of digits can be computed exactly by an algorithm. Therefore, appearance of randomness does not necessarily mean an absence of organization. In this case, "random" is another word for "mysterious."

Thus, taking the statistics of connectivity to be the "bottom" is not empirically grounded. It is a particular theoretical commitment, that the variability of synaptic connections and neural properties is noise around an average, rather than organization. Is it plausible? We will examine this question in more detail in the next section, but we can already note that there is a theoretical tension between the idea that the structure of the brain determines the mind and the assumption that only statistics of brain structure matter. What is the mind of the average brain that is being modeled and simulated? What is its personality, and what are its memories? Does the average brain remember its average childhood, and what is its profession? Letting aside the problem of minds, what behavior would the average brain display? Could it ride a bike? It seems obvious that, whether we are talking about mind or behavior, the individual history of the organism reflected in the structure of its brain matters, and this history disappears if we take inter-individual statistics.

Radical Connectionism Is False

For these reasons, other approaches focus on the measurement of the entire set of individual connections between all neurons, the *connectome* (Seung, 2012). But why focus on this level, rather than, say, the individual morphology of neurons? Again, this is because of a particular theoretical commitment—namely, *connectionism*. As we have seen in chapter 1, connectionism is one of the major theoretical frameworks in neuroscience. It stipulates that what determines brain function is the connections between essentially stereotypical neurons. Thus, the physiological basis of learning is changes in connection strength. In its most radical form, behavior, memory, and experience are all determined by the connectome. Indeed, when the connectome of *C. elegans* was published several decades ago, the running head was "The Mind of a Worm." Yet, as we have mentioned, the relation between the neuroanatomy of *C. elegans* and its behavior (let alone mind) remains largely unknown. Why?

Beyond the technical difficulties, such as the issue of identifying the polarity of synapses (excitatory or inhibitory), we know that radical connectionism is false. That is, behavior is not in fact specified by the connectome alone. We know from invertebrate studies that the behavior of a nervous system is fundamentally changed by neuromodulators, which also exist in mammals including humans (Brezina, 2010). Comparative studies show that structurally very similar circuits can serve different functions in different related species (Katz, 2007). Conversely, similar behaviors in related species can engage different patterns of neural activity (Sakurai et al., 2011). In fact, some protists such as the ciliate *Paramecium* control their motility with action potentials, with molecular and genetic mechanisms similar to neurons (Brette, 2021), and can display a variety of behaviors despite having no connectome at all, being unicellular organisms. These facts should not come as a surprise, since we know that many structural aspects other than the strength of synaptic connections have an effect on neural dynamics: the expression of ionic channels, their phosphorylation, conduction delays, morphological changes in dendrites and axons, and so on. There are also ephaptic interactions, which mediate electrical interactions between neurons without any synapses at all (Anastassiou et al., 2011). Glial cells, once thought to be involved only in metabolic support, might also be involved in cognitive aspects of brain function (Araque and Navarrete, 2010).

In brief, connectomics-based modeling is not data-driven, but driven by connectionist theory, and radical connectionism is false. But, in principle, we could imagine that the structure of brains could be measured at the molecular level and then simulated with advanced technologies—this is the idea behind mind uploading fantasies. Thus, a bottom-up approach is at least theoretically possible, if not practically achievable. Or is it?

Measurements Are Interactions, Not Abstract Mappings

The idea that one could measure the state of all molecules of an object and then apply the laws of physics to predict their future state comes from Newtonian mythology: if the laws of all objects are described and their initial conditions are measured, then it should be possible to deduce the behavior of the system. Take the example of a gas, a collection of molecules that bump into each other. If one knows the position and moment of each molecule, then in principle it is possible to deduce the future trajectories of all molecules and their collisions, in other words, the complete behavior of the system.

However, this billiard ball model is a seriously outdated view dating back to the seventeenth century, which has been shattered in the early twentieth century by quantum physics, as clearly explained by physicist Arthur March in *The New World of Physics* (March, 1962). This might have been quite new in 1962, but not in the twenty-first century.

It is a mythological view because no one has ever measured the positions and moments of the particles of any gas. For a good reason: this is impossible, not just technically, but in principle. Heisenberg's uncertainty principle states that the position and moment of a particle cannot be simultaneously measured with high precision. Fundamentally, this is because a measurement is an interaction between two physical systems, not just a formal mapping from reality to model. When we immerse a thermometer in a bath, the reading is slightly biased by the initial temperature of the thermometer. We usually ignore this small bias when measuring macroscopic systems, but at the molecular level, the measurement is a strong interaction, to the point that it both limits precision and changes the measured system. In several texts addressed to biologists, Niels Bohr pointed out an important epistemological consequence of his discoveries in quantum physics, namely that the knowledge we can obtain about living beings is fundamentally limited by the fact that the act of atomic-scale measurement is incompatible with life—it kills the subject (Bohr, 1958).

Indeed, electrophysiological properties of neurons are measured in brain slices, usually at a temperature lower than the body temperature; detailed connectomes are measured by electron microscopy after fixing the brain with chemicals and cutting it into thin slices. Of course, there is considerable progress in measurement techniques, and one can expect that less intrusive techniques will be developed in the future. But there is a limit that is not practical but fundamental: the "bottom" cannot be precisely measured, which is why there is no such thing as purely "data-driven" science.

Models Are Laws Plus Constraints

Theories are not *driven* by data. When we say that the behavior of a system can be inferred from the current state of all its elements, we are merely describing the properties of a formal model, not the possibilities of prediction enabled by an act of measurement. Historically, the macroscopic properties of gases (the "top") were established before Boltzmann worked out a way to deduce them from microscopic properties, with statistical mechanics. Importantly, statistical mechanics does not rely on *measuring* the positions and moments of particles, since this is impossible. In reality, statistical mechanics starts from a theoretical description of a gas—that is, assumptions about the distribution of particle properties and their independence.

More generally, in condensed matter physics, macroscopic properties are not directly deduced from microscopic theories. Rather, macroscopic theories are made *consistent* with microscopic theories, by the addition of auxiliary assumptions—that is, of models (Drossel, 2021). Thus, even though the models might be expressed in terms of microscopic objects and laws, they are built by making the "bottom" and the "top" meet.

Thus, in scientific practice, models are established from two types of constraints: the general laws of physics, the fundamental ones being microscopic, and constraints about the object of study, which operate at the scale of the object. Therefore, even when a macroscopic phenomenon is said to be "reduced" to fundamental laws, modeling practice is never entirely bottom-up: there are fundamental microscopic laws as well as macroscopic constraints. The very idea of reductionism is itself highly fallacious: of course, all physical phenomena cannot be reduced to the same few fundamental laws, since they are different phenomena. They can be reduced to the same general laws plus specific constraints. Often, these constraints are inferred from theories established at the scale of the object of interest.

In statistical approaches to brain modeling, the constraints are statistical measurements together with the theoretical assumption that neurons are random variations around templates. In connectomics-based approaches, the constraints are neuroanatomical measurements together with the theoretical assumption of radical connectionism. But these theoretical assumptions do not come from empirical observations. In fact, they contradict a fundamental macroscopic constraint that we know from observing organisms, which is the subject of physiology: living organisms are organized.

Living Beings Are Organized

How to Model a Neuron

Organisms are organized, and a single neuron is organized too. By this, I mean that the microscopic components of the cell are arranged in such a way that it ensures certain macroscopic properties of importance, at the cellular level. For example, the arrangement of ionic channels in the membrane is such that it ensures the generation and propagation of action potentials. To see what this implies for neural modeling, we will now look at how neuron models are built in practice. Cellular neurophysiology is certainly the field of neuroscience where quantitative models have been most successful, starting with the pioneer work of Hodgkin, Huxley, and colleagues (Hodgkin and Huxley, 1952), who established the biophysical basis of action potentials. They also built a

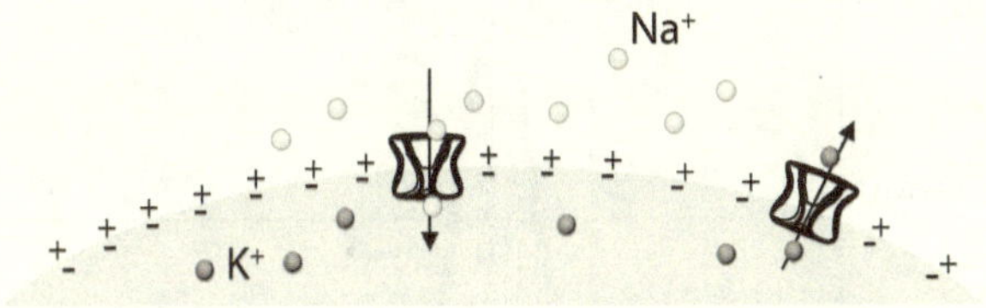

FIGURE 3.3. Ionic basis of neuronal electricity. The left channel lets sodium (Na^+) ions enter the cell (gray); the right channel lets potassium (K^+) ions exit. As the membrane is normally mostly permeable to K^+ ions, a charge imbalance located at the insulating membrane develops, constituting the negative resting potential.

quantitative model of action potentials, taking the form of a set of differential equations.

Neurons produce and transmit electrical signals by the opening and closing of various kinds of ionic channels in their membrane (figure 3.3). Ionic channels are proteins in the membrane that can let specific types of ions pass through a pore when they are open. The main types of channels are sodium channels, potassium channels, and calcium channels. The extracellular medium has roughly the ionic composition of seawater—that is, mainly sodium and chloride ions. The intracellular medium has mainly potassium ions. Thus, when a sodium channel opens, sodium ions flow into the cell by diffusion, creating a positive current. When a potassium channel opens, potassium ions flow outside the cell, creating a negative current. As ions accumulate, a misbalance of charge builds up between the two sides of the membrane, creating a difference in electrical potential called the *membrane potential*, which counteracts further ionic diffusion. At rest, the membrane is mainly permeable to potassium, which creates a negative membrane potential (typically about −70 mV in neurons).

The action potential is a transient change in the membrane potential due to the opening of sodium and potassium channels (figure 3.4). These channels are *voltage-gated*: they open when the membrane potential is high enough. Thus, when the neuron is stimulated by various means and its membrane potential increases (e.g., when it receives an input from another neuron), sodium channels open, letting sodium ions flow into the cell. The membrane potential then increases further, which opens more sodium channels, leading to further potential increase: this is a positive feedback loop, explaining the explosive nature of the action potential. Then potassium channels open, but with a delay, letting potassium ions leave the cell, and this restores the resting potential. The

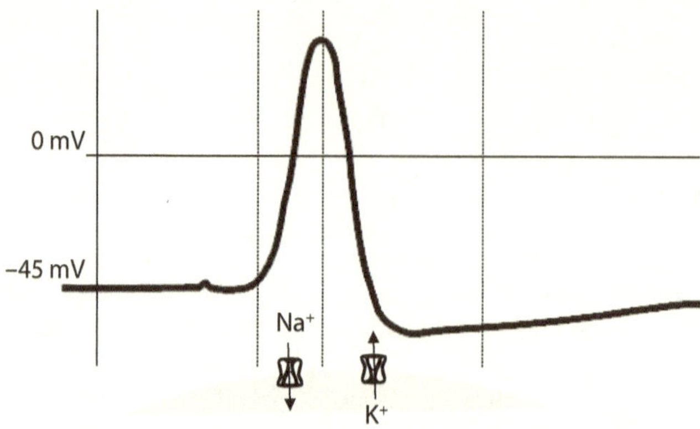

FIGURE 3.4. Action potential recorded in the squid giant axon (membrane potential as a function of time; adapted from Hodgkin and Huxley, 1952). The entire trace lasts about 5 ms.

result is a transient electrical spike that lasts for a couple of milliseconds. In axons, this spike propagates by the progressive opening of neighboring sodium channels. Thus, proper electrical function depends on the spatial arrangement as well as the detailed kinetic properties of channels.

Based on electrophysiological measurements in the squid giant axon, Hodgkin and Huxley built a mathematical model of the action potential, including sodium and potassium currents, which was able to reproduce the electrical response of the membrane to currents, and to correctly predict the propagation velocity of the action potential along the axon. Since then, the general biophysical basis of excitability was found to be universal across species (Hille, 2001), including plants and unicellular organisms, but with diverse structural variations in the ions and channels involved. Many studies have then characterized the structure and properties of single ionic channels.

How does one build a functional neuron model? A naïve mathematician starting in theoretical neuroscience might think that parameters of a model are directly measured, like we measure temperature and geometry. But parameter measurement is generally much more indirect. Hodgkin and Huxley used a very particular preparation, the squid giant axon (figure 3.5), because it is so large (about 1 mm in diameter) that one can insert a wire in it, making the axon isopotential (no spatial variation in membrane potential). The squid giant axon is in fact one of the exceptions to the neuron doctrine, according to which the nervous system is made of discrete individual cells. The giant axon is a syncytium resulting from the fusion of hundreds of cells (Young, 1939). They could

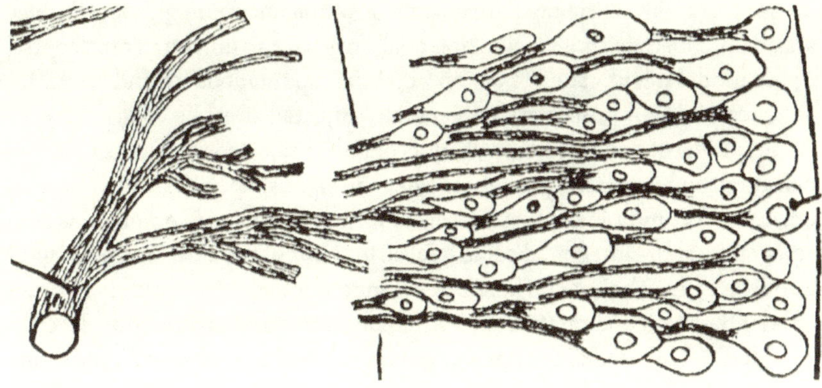

FIGURE 3.5. The squid giant axon of Sepia, a cephalopod, results from the fusion of many cells. Used with permission of the Royal Society (UK) from Young JZ. 1939. Fused neurons and synaptic contacts in the giant nerve fibres of cephalopods. *Philos Trans R Soc Lond B Biol Sci* **229**(564). Permission conveyed through Copyright Clearance Center, Inc.

then manipulate the ionic solutions both inside and outside the axon, while recording the electrical currents through the membrane. For example, they could remove sodium from the extracellular medium, so as to isolate the potassium current. Needless to say, this is a highly invasive procedure.

In contrast with the squid giant axon, neurons are complex cells with treelike morphology and dozens of ionic channels of various types expressed at various densities depending on their subcellular location. Measuring the properties of a single channel type typically involves replacing the liquid content of the cell (the cytosol) with a solution prepared by the experimenter, and blocking the other channels with a cocktail of specific molecules. Since this is a highly invasive procedure, this is normally done not in a living animal but in a brain slice, usually not at physiological temperature, but at lower temperature.

Naturally, this cannot be done for each neuron of the same brain, and therefore statistics are collected from a few representative neurons. But even for a single neuron, it is not possible to measure the properties of each type of channel at each subcellular position in the same neuron, because the measurement of a given channel type involves pharmacologically blocking other types. For these reasons, detailed biophysical neuron models used in theoretical neuroscience include ion channel models collated from a variety of studies, often in different species, different cell types and brain areas, and different experimental conditions—which we might call "Frankenstein models."

The interesting fact is that these Frankenstein models are not simply collages of models of the various components, because in general, such collages do not work "out of the box." That is, the resulting model might not produce action potentials, or those action potentials may have an unnatural time course, or the model might generate action potentials spontaneously, and so on. Instead, component properties are generally adjusted after the collage. For example, properties of the sodium channel might be adjusted so that the modeled neuron produces action potentials, with the right shape. Thus, the model is obtained by combining measurements of various subcellular components with constraints at the integrated cellular level, in terms of the electrophysiological response of the whole cell. Therefore, the methodology is not entirely bottom-up: it integrates constraints from both microscopic and macroscopic levels.

Other neural modeling approaches start from a parameterized model that includes the known channel types, and the parameters are tuned so that the model correctly predicts the electrical responses to currents. This procedure can lead to satisfying results (Naud et al., 2014; Rossant et al., 2010), but again in this case, parameters are not measured but constrained by the integrated level—in fact, the most successful models are phenomenological models, meaning that parameters have no direct biophysical meaning. Whether using parameter fitting techniques or adjusting measurements to make the model "work," these methods of building functional neural models attest the failure of a purely bottom-up strategy. The reason for this failure is that the electrical behavior of the neuron depends on relations between the different components (properties of different channels, neuron morphology), which are not directly measured. These relations cannot be captured by measuring the various components independently in different neurons, because neurons are structurally different: they are individuals.

Neurons Are Different Individuals

Scientists trained in mathematics, computer science, or physics tend to approach biology with a platonic view of cells: cells are instances of an ideal form, like atoms or molecules. This view is reinforced by the fact that cells of animals are clones (with some differentiation into cell types). But even twins are different individuals, with their own thoughts, tastes, and lives, especially if they grow up separated.

This individuality is obvious when looking at the morphology of neurons. Figure 3.6 shows motoneurons in the ray spinal cord stained by Ramón y Cajal (1899) using the Golgi method. Although they are morphologically similar, their detailed shape is diverse. It is easy to underestimate the complexity of a cell when wrongly thinking of its components as electronic components with fixed specifications. Consider for example the sodium channel, the "trigger"

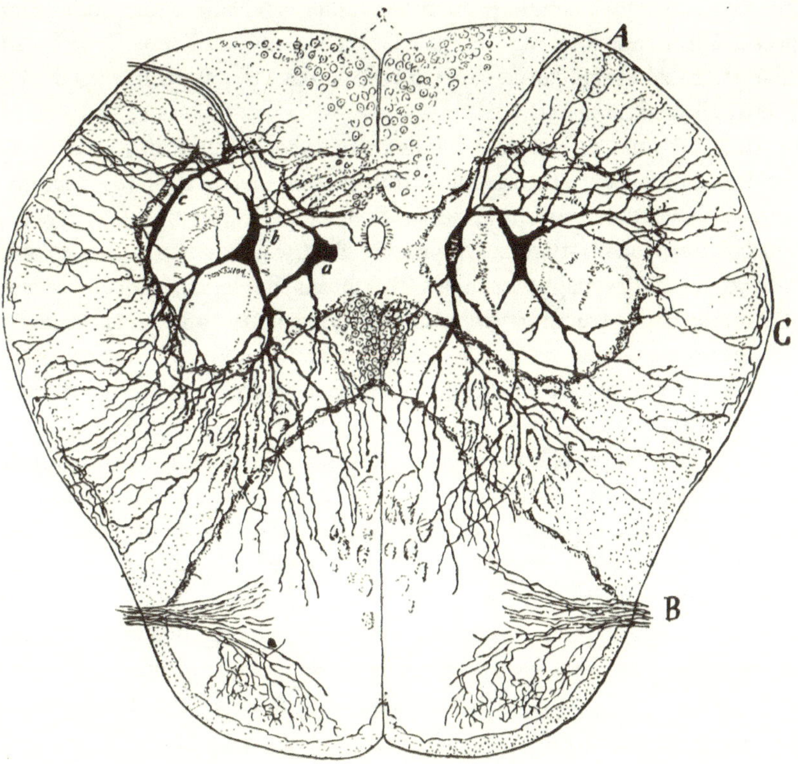

FIGURE 3.6. Transverse section of the ray spinal cord (from Ramón y Cajal, 1899).

of action potentials. The sodium channel is in fact not a single protein, but an assemblage of different proteins: the main unit, called the alpha subunit, of which there are nine types and many more subtypes, and a number of beta-units. Those beta-units may or may not be present, changing the properties of the channel. The alpha subunits can also take different forms, by alternate splicing. More importantly, their properties can be modulated by many cell processes—for example, they can be phosphorylated by enzymes, which can change every aspect of their kinetics. All these things depend not only on cell type (species, brain area, etc.) but on the individual cell, and they can vary both over time and with subcellular location. This variation arises from plasticity processes (gene expression, modulation by enzymes, etc.), and from developmental processes. For example, neural properties may depend on the cell's position on the cortex (e.g., in relation with tonotopy in the auditory system [Kuba et al., 2005]). Variations in different neural properties are not independent—for example, properties of different channels (e.g., sodium and potassium channels) co-vary across cells (McAnelly and Zakon, 2000)—and channel distributions can co-vary with subtle aspects of the cell's morphology (Goethals et al., 2021).

Thus, neurons have diverse detailed structures because they have diverse histories, not because the mapping from genetic specification to structure is noisy—this detailed specification does not exist. Despite this diversity, neurons are functional—they can respond to electrical signals and propagate action potentials to synaptic terminals. This means that cellular processes ensure that microscopic structures vary while satisfying global constraints. Therefore, in order to model neurons, it is necessary to pay attention to those global constraints, and perhaps to the processes that enforce them, and not just to the structural aspects.

A simple example of processes ensuring global constraints is homeostasis. In neurons, the resting potential as well as the firing rate are regulated to remain within certain bounds. There is a significant body of theoretical work on this subject, taking inspiration from control theory (LeMasson et al., 1993; Marder and Goaillard, 2006; O'Leary et al., 2014). Although it is fair to say that these homeostatic processes remain poorly known, this line of work shows that such constraints imply relations between the different structural elements (in particular, ionic channels), while allowing for large individual variations among neurons.

A more complex example is the initiation of action potentials. In most vertebrate neurons, action potentials initiate in a small structure on the axon called the axonal initial segment, near the cell body. It turns out that this structure can move and shrink, with development and activity (Grubb et al., 2011). The action potential propagates from the initial segment in both directions, forward to the axonal terminals and backward to the cell body and dendrites. Theory shows that, in order for backpropagation to be possible, the initial segment must be closer to the cell body if the neuron is larger. Empirically, this relation is indeed observed (Goethals et al., 2021; Hamada et al., 2016). The cellular process that ensures this constraint is unknown.

Thus, there is a global logic in the relations between structural elements, which disappears when we examine the elements individually. The idea that there are global properties that cannot be captured by looking at local properties is neither new nor specific to biology. In mathematics, it is the subject of topology. For example, a Möbius strip is a surface with a single side (figure 3.7). Locally, it is indistinguishable from a two-sided infinite strip. But it has the property that any two points on the surface can be connected with a curve that does not cross the edge.

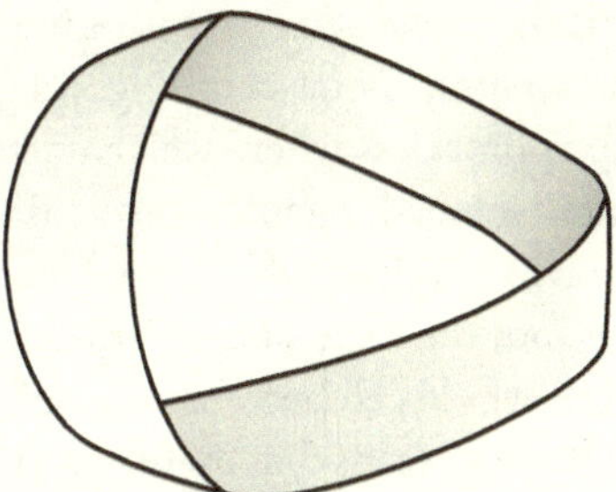

FIGURE 3.7. The Möbius strip is defined by a global topological property, not by local properties.

Individual neurons are organized, and brains are organized too. If a given behavior engages a large part of the brain, then there must be certain structural relations between neurons on a global scale. To capture these relations, one must then consider the individual behavior as a modeling constraint, which again bars a purely bottom-up approach.

Where do these global properties come from? As we have seen in chapter 2, the organization of a living being is the expression of its autonomy: each individual has its own way of working, which is constrained by genetics but also depends on its own individual history and environment—what neurobiology calls "plasticity." In the same way, each brain has its own organization. This is why the logic of living cannot be captured by statistics of structural measurements.

In this respect, living organisms differ from machines. Living beings are not just organized: they are self-organized.

Living Organisms Are Self-Organized

Machines are made from parts that have an independent existence and specifications, and the parts are then assembled into a functioning machine. But the parts of a living being have no independent existence; they are not assembled into living organisms: they develop. As we have seen in chapter 2, each living cell is continuous with a previous cell, recursively over 4 billion years: cells are not assembled, and neither are multicellular organisms. A living organization develops, rather than being designed and assembled. The fact of development implies that parts do not preexist to the whole: "top" and "bottom" are codetermined. This is another great challenge to bottom-up approaches: the bottom does not exist independently of the top.

When building a machine, we start by gathering its parts, according to specifications, then we assemble them according to a global plan. This logic is mirrored in bottom-up approaches where we characterize neurons, then we look at how neurons are arranged together. It seems legitimate to start with parts and proceed with wholes, since wholes are composed of parts. But this is the logic of machines, not the logic of life. A multicellular organism is not assembled: it develops. A whole cell becomes a whole embryo, then a whole child, then a whole adult. This means that there is always a whole before there can be parts.

One particular implication of development is that, since parts are produced by wholes, properties of parts depend on properties of the whole. That is, properties of microscopic elements depend on events and properties defined at a more macroscopic level. This phenomenon is a fundamental challenge to bottom-up approaches known as "downward causation" (Campbell, 1974; Noble, 2012), which we have discussed in chapter 2. Thus, there is bidirectional causality between cellular and organismic levels. It follows that properties of the "bottom" depend on properties of the "top." But what is the "top" exactly?

Brain, Body, and Environment

Brain Structure Does Not Determine Behavior or Cognition

In bottom-up approaches, the bottom is the microscopic structure of brains while the top is cognition and behavior. The problem with this terminology is that cognition and behavior are not functions of brain structure, but rather properties of the coupling of the brain with the body and environment. In other words, cognition is embodied (Gallagher, 2006). For this reason, one cannot go "up" from the structure of brains to cognition and behavior, because the "top" does not actually sit on the "bottom."

Of course, no one denies that brains are coupled with bodies that act in environments. But the consequences for brain modeling tend to be underappreciated. I will start by examining the problem of determining the function of a neural circuit from its connectome. If we are to believe radical connectionists, the connectome determines who we are (Seung, 2012). For example, can we understand how we identify faces from the detailed wiring of the visual cortex? This perceptual ability is generally framed as the transformation of an image into a face label. The first issue we face is that we need to relate the activity of neurons to images and face labels, none of which are in the domain of spikes. The way this is typically done in computational neuroscience is to define encodings: there is a mapping E (encoding) from the image to the neural network input, and another mapping D (decoding) from the state of some

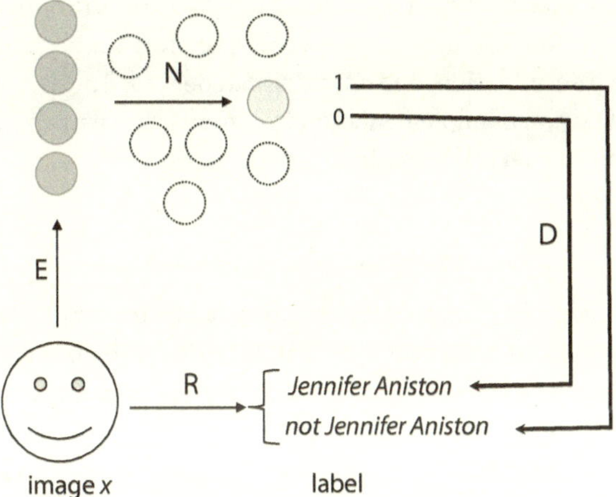

FIGURE 3.8. Implementing a Jennifer Aniston recognizer with an arbitrary connectome.

neurons (the "output neurons") to the face label. In this way, the neural network can be seen to implement a transformation of the image into a face label. Crucially, the encoding and decoding mappings are not in the connectome. The problem is that almost any connectome is compatible with almost any input–output transformation, provided we define the right encodings.

To see this, suppose that we have mapped a nervous system and assigned a subset of its neurons to inputs, and we look at the output of one neuron (figure 3.8). To simplify, we assume that the neurons have a binary activity (0/1). This defines a binary neural mapping N from inputs to output. Unless this mapping is trivial (always the same output), it can be seen to implement the recognition of an image of Jennifer Aniston.

Here is a mathematical proof. We consider that the network recognizes Jennifer Aniston when its output is 1 ($D(1)$ = Jennifer Aniston; $D(0)$ = not Jennifer Aniston). Let R be an image recognition system that maps an image to the binary label Jennifer Aniston/not Jennifer Aniston. We then define the encoding of images to the neural network as follows. We choose an input I_1 such that $N(I_1) = 1$—that is, identified to Jennifer Aniston—and an input I_0 such that $N(I_0) = 0$. Then for every image x, we define $E(x) = I_1$ if $R(x)$ = Jennifer Aniston; otherwise, $E(x) = I_0$. We have chosen an encoding and a decoding such that $R(x) = D(N(E(x)))$—that is, such that the neural network implements the recognition of Jennifer Aniston.

In a way, we have "cheated": all the work is actually done by the encoding E, rather than by the neural network. But this is precisely the issue: unless we consider the actual coupling between the network and the body and environment, and therefore not just brain structure, the function of the network depends on an interpretation, and anything can be said. This is a critical flaw of the standard notion of implementation in computational neuroscience, which we will encounter several times in this book.

The Brain Does Not Receive Images

At first sight, it might seem that the problem can be trivially solved: the retina receives images from the world, and thus, there is an obvious encoding of images in the activity of retinal ganglion cells (the spiking output of the retina), like a camera. Unfortunately, this is false.

Connectionist models of image recognition mislead us into believing that the visual system processes images, encoded in patterns of pixels given as inputs. But images are the result of perception, not their input. In reality, the eyes actively scan the visual scene with movements on different timescales (from ~10 ms to seconds) that we are mostly unaware of, except for the largest abrupt movements or *saccades* (Martinez-Conde et al., 2004). The arrangement of photoreceptors on the retina is inhomogeneous, both in spatial resolution and color properties, and the eye has optical aberrations, also unperceived (O'Regan and Noë, 2001). Thus, the "input" to the visual system is not simple, and in particular it is not specified by the structure of the nervous system. More deeply, this "input" is not exactly an input: seeing is an activity that involves a closed-loop interaction between the nervous system and the visual scene, through the body (Ahissar and Assa, 2016).

Similarly, the brain does not have tactile "inputs." Rather, the person touches objects and it is the interaction between body and surfaces that is informative, as described by Gibson (1962). Even when the person is being touched, the mechanical stimulation is not local but propagated across a large area by the skin. Again, the "input" depends on nontrivial properties of the body, which are not included in brain maps.

Motoneuron Spikes Are Not Rigid Commands

In the same way, it is not at all obvious how neural activity results in an organism's motion. Muscle control is not simple at all. For example, muscles adapt to the neural stimulation on different timescales, and this is a very familiar phenomenon: on the timescale of weeks, this is called "exercising." Therefore, there is no fixed relation between a motoneuron's activity and its behavioral effect. In addition, as we have previously discussed in this chapter, the body mediates nonlocal interactions between distant neurons. For example, contracting a muscle stretches the antagonist muscle, so that sensory neurons innervating both muscles are activated. Therefore, there are interactions between neurons that are not captured by the connectome, or by any other structural map of the nervous system, because they are mediated by the body.

Behavior depends not only on the physiological properties of the body, but also on its posture. Clearly, the contraction of a leg muscle does not have the same effect if the person is standing or if it is sitting. Besides, this example shows that it is not only the position of the body that matters, but also its relation to the environment (e.g., the ground or the chair). Conversely, posture is changed by the activity of the nervous system. Specifically, posture is adjusted in anticipation of one's later movements (Le Mouel and Brette, 2017). For example, a runner places herself in starting blocks, so as to propel the body forward as fast as possible when the race starts. Another example: when facing a cliff, the body leans backward so as to avoid falling. But neither "leaning backward" nor "starting blocks" can be read in a connectome. More broadly, it is impossible to understand anticipation, a fundamental property of behavior, unless one explicitly considers the environment.

The Fundamental Coupling between Brains and Environment

As we have seen in chapter 2, there is a fundamental reason why brains cannot be studied as isolated objects. Living beings are precarious organizations that maintain themselves out of thermodynamic equilibrium by interacting in particular ways with the environment. This fact underlies all normative properties of living beings—in particular, behavior. Thus, organism and environment are fundamentally entangled: a living organization is relative to the environment, and what we call "the" environment depends on the organism. The organization of living beings and brains is to be understood with respect to this particular coupling, not as isolated objects.

Therefore, proper understanding of living phenomena requires taking a systemic approach (Krakauer et al., 2017; Noble, 2008), where the fundamental entities are systems rather than isolated objects, and these systems have normative constraints: they are organized so as to achieve something (namely, their continued existence). Yet, there is a widespread belief that scientific explanation is by nature reductionist, that natural phenomena must be ultimately explained by the local interactions of elementary objects. This is a very different kind of explanation. But as I mentioned, this belief is largely mythological. Scientific explanations, especially in physics, generally explain phenomena in terms of the operation of various kinds of constraints. (See in particular Juarrero [1998, 2002, 2023] for a discussion of causal constraints.)

Science Beyond Billiard-Ball Causality

Aristotle's Four Causes

The classical analysis of causality dates back to Aristotle, who categorized explanations of phenomena into four different "causes": material, efficient, formal, and final. For example, to explain how a chair comes to be, one might describe the chair as an object made of wood, which has been cut and assembled according to some technical processes, and arranged into some particular shape such that one can sit on it. Here, the material cause of the chair is wood: what the object is made of, the physical substrate of the phenomenon. The efficient cause is the processes underlying the phenomenon—in this case, the cutting and assembling processes. The formal cause is the pattern responsible for the phenomenon: the shape of the chair. The final cause is the goal: an object that allows one to sit.

Each of these causes is important for a theory of chairs. A chair cannot be made of water, or of fire: only some types of substances can be turned into chairs (material cause). But wood is not a chair, it only becomes a chair once it has been subjected to a number of processes, such as cutting and assembling (efficient cause). These processes must be such that they produce an object one can sit on (final cause). To this end, they must be organized in a particular way (formal cause). For example, there should be at least three legs, and a horizontal piece of a certain size; that piece should be on top of the legs, and not below. These causes are interrelated: processes (efficient cause) operate on certain kinds of substance (material cause), and are arranged in certain ways (formal cause), constrained by the function of the system (final cause).

Strikingly, this analysis appears most relevant for characterizing human artifacts. But a prominent feature of Aristotle's worldview was the emphasis he put on final cause to also explain all natural phenomena. For example, he believed that planets have a circular motion because of the perfection of circles. It is this distinctly teleological feature that has been subsequently rejected by the scientific worldview. Similarly, formal cause has been a controversial notion outside of engineering, because forms are mental objects, so to explain natural phenomena in terms of formal cause is to suppose they have been designed.

In ordinary conversations, efficient and final causes are especially prominent. If I say, "I pushed him and he fell," I can distinguish two causes. The efficient cause of his falling is the transfer of kinetic energy from my hand to his back. The final cause of his falling is my intention of making him fall. The scientific worldview rejects explanations in terms of intentions, especially when applied to nonmental phenomena (such as planets).

But science is not an ordinary conversation, and there is no reason why scientific explanations should follow the demands of folk causality. We do not find teleological explanations in science, but scientific explanations also rarely take the form of a narrative describing a sequence of interactions between objects. In fact, the kind of explanations we find in physics are often close to formal and even final causes, which are not particularly problematic once rephrased in terms of constraints, a reformulation that avoids the anthropomorphic aspects of Aristotle's original analysis.

Formal Cause as Constraint

What is a scientific law? Take for example the ideal gas law. It stipulates a particular quantitative relation between pressure, volume, and temperature, expressed by an equation ($PV = nRT$). The equation is formal: it exists in the mind of the scientist, not in the object. However, the equation is the formalization of a constraint, and this constraint is realized in the real world. Thus, Aristotle's formal cause becomes acceptable once we amend it appropriately: it is not the form that causes the phenomenon, but the constraint that the form expresses. Constraints do not require designers or minds.

When the scientist wants to predict what happens when you reduce the volume of a container at fixed temperature, she uses the constraint of the ideal gas law to infer that the pressure will increase. Of course, we might want to explain the phenomenon in terms of the collisions of atoms, but these are narratives rather than science.

Newton's laws are also formal expressions of constraints. For example, the second law says that the acceleration of an object times its mass equals the total force exerted on that object. From this constraint, we can infer the trajectory of an object, for example in a gravity field. Dynamics is secondary to the constraint on acceleration. The various conservation laws of physics are examples of general constraints.

To know whether an object would float or sink, we do not examine the motion of individual molecules in the fluid. Instead, we use Archimedes's principle, which stipulates that an immersed object is subjected to a force opposite to the weight of the fluid that the body displaces. This is a global relational constraint (relation between the object and its environment), and yet, physicists are generally at ease claiming that the buoyant force causes the object to move up or down in the water.

There certainly was, at some point, a hope that everything could be explained in terms of local mechanical events—the "materialist-mechanistic" program dating back to the seventeenth-century view of Huygens, as related by Arthur March (March, 1962). But this program failed, in particular with the development of electromagnetism and then of quantum physics. It is often said that quantum physics defies understanding, in particular that it contradicts our intuitions about causality, and yet the equations work perfectly and predict real phenomena. What this means is that the theory of quantum physics expresses constraints between measurements that cannot be fit into the narrative structure of folk causality. Sometimes scientific explanations can be cast as stories, but often they simply cannot.

Furthermore, as we have seen, it is inaccurate that all physical phenomena can be reduced to fundamental laws. In order for the same fundamental laws to produce diverse phenomena, they must be augmented with specific constraints. These constraints can be local, such as the initial position and velocity of a particle, or they can be global, such as the structure of an object (solid, liquid, etc.) or its global topology (e.g., whether it is a closed surface). Some of those constraints are even expressed in terms of the future.

Final Cause in Science

The problem with Aristotle's final cause is that it attributes mental properties, such as intentions, to entities that do not have minds, such as planets. But if we understand final cause more broadly, in terms of what phenomena tend to, one must realize that they are also prominent kinds of explanation in science, and not only in biology. For example, if we release a ball on top of a valley, we will say that its motion will tend toward a final state of equilibrium, which is at the bottom of the valley. To reach this conclusion, we might say that the ball should ultimately reach a minimum of free energy because of the second law of thermodynamics. In this case, it corresponds to the minimum kinetic energy (the ball is at rest) and the minimum potential energy of gravity (the bottom of the valley). The ball obviously has no intention, but the physical processes that govern the ball's motion are directed toward a final state of minimal free energy.

Of course, we could have calculated the full trajectory of the ball using Newton's laws, but this is clearly not the most convenient way, nor is it the most insightful explanation—"because it turns out to be the asymptotic state of the calculated trajectory" versus "because it is the lowest point." In fact, in realistic cases, this calculation would require a knowledge of friction with both air and ground, which one may not possess.

A more sophisticated example is a wire frame dipped into a soap solution. To determine the shape taken by the soap film, we do not simulate or analyze the behavior of all individual soap molecules as they come in contact with the wire. Rather, we use the fact that the soap film must minimize free energy, so that its surface has the smallest possible area with the constraint that it is bounded by the wire frame.

Thus, it is true that certain physical phenomena are best explained in terms of their final states, what they tend to. This is clearly the case also in biology. For example, behavior is best explained in terms of what it tends to achieve, rather than by a particular sequence of muscle twitches. In physiology, we explain systems by the result of their normal operation: the digestive system is organized in such a way that food is transformed into more usable forms of energy. This causal explanation works across species with very different digestive systems, from protists to humans.

But while explanations in terms of final causes are clearly useful, what is disturbing is that we explain phenomena in terms of events that are posterior to them. There are at least three ways to address this problem in the context of biology.

Final Cause in Biology

The first way, of course, is the notion of goals, taken literally: a cognitive entity makes a plan to achieve a certain outcome, consisting of a sequence of actions. In this case, the final cause of the actions is the outcome, and this outcome is ensured by the planning activity. Planning may apply to (part of) behavior, but not to, say, the function of the digestive system: no cognitive entity has designed the digestive system in such a way that it turns food into a usable source of chemical energy. Planning is the basis of teleology in machines (there is an intelligent designer), but this cannot apply to the organization of living organisms.

The second way is Darwin's theory of evolution. This is essentially what Monod referred to when he claimed that the "teleonomic project" of living organisms is invariant reproduction. Since feeding is necessary for organisms to survive, organisms with efficient digestive systems are selected. Therefore, the function of digestive systems is a final cause of their organization, which acts through the mechanisms of Darwinian evolution.

The third way is more subtle and derives from the closure of a living organization. A living being is an organization of out-of-equilibrium processes that maintains itself. In systems that satisfy this constraint, the end result of a process is also a condition for the continuation of the organization to which it belongs. In this sense, we can say that a living organization is oriented toward the future, namely its future existence. This point was defended in particular by theoretical biologist Robert Rosen (2005), who analyzed living organizations in terms of Aristotle's four causes and concluded that living beings are closed to efficient causation. It is in this sense that a part of an organism can be said to have a "function" for the organism: the part contributes in a specific way to the operational closure of the organism's organization. These are subtle concepts that we will examine further in chapter 7 when we discuss anticipation.

As we have noted, bottom-up approaches tend to present themselves as "data-driven," but with the reductionist mindset, what is considered as "data" is a set of measurements of various parts of the brain. This leaves out a crucial datum: a living organism lives. A living organism is an organization of processes—that is, a set of processes arranged in such a way that it "works." It is this normative property that allows one to assign functions to subsets of the living organization, to speak of pathologies, to attribute goals to the organism, and to speak of behavior—in contrast, the climate does not "behave" or "act," except metaphorically.

Bottom-up approaches face the difficult problem of specifying what "the bottom" should be: neural populations, single neurons, single molecules? Now we know why this is so difficult: when dealing with organizations, the right question is not which elements matter, but which *relations* and *processes* matter—that is, ensure that the system "works," and some of these relations involve more than the brain: they involve the organism in its environment.

Understanding brains requires caring about their organization, in particular what it means for the brain to "work." It is this realization that is at the origin of the most popular conceptual framework to think about brains: brains as biological computers.

4

Biological Computers

IN POPULAR culture as well as in the scientific and philosophical literature, the brain is commonly described as a sort of biological computer shaped by evolution. The brain processes information, it implements algorithms, it computes. One speaks of neural codes, hardware and software, reverse engineering the brain, and so on. However, this lexical field covers a constellation of related ideas and theories. In many cases, these words are used metaphorically, as in "reading and writing the neural code," where "writing" is shorthand for "stimulating" (Stanley, 2013). In other cases, they refer to precise theoretical concepts, but there are different computational theories of the brain. For example, classical computationalism at the root of cognitivism sees cognition as the manipulation of symbols, in line with rule-based artificial intelligence systems ("good old-fashioned artificial intelligence," or GOFAI), whereas connectionism sees the dynamics of neurons as calculations, but not necessarily attached to mental symbols, except for inputs and/or outputs.

Thus, there are different computational concepts that need to be disentangled and discussed separately. First, a computer is a programmable machine (Brette, 2022a). As we have seen, living organisms are organized, but otherwise they are not machine-like. There is also no useful sense in which brains are programmable (*Programmable Machines*). But do brains compute? Here we need to be more explicit about what we mean by "compute" (figure 4.1; *Programs, Algorithms, Computations, and Beyond*). If we mean that some behavior can be described as programs in the broad sense of a sequence of predetermined actions, as in a "concert program," then that is certainly often the case. If we mean that we can follow algorithms, in the broad sense of a sequence of fully specified elementary actions to solve a problem, then that is sometimes the case, as when we follow a cooking recipe. But if we mean computation in the sense of computer science, then we are referring to something much more particular, which operates in the formal domain with a finite set of rules. What we call "information processing," the transformation of input data into output data, includes computation but also noncomputational ways of solving data problems. When

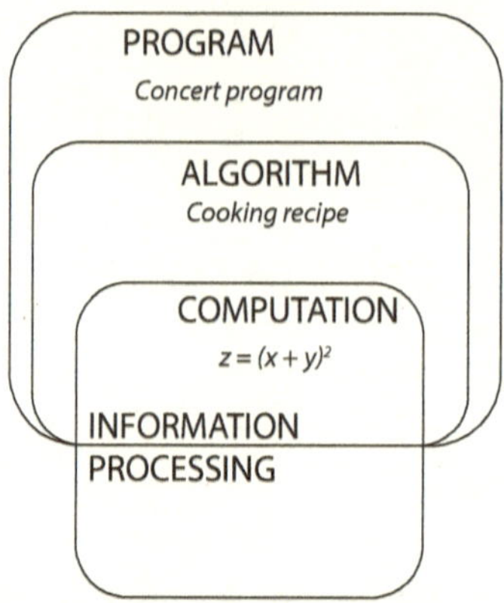

FIGURE 4.1. Program, algorithm, computation, and information processing in the common sense of the terms, all underlying different meanings of "compute" as used in brain theory.

we examine the normal behavioral repertoire of organisms, we find that a large part is interactional and dynamic, including perceptual processes (*Behavior as Interaction*). Neither dynamics (*Why Dynamics Is Not Computation*) nor interaction (*Why Interaction Is Not Computation*) can be meaningfully characterized as computation, unless by "computation" we mean "anything that solves a problem." This undermines computational reductionism, the claim that all of cognition is ultimately decomposable into cognitive atoms, which are elementary computations. Nonetheless, an important part of behavior does not rely on interaction, and it is especially in those cases that neurons are said to "compute." Unfortunately, we will see that the classical concept of "neural computation," by which brain processes are said to implement a computational model, is flawed beyond repair (*Neural Computation*).

Programmable Machines

Brains as Machines

We start with the concept of machine, since a computer is a kind of machine. What is it about brains that is machine-like? As we have seen in chapters 2 and 3, living organisms and machines are organized, which means that their structure is arranged in such a way as to fulfil a goal.

Thus, a desirable feature of the machine view is that it acknowledges the normativity of living entities. A machine can be broken; a brain can be pathological. A machine can have a bug or make errors; actions of an organism can fail or succeed. Breaking a rock in two parts makes two rocks, but breaking a machine or a brain in two parts gives a broken machine or a dead (or at least dysfunctional) brain. Both machines and brains are organizations: arrangement of processes together with norms.

A machine is primarily defined by its final cause (function) and its formal cause (abstract organization), and only secondarily by the specific processes and substrates (efficient/material causes) that *implement* the functional specification. Thus, in contrast with the reductionism of bottom-up approaches, the machine view offers an account of all four causes of Aristotle, and a particular way to articulate them—namely, the concept of *implementation*. Thinking of brain structure in terms of implementation means that to understand brains, one should start by describing cognitive function at a detailed conceptual level, then work out how this function might be "implemented" by circuits of neurons. This approach is often described as "top-down" or "principle-based," and was famously defended by David Marr (Marr, 1982), which we will discuss further later.

Indeed, machines fit very well with Aristotle's account of causality. However, this is an issue because that account is teleological and anthropomorphic, as we have seen in the previous chapter. Final and formal causes are primary in Aristotle's worldview because that view is modeled on how humans build artifacts: first define the function of the object (final), then think of the logic of the object (formal), then build it (efficient/material). That is not how a living organism is made—that is, it is not made.

This anthropomorphic view tends to introduce prejudices about what brains are supposed to achieve—what David Marr called the "computational level." For example, the visual system might try to compute the three-dimensional layout of objects from the two-dimensional images projected on the retinae. This way of phrasing the problem of perception is typical of what Thompson et al. (1992) called "computational objectivism." The outside world can be described objectively, independently of the organism, for example with the laws of physics, and the organism receives imperfect projections of this world on its sensory organs—much like the shadows in Plato's cave. According to computational objectivism, perception is then an inverse problem (or an inference problem), which consists in recovering the objective description from the sensory inputs. But this objective description is only primary for the engineer who makes the machine, not for the organism. Given that an organism follows its own goals and not those of its maker, there is no reason a priori why its actions should be based on the objective description of the engineer, rather than on an adequate abstraction of the organism's interactions with the world. These considerations will lead us to reformulate the "computational" problem of perception in a rather different way (see chapter 6).

Machines are defined by their functional specification, the relations between their parts, independently of their material existence. This is a key aspect of computationalism: cognition and consciousness are determined by the relations between functional states, at an abstract level, independently of the material "implementation" of those relations. In other words, the "hardware" plays no role in the specification of the "software," it only supports its execution. But as we have seen in chapter 2, this is not the case of living systems, because the normativity of living systems is grounded in the physics of this world: it is because of thermodynamics that living systems need to feed, and therefore that they have goals, that they take actions, that they anticipate; in other words, that there is such a thing as "behavior." Therefore, Marr's "computational level" is in fact grounded in the material level, which is then improperly called "implementation."

Living systems are not just organized, like machines; they are self-organized. In particular, parts do not preexist to the whole, and do not even exist as stable material entities: they are better thought of as attractors of self-organizing processes, such that there is no clear distinction between "hardware" and "software"—everything is "soft." The properties of a living part depend on its context within the organism: higher levels of integration modulate the properties of lower levels, down to the molecular content (downward causation; see chapter 2). Parts are not necessarily lawful: in many cases, living processes are constitutively stochastic, harnessing randomness rather than resisting noise. Parts of living systems are not assembled: the organism grows, while remaining functional (i.e., while living). They are not designed: there is no plan of the body, or of the brain, and no entity to do the planning.

These are fundamental differences with machines, which makes the classical neurocomputational concept of "implementation" dubious. Thus, brains are not programmable machines. But are they programmable living systems?

Programmable Things

Let us leave the machine concept aside and simply consider that a computer is a programmable thing. A computer is not just something that implements a computation. A simple finite-state automaton runs a computation, it is not a computer. A Turing machine, in contrast, runs a computation specified by a series of instructions on a tape, which can be modified arbitrarily by the operator. Thus, something is a computer not just if it implements computations, but if it can be *programmed*.

Is the brain programmable? Many philosophers of mind realize that the brain is probably not *literally* programmable, and therefore discuss the idea that the brain computes many things, rather than the brain is a computer

(Chalmers, 2011). However, this distinction deserves some discussion, since the two notions are often conflated (Shagrir, 2006).

What do we mean by "programmable"? Broadly speaking, it means that the machine can execute instructions given in advance by a user. For example, a programmable heater can be set to work at predetermined times, with predetermined target temperatures. Obviously, the programs that computers can run are quite different from those of programmable heaters, but we focus first on the notion of programmability.

Are We Programmable?

Consider the following instructions to ride a bike: push with your left foot, then push with your right foot, and repeat. These words describe someone riding a bike, and they even superficially look like a program, with discrete elementary steps and recursion. However, when we teach a child how to ride a bike, we might use this description as "instructions," but the child still needs to practice before it can successfully execute them. This is because there is an implicit prescription that comes with these instructions: push with your left foot, then push with your right foot, *while trying to ride the bike*. Thus, the instructions *guide* the behavior, but they do not specify it: they are not a program.

Thus, teaching a child how to ride a bike is not a case of programming. It is in fact rarely the case that behavior follows explicit rules, as Dreyfus explained in *What Computers Can't Do* (Dreyfus, 1978) when he criticized rule-based expert systems. People may learn by initially following guiding rules, but rules are for beginners: experts normally go well beyond the rules.

However, there are obviously cases when people can execute instructions. For example, a cook can follow a recipe, or a lab technician can follow a protocol. Closer to the notion of a computer, an educated person with a pen and paper can execute an algorithm. In fact, in the seventeenth century, a "computer" was a person who did calculations (Hutto et al., 2018). Indeed, it is the words that we use to describe computers that come from the mental domain, not the other way around. It is also no coincidence that human languages and programming languages have features in common, such as discreteness and recursion: programming languages are inspired by human language, and computer science borrows linguistic concepts such as semantics and syntax. But the fact that we talk about computer "memory" does not mean that the computer literally remembers what you wrote when you open a text file, and the fact that we talk about program "semantics" does not mean that program variables actually mean something to the computer. Therefore, this superficial resemblance cannot possibly be taken as evidence that brains are computers:

it is computers that have been designed so as to mimic the ability of humans to carry out calculations, one ability among many.

Nonetheless, it is the case that people can execute instructions. Does this mean that people are programmable, like computers? There is a subtle but important distinction to make here. Something is programmable not just if it *can* follow instructions, but if it *will* follow instructions. In our examples, we have simply assumed that the person would agree to execute the instructions and behave as a machine. In reality, someone can be instructed to do something, but that person might very well decide to ignore those instructions. Most human behavior and mental life is not the result of instructions. Even enslaved persons perceive, feel, think, and dream, and these mental activities are not instructed.

When we say that the computer can execute arbitrary instructions, "can" and "arbitrary" refer to the agency of the user, not of the computer: it is the user who decides to run certain instructions rather than others, not the computer. Thus, the concept of programmability clashes with a key feature of the living that we have discussed in chapter 2: autonomy. A living being is autonomous—that is, it is an entity whose rules are specified by its own organization. This is the exact opposite of a computer, whose rules are programmed. The relation between environment and organism is usually not one of command, but one of coupling. The organism has an autonomous activity, even in the absence of any "stimulus" or "input," and interacts with the environment when it is useful to maintain its own activity.

Therefore, the ability of people to follow instructions does not make them programmable. In any case, this would exclude animals with nonlinguistic abilities. But *neurocomputationalism* does not exactly consider that people are computers. Rather, the claim is that *brains* are computers.

Instructions in the Brain

Perhaps the brain can be programmed by receiving instructions not from an external source, but from an internal source. There is a distinctly dualist ring to this idea. Indeed, Wilder Penfield, who discovered the cortical homunculi (sensory and motor "maps" of the body on the cortex), claimed that the brain is literally a computer because he was a dualist (Penfield, 1975). Penfield was a neurosurgeon, and in the 1940s, he electrically stimulated the brain of conscious patients at different places. He observed for example that stimulating the cortex at specific positions would trigger movements of particular parts of the body, or elicit specific sensory experiences. But he never elicited a sense of agency: patients would always report that it was Penfield who made their arm move, not themselves. Penfield concluded that the sense of agency must be possessed by something else than the brain, and that something was programming the brain-computer, in the same way as he was triggering various sophisticated but mindless behavior.

Of course, modern neurocomputationalism is not founded on brain–mind dualism, at least not admittedly so—we will see in the next chapter that the concept of neural codes is still very much influenced by Cartesian dualism. However, it still relies on some nonmental form of dualism—namely, the distinction between hardware and software. The concept of programmability is committed to a distinction between *software*, a set of modifiable elements, and *hardware*, a fixed set of processes that act as a function of those elements—the program and the machine that executes it. It is crucial that the hardware is fixed, for otherwise the "program" would not specify what it does—that is, would not be a program at all. The problem is that there is no such distinction in the brain, because as we have seen in chapter 2, everything in the brain and more broadly in biological organisms is "soft." In particular, classical connectionism claims that brain function is determined by the set of synaptic weights, but synaptic weights are certainly not the only modifiable elements in the brain. For example, axonal conduction delays, morphology, and electrophysiological properties of neurons all change with development and experience.

Are Brains Programmable?

To say that the brain is programmable means that there are elements (e.g., synaptic weights), playing the role of programs, that can be modified arbitrarily in view of some goal. But what does it mean that these can be modified "arbitrarily"? Who or what modifies those weights?

One might say that evolution modifies those weights, that is, the brain is a computer programmed by evolution—perhaps as it concerns so-called instincts. This claim raises two issues. First, as we have seen in chapter 2, evolution does not modify synaptic weights directly, because synaptic weights are not "encoded" in the genome. There is no map of the brain in the genome, even at the coarser level of brain areas. The genome specifies proteins and constrains developmental processes. Second, to program is to lay out a set of instructions in view of a goal. The view that evolution is a case of programming is therefore intelligent design, not Darwinism. Evolution is not literally a case of programming but rather the natural selection of random structural changes, which are generally nonspecific (not "add a synaptic connection between these two neurons" but perhaps changes that induce growing a bigger brain, inducing particular folds). It is therefore highly misleading to claim that the brain is a computer programmed by evolution.

Another view is that the brain is a "self-programmed" computer. If this claim refers to synaptic plasticity, then the modifications are not arbitrary either, but are governed by some particular processes. It is not the case that a cat can "self-program" itself into playing chess. Synaptic plasticity is not a case of programming either.

On the Universality of Neural Network Models

One reason behind the misconception that brains are programmable is that models of the nervous system can be rearranged to implement many different things. One of the founding results of computational neuroscience is the demonstration by McCulloch and Pitts (1943) that a binary neural network model can implement any logical function: the connection weights between formal neurons can be chosen such that the input–output mapping implemented by the model matches any specified logical function (e.g., the XOR operation; see figure 1.1). But of course, the network *can* implement any logical function *if* one modifies the parameters of the network. Thus, the universality of these networks relies on a distinction between parameters and dynamical variables of the model.

But the concept of parameter is a very peculiar one: a parameter is a quantity that can change, since it can be modified, so what is the difference with a dynamical variable? The answer is that the difference is drawn by the modeler. Here, the dynamical variables are the activation values of the binary neurons, while the parameters are those variables that are fixed during the execution of the model, but that can be modified by the modeler. This is again a dualistic distinction, and this distinction does not exist in the brain: the "synaptic weights" are not parameters, but also dynamical variables.

The distinction between a dynamical variable and an arbitrarily modifiable parameter is a distinction between two sources of change of a very different nature: on one hand, the process under study whose variables undergo lawful changes; on the other hand, an entity that modifies parameters in a way that is neither lawful nor random, but *arbitrary*. The former is the computer and the dynamical variables are the computational states; the latter is the user and the parameters are the program.

Concretely, what it means for some system to be programmable is that there is a way by which a user can provide instructions to that system, such that the system behaves in a certain way. This should be rather obvious, for a disorganized pile of transistors "can" also implement any computable function if the transistors are wired appropriately, just as the formal networks of McCulloch and Pitts, and yet, a disorganized pile of transistors is not a computer (Brette, 2022a). The universality of neural network models does not turn the brain into a computer, any more than it turns a pile of transistors into a computer.

Therefore, neither the brain nor a person can possibly be a computer in any literal sense, because they are not programmable. This is why discussions of computationalism generally focus on whether the brain computes, rather than on whether it is a computer. Even David Marr, a pioneer of computational neuroscience, did not insist that the brain is a computer: "In a sense, of course, the brain is a computer, but to say this without qualification is misleading" (Marr, 1982). Instead, he proposed to see the brain as an "information-processing" device, something that computes many things. We will now look at what it means for something to compute. What is a program, an algorithm or a computation?

Programs, Algorithms, Computations, and Beyond

Program

Do brains compute? Do they implement algorithms? Is there a genetic program? To answer these questions, we must clarify what we mean by "program," "algorithm," and "computation." In computer science, these terms are essentially synonyms. But in broader use, including the way they are used in biology and neuroscience, they refer to slightly different concepts, which is a source of confusion. Specifically, the concept of "program" is broader than "algorithm," which is broader than "computation."

A *program* is a set of explicit instructions that specify the behavior of the system in advance (*pro-*: before; *-gram*: write). There are many kinds of program. For example, a concert program specifies in advance the sequence of musical pieces that will be played. However, the concert program is not an algorithm. Two important elements are missing: the steps of an algorithm should unambiguously specify the behavior of the system, and an algorithm should solve a problem—that is, it is directed toward some goal, such that it makes sense to say that the algorithm succeeds or fails. Neither element is present in the concert program. The concert program is a program in the broad sense of a predetermined schedule of events, but not in the sense of computer science.

What about the genetic program? This is a controversial notion (Peluffo, 2015), precisely because of the ambiguity between the broad sense of program in colloquial use, and the narrower sense of program in computer science (i.e., algorithm). The development of an organism is generally lawful and largely constrained by its genetic material, although not at all in full detail. But the same could be said of a concert program. However, in contrast with a concert program, it can be said that the genetic program is oriented toward the formation of a reproducing adult. It has the teleological feature of programs in the sense of computer science. But would we say that it is an algorithm? This is more questionable, because one cannot follow the "genetic code" and infer the adult organism. The relation between genes and development is of the kind: "when gene A is expressed in the usual context, legs grow." This is not a description of how legs grow, but of the conditions for growing legs. In fact, as we have seen in chapter 2, the adult organism depends not only on the genetic material, but also on the cell that material belongs to, and on various environmental factors. Thus, the phrase "when gene A is expressed, legs grow" is contextual to the rest of the genome, to the rest of the organism, and to the environmental factors. Therefore, the genetic material is not an algorithm, because it does not fully specify its operation, and because it is not made of unambiguous elementary steps. The genetic "program" has some features of programs in computer science, but clearly not all. It lies somewhere between concert programs and programs in computer science.

Algorithm and Effective Method

The term "algorithm" is also used outside of computer science. For example, in the medical domain, there are algorithms used by physicians (figure 4.2). The goal of a medical algorithm is to diagnose a disease or to treat the patient. It consists of clearly defined steps, such as detecting a potential infection or providing some drug, and measurements may condition transitions between steps. This is similar to lab protocols or cooking recipes, although the term "algorithm" is rarely employed in those cases. For example, consider the following recipe: fill a pot with water, bring it to a boil, put a handful of pasta in the pot, wait until the pasta is cooked. A person who has never cooked pasta could follow the recipe step by step and obtain the expected result.

This is close to the notion of "effective method" in computer science. An effective method is a sequence of elementary instructions to solve a problem, which can be carried out "mechanically" by humans, without any "insight" or "intuition" (Copeland, 2020). There is obviously some ambiguity in this definition, since it entirely depends on what we consider as an elementary instruction, which can be carried out "mechanically" by humans. The pasta recipe described here is an effective method for cooking pasta only if we consider "put a handful of pasta in the pot" as an elementary instruction—that is, if we take perception and sensorimotor skills for granted. Thus, the fact that people can follow cooking recipes does not make perception and action the result of algorithms; in this case, they are the "elementary instructions" of the algorithm.

Therefore, when it is stated that behavior or mental activities are "computational" or "algorithmic," an important question to ask is: with respect to what

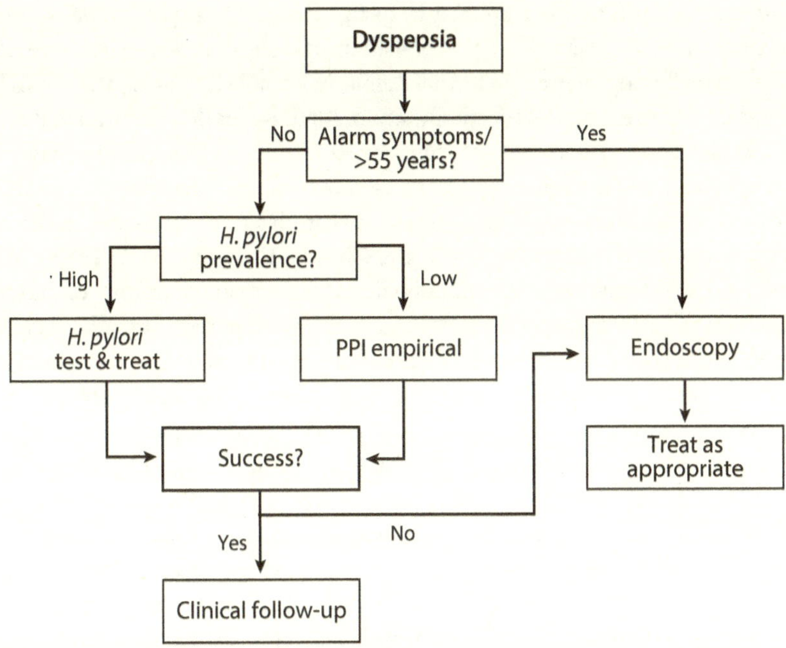

FIGURE 4.2. Diagnostic algorithm for dyspepsia (adapted from Aabakken, 2011). PPI is a proton-pump inhibitor (a prescription drug).

set of elementary instructions? It must be realized that the notion of effective method is too broad to found a general theory of cognition. Indeed, an effective method is defined as a sequence of simple steps than can be carried out by humans, so to state that human cognition is made of effective methods is simply to say that it can be decomposed in simpler steps. This is rather vague: one must specify the nature of these simpler steps.

Computation

This leads us to the notion of computation. Much confusion has been generated in the literature by the colloquial use of the term *computation*, which is often used in a very broad sense to designate a process by which some data is mapped to some other kind of data, with a norm (the output may be correct or not). But this is more accurately termed "data processing." In computer science, computation takes a much more precise and limited sense.

A computation is essentially an effective method in a formal domain (as opposed to a domain of interactions with the world), generally in the domain of numbers, which can be applied by a human with just pen and paper—we will see in chapter 8 the significance of this detail. Clearly, a recipe is not a computation. There have been several attempts to formalize the intuitive notion of an effective method in formal domains, notably by Turing and Church, leading to different models of computation. All these models feature a finite set of instructions and a property of recursion. It turned out that these formalisms are equivalent, in the sense that everything that can be calculated by one kind of effective method (e.g., a Turing machine) can also be calculated by the other kind (e.g., lambda calculus). This is the Church-Turing thesis: "[Turing machines] can do anything that could be described as 'rule of thumb' or 'purely mechanical'" (Turing, 1948). This is obviously not a theorem, but rather a claim that what we usually mean by "calculation" or "algorithm" in the mathematical domain can be captured by a formalism equivalent to Turing machines.

A computation operates on data (not pasta). It has an input and an output, and it implements a particular mapping from input to output by a sequence of steps that belong to a finite set of elementary operations on symbols. The latter restriction is crucial for the notion of computation. Otherwise, the concept of computability would be totally meaningless. For example, Alan Turing demonstrated that there is no program that can take another program as input and calculate whether that program eventually stops: the halting problem is not computable. Obviously, this theorem would be false if we were allowed to choose any kind of instruction (choose as an instruction the mapping that assigns a stopping/nonstopping label to programs). Therefore, a process can only be meaningfully called "computation" when it is composed of a finite set of elementary instructions. Otherwise, the process is just a mapping between different kinds of data, not a computation.

We are familiar with computations: an algorithm to compute a square root, for example, or to sort words in alphabetical order. Square root algorithms typically use additions, multiplications and divisions, recursively iterated. For example, the Babylonian method calculates $\sqrt{a}$ by iterating the following formula: $x \leftarrow \frac{1}{2}(x + a / x)$. Sorting algorithms recursively combine comparisons and permutations, and words can be compared letter by letter with a table, recursively. But are there noncomputational ways to solve data problems?

Solving Data Problems by Setting Constraints

There is another way to calculate a square root. Take an apple, place it at a certain height, and let it fall. Measure how much time it takes before the apple reaches the ground: it is proportional to the square root of the height. Here, we did not do a calculation: we applied constraints to a physical system whose measurable properties follow the relation we are interested in.

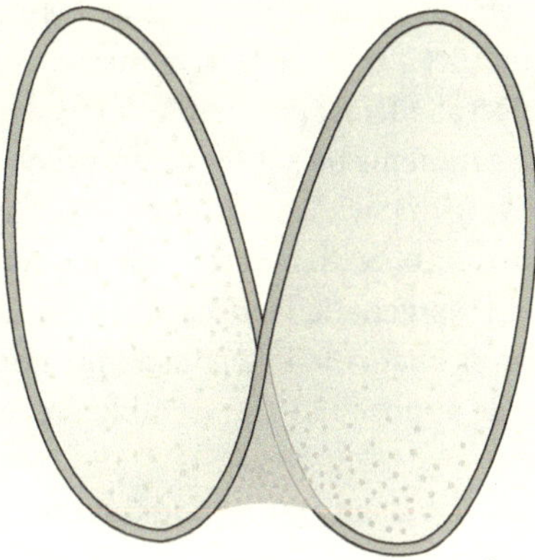

FIGURE 4.3. Minimal surface formed by soap film on a wire frame (inspired by Morgan, 1990).

Let us give a more sophisticated example of problem-solving activity that is *not* a computation. If you immerse a wire frame into soapy water and then remove it, a soap film will form in such a way that its surface will be the smallest possible surface with the wire as boundary conditions (Morgan, 1990; figure 4.3). This operation implements a mapping from the wire frame to the associated minimal surface, but it is not a computation: the result is not obtained by a sequence of elementary calculations like the Babylonian square root algorithm, but by harnessing the property of closed physical systems to settle in configurations with minimal thermodynamic free energy.

Of course, you *could* compute the minimal surface from a mathematical representation of the wire frame, but that is not how you did it. It must be stressed that a computation is not just a mapping with certain properties (such as being computable), but a particular sequence of elementary symbolic steps that implements that mapping. This means that, when applied to an input, the computation stops after a finite number of steps and outputs a number or symbol that matches the value that the mapping associates to the input. Immersing the wire frame does not fit this definition. This procedure solves a problem that can be expressed in terms of data, but not by means of computation.

Thus, one particular way to solve a data problem is to map the data to constraints on a physical system (e.g., to build or distort a wire frame), then to let the system settle into some state, which is then interpreted as the output. In this way, we implement a mapping from input data to output. This way of solving problems is sometimes called "analog computation," but these are very misleading terms, for what is called "computation" here has absolutely no connection with computability theory, effective methods, or Turing machines. There is no computation at all, just problem solving.

The idea of solving problems by setting appropriate constraints resonates with concepts of the living that we have discussed in chapter 2: a living system produces its own constraints, which determine the way its processes work. Similarly, Waddington's "epigenetic landscape" (Waddington, 1942) pictures development as a process channeled by constraints imposed by the genome and the environment, as opposed to an assembly process based on a map of the adult organism.

This kind of idea has also been explored in theoretical neuroscience, typically in models inspired by statistical mechanics. For example, in the Hopfield model (Hopfield, 1982) and in the Boltzmann machine (Ackley et al., 1985), which are inspired by spin glasses (some kind of magnetic alloy), each possible state of the neural network has a corresponding energy and the network tends to settle at the lowest energy state—there is much to say about the notion of state in these and other formal neural models, but we will leave this issue for chapter 8. In these spin glass models, the energy depends on the strength of interactions between nodes—that is, on the synaptic weights—and therefore the weights act as constraints on the dynamics of the model, especially on the model's attractors.

In practical implementation (in particular on computers), spin glass models might take the form of algorithms, with discrete iterations, but this is not a defining feature. Actual spin glasses do not update themselves in that way. What matters for their properties is the thermodynamic equilibrium. Variations of spin glass models include stochastic and asynchronous dynamics, or continuous-time dynamics. These models may implement a mapping from an initial state to a final state, or from a set of weights to an equilibrium state, but in either case the path from input to output is not predetermined. This is quite different from a multilayer perceptron, which implements a mapping from input to output by a series of deterministic synchronous iterations.

Human Computation

Despite these subtleties, it is clear that adult humans can perform computations in the usual sense, such as adding two large numbers. This is obviously the case, since computations are precisely defined as those activities that can be carried out by humans, with the potential help of a pen and paper. However, consciously performed computations represent a rather narrow domain of cognition: it excludes cats, young children, and a fair proportion of adult humans; it excludes perception, action, imagination, memory; it excludes many cognitive activities involved in carrying out a computation: hearing, understanding the words, memorizing them, using a pen and paper, uttering the result.

Are there computational features in behavior, beyond actually performing calculations? Another example is human language. Chomsky has shown how well-formed sentences can be produced by effective methods (generative syntax), which are recursive (Chomsky, 2002). The similarity with programming languages is of course not a coincidence. Just as computers have been designed on the model of human computers (people performing calculations), programming languages have been inspired by human language. Programming languages have "syntax" and "semantics" and consist of "instructions." However, a generative grammar is not an effective method to speak, nor does it convey the meaning of words. Thus, Chomsky's insights do not imply that speaking is a kind of computation.

More broadly, as we have seen in chapter 2, behavior is made of actions, which are oriented toward the achievement of something desirable for the organism. Thus, behavior is anticipatory: this fits the concept of "program" in the broadest sense. Actions are goal-driven, so that they depend on abstract properties of situations, rather than on proximal stimuli. But again, this is not specifically algorithmic or computational.

One computational feature of behavior is that in many cases, it is recursive. This is again a side product of the anticipatory nature of behavior: a situation can be considered favorable by anticipation—that is, because further actions would then lead to a favorable situation. It follows that a distant goal may be reached by a series of actions, each leading to a new situation that has no immediate value, except for the last one. Riding a bike is a simple example of a recursive behavior: push on your left foot, push on your right foot, repeat. This gives an algorithmic flavor to behavior, but again as we have seen, only in the broad sense of "algorithm"—in reality, this is actually not an effective method to ride a bike.

Thus, behavior clearly has a number of computational features. Namely, behavior is goal-driven, it can be decomposed into actions, and it is recursive. However, the term *computational* is misleading here, because the features in question are those of "programs" and "algorithms" understood in a broad sense, not in the more specific sense of computer science. The discrete recursive structure of bike riding is not an effective method to ride a bike. More generally, the "elementary" steps that compose a complex behavior are not generally elementary calculations, at least not the steps than we can observe in behavior. In particular, perception cannot be further decomposed at the behavioral level. Rather, it is a hypothesis of computationalism that perception involves a sequence of unconscious and invisible elementary steps that lead to the production of a symbol. In fact, perception itself is not observable as such, only perceptually guided actions are. Thus, while behavior might be decomposable into sequences of actions, there is no obvious indication that an action can be further decomposed into a perceptual calculation, followed by symbol manipulation, followed by motor command (what Hurley [2001] called "the sandwich model" of cognition). In other words, behavior may be (at least in some cases) seen as algorithmic in the sense of recipe-following—wait for the water to boil, then put the pasta in the pot, wait until the pasta is cooked—but not necessarily in the scientific sense of algorithmic theory, that is, something that can be decomposed into a finite set of predefined elementary operations on data. This was in fact one of the key difficulties that expert systems (or GOFAI, for "good old-fashioned artificial intelligence") faced in the 1980s—it turns out to be quite difficult to recognize a box of pasta and to put its contents in a pot.

Computationalism is the claim that *all* cognitive activity is a sort of unconscious calculation. In other words, computationalism is a form of cognitive reductionism: all cognitive activity in all the animal kingdom (perception, decision, motor control, etc.) is actually composed of unconscious atoms of cognition, which are calculations. This is not a trivial claim. However, behavior can also be based on interaction rather than computation.

Behavior as Interaction

Spatial Navigation

Spatial navigation is often presented as a classic example of computation, which is displayed by many species. For example, desert ants can venture far from their nest looking for food (Wehner and Wehner, 1990), and when they find food, they return to the nest by the shortest path, in a straight line, without using visual or chemical landmarks (figure 4.4). Thus, the homing vector depends only on food position and nest position, independently of the tortuous path that led to the food.

There is a computational structure in this behavior, which can be seen by decomposing the outward path into small displacements. We call *homing vector* $\boldsymbol{x}(t)$ the straight route that the ant would take if it found food at time t. The homing vectors at two nearby instants t and $t+dt$ are related by the ant's displacement, $\boldsymbol{v}(t).dt$, where $\boldsymbol{v}$ is the ant's velocity vector. Thus, one can define a mapping that takes the homing vector at a given instant, and map it to the homing vector at the next instant, and it turns out that the result can be written as a simple calculation based on two behavioral quantities, the homing vector and the velocity vector:

$$\boldsymbol{x}(t) \overset{\varphi}{\mapsto} \boldsymbol{x}(t+dt) = \boldsymbol{x}(t) + \boldsymbol{v}(t)dt$$

FIGURE 4.4. The desert ant returns to its nest along a straight path after wandering until it reaches food (adapted from Wehner and Wehner, 1990).

This mapping can be iterated from the nest to the final food position to obtain $\boldsymbol{x} = \int \boldsymbol{v}$. This is a computational model of behavior: the input is the velocity vector $\boldsymbol{v}$, the output is the final homing vector $\boldsymbol{x}$, and the computation is path integration. It can be decomposed into small steps that are elementary operations on data.

Thus, the mere fact that the ant returns to its nest along a straight path implies that the ant's homing behavior has a computational structure.

Computational Structure in the Eyes of the Observer

The first point I want to make is that the computational structure is in the observations that we make, as external witnesses of the situation. It does not follow that the organism is itself engaged in computing something.

To see this, consider a completely different way to solve the path integration problem, which does not rely on calculating vectors, suggested by Wehner and

Wehner (1990). In Greek mythology, Theseus managed to find his way out of the Minotaur's labyrinth thanks to Ariadne, who gave him a thread that he attached to the door and then unspooled along his path. To escape the labyrinth, he just had to follow the thread back to the entrance.

Out of the labyrinth, one could solve the path integration problem by tightening the thread, so that the thread takes the shape of the shortest path to the starting point, a straight line. Solving this problem involves no computation. Whatever Theseus's position, the action that solves the problem is always the same: pulling on the thread. The procedure is neither computation nor information processing. Instead, it relies on interacting with the environment. In neuroscience and cognitive science, this is related to the concept of *situated* or *embedded cognition.*

The interesting point about this example is that everything we have said earlier about the computational structure of behavior remains valid. It is still the case that the homing vector is obtained by iterating the operation $\boldsymbol{x}(t+dt)=\boldsymbol{x}(t)+\boldsymbol{v}(t)dt$, and yet neither the brain nor the organism computes the homing vector. This is not just a question of boundary—that is, of whether cognition occurs within the boundaries of the skull or is extended (Clark and Chalmers, 1998)—because the thread does not implement the vector update operation either.

The paradox arises from our loose use of the term "behavior": when we analyze the properties of the behavioral measurement $\boldsymbol{x}$, we do not describe the actions of the organism, but the results of these actions in the environment, as observed by us. In fact, $\boldsymbol{x}$ is not exactly a behavioral measurement, but a *hypothetical* one: it would be the homing vector *if* the ant, or Theseus, decided to come back home. But as long as Theseus is not pulling on the thread, the thread is loose and does not point toward the nest. Thus, the thread is not computing anything at all. And when the thread is tightened, it does not compute the homing vector any more than soap computes minimal surfaces.

This is important, because in the case of Ariadne's thread, it is obvious that the brain is not implementing a computation, namely path integration, despite the computational structure that we can see when the problem is formulated in a certain way.

Of course, the ant does not use a thread. In fact, there is good evidence that it does not solve the path integration problem by interacting with the environment, because when the ant is moved to a different place, its homing vector is unchanged, so that it runs back as if the nest had been displaced by the same amount. This is an example of what philosophers call "representation-hungry" behavior (Clark and Toribio, 1994), which we will discuss shortly.

Nonetheless, the thread example makes the point that not all problems are intrinsically computational problems. Spatial navigation is not a computational problem: it is a problem that has computational solutions, but also noncomputational solutions. A problem becomes a computational problem only when we ask how to solve it with a computation. In the same way, spatial navigation is not an "information processing" problem. It becomes such a problem when one frames it as the problem of implementing a mapping from input data (the time series of velocity vectors) to output data (the homing vector). But many problems can be solved partly or entirely by interacting with the environment: this is the insight of theories of embodied and situated cognition (Pfeifer and Bongard, 2006). Not all behavior, including problem-solving activity, is computation.

Generically, the problems of organisms are not *intrinsically* computational. With the exception of humans performing calculations, animals do not normally deal with data, and the goal of any behavior is not to output data: it is to change the organism's situation by acting in the world. Thus, the computational nature of a behavioral problem is not intrinsic, but rather derives from the way it is framed. Therefore, it is misleading to call the properties of some particular behavior or cognitive activity the "computational level," as Marr did (Marr, 1982), for this already assumes that the organism solves the problem exclusively by means of computation. As we have seen, this is not necessarily the case, whether computation is understood as conscious or unconscious. A better terminology is simply the *problem level.*

But Is It Cognition?

Of course, one might be reluctant to call the thread trick "cognition." But as Ramsey argued (Ramsey, 2017), what we decide to call cognition is not such an urgent matter. If we want to explain spatial navigation, and it turns out that this ability is not based on updating a representation of one's position by a calculation, then so be it: spatial navigation is not computational, and the brain does not compute position.

In any case, if we want to understand what cognition is, and not just behavior, then we have to be careful to define cognition not in terms of how it is supposed to work, which would be prejudiced, but in terms of the phenomena we want to explain (e.g., perception, memory, etc.).

Take for example visual perception, which is traditionally considered part of cognition. Marr framed vision as a case of "information processing." Artificial neural networks frame image recognition as a mapping from a pattern of pixels to object labels, implemented by a sequence of operations. Most neurocomputational models depict visual processes in a similar fashion. I will leave aside for now the questionable identification between output symbols and percepts, and focus on the fact that perception is framed as an input–output mapping.

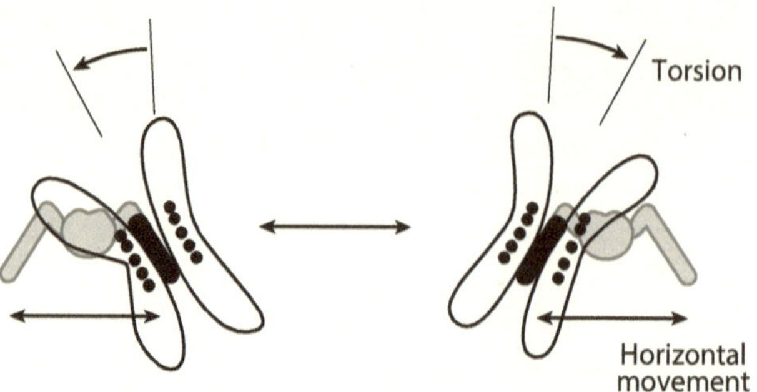

FIGURE 4.5. Jumping spiders recognize mates by aligning their linear retinae with the mate's legs (adapted from Land, 1969).

As far as we know, this is not how vision normally works. Instead, the eyes actively scan the visual scene with both fast and slower movements (from ~10 ms to seconds) (Martinez-Conde et al., 2004), and the miniature unconscious eye movements that we make when fixating an object are perceptually important (Rucci et al., 2007). In fact, if the retinal image is stabilized, visual perception quickly fades. Even *Drosophila* makes ocular saccades (Fenk et al., 2022), and jumping spiders recognize mates or prey by scanning the visual field with two linear retinae (Land, 1969; figure 4.5). Thus, seeing involves a closed-loop interaction between the nervous system and the visual scene (Ahissar and Assa, 2016). The sensorimotor theory of perception (O'Regan and Noë, 2001) and Gibson's ecological theory of perception (Gibson, 1979) consider vision as constitutively exploratory. Although these theories are debated, the least we can say is that vision includes an exploratory component, an interaction with the environment. The visual system does not simply map images to symbols, and a fortiori, it does not compute symbols from images.

Against the Stimulus

The fundamental insight of the interactional perspective is that behavior does not necessarily solve input–output problems, because the sensory "input" is itself the result of the organism's actions. As Dewey (1896) put it: "the motor response determines the stimulus, just as truly as sensory stimulus determines the movement." In the modern sense, the concept of stimulus is an experimental concept. It was inherited from the vitalistic physiology of the eighteenth century, which placed the source of animate motion in the "irritability" of living tissues, then morphed into the mechanistic concept of reflex in the nineteenth century (Danziger, 1997). But in an ecological setting, there is an

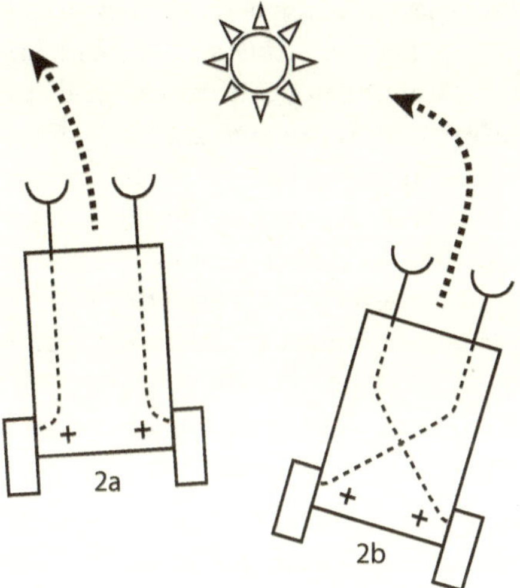

FIGURE 4.6. A vehicle that avoids (2a) or seeks (2b) light with elementary feedback circuits (Braitenberg, 1984). Original figure by Schoch T. CC BY https://creativecommons.org/licenses/by-sa/3.0/deed.en.

environment with which the organism interacts, rather than a stimulus that commands the organism's responses.

An influential book in this theme was *Vehicles* (Braitenberg, 1984), where neuroscientist Braitenberg described fictional vehicles implementing simple reactive schemes, showing that simple interaction rules can result in complex behavior. One of the simplest vehicles has two wheels, each connected to a light sensor that directly commands the wheel speed (figure 4.6). If each wheel is connected to the sensor on the same side, then the vehicle avoids sources of light. If connections are crossed, the vehicle is attracted by sources of light. Note that the vehicle does not calculate the steering direction. Instead, that direction results from the speed difference between the two wheels. There is also no measurement of light intensity. In fact, the two wheels are structurally independent if we consider only the vehicle, and they become coupled only through the interaction of the vehicle with the environment. The vehicle itself does not compute source position, steering direction or light gradient.

In robotics, Rodney Brooks pioneered an influential approach named *behavior-based robotics*, which emphasizes interaction with the world rather than by planning (Brooks, 1991). Complex behavior is then obtained not by decomposing a difficult problem into simpler components that are implemented as independent modules exchanging symbols (a perception module, a motor module, etc.), but by starting with a system implementing a simple behavior, then building layers on top of that system, which control lower layers. This was called a *subsumption architecture* (Brooks, 1986).

To a biologist, the subsumption architecture immediately evokes evolution. It is certainly the case that brain structure did not evolve component by component, but from simple functional systems to more complex ones—the same remark could be made with development in mind, since brains are not assembled but grown. Thus, an idea that has gained momentum is that brain function should be understood with an evolutionary lens, starting not from computation but from elementary feedback interactions with the world (Cisek, 2019).

We will now turn our attention to a particular way in which behavior can be based on interaction rather than computation: the paradigm of control.

Behavior as Control

Many critics of computationalism have focused on *feedback control* as an alternative to computation, such as van Gelder (1995), who asked "What might cognition be, if not computation?" Ariadne's thread is not a case of control, and many interactions with the world are not either. But the conceptual appeal of control is that it provides a generic way to couple agent and environment in a normative framework, which is why it has received significant attention in the literature.

Consider the problem of sound localization. Many animals can localize the source of sounds based on tiny differences between the acoustical waves arriving at the two ears—interaural time differences or ITD, and interaural intensity differences or IID (figure 4.7). Typically, a sound coming from the left will arrive earlier and louder at the left ear than at the right ear. However, the problem is complicated by the fact that the sounds themselves can be very diverse, the interaural differences depend on frequency because of sound diffraction by the head (Benichoux et al., 2016), and the environment introduces sources of confusion such as noise and reflections (Gourévitch and Brette, 2012). Thus, sound localization is determined by an abstract feature of sounds, namely the position of their source, which is not obviously present in the acoustical waves captured by the ears. In particular, unlike in vision, there is no topographic relation between the position of the sound source and the activation of the cochlea.

Consider the behavior of turning one's head toward the source of a sound. The mapping from the two acoustical waves to the head's position has the distinctive property that it is invariant to the sound produced by the source (assuming the sound is broadband). Seen as an information processing problem, that of implementing this mapping between input waves and output

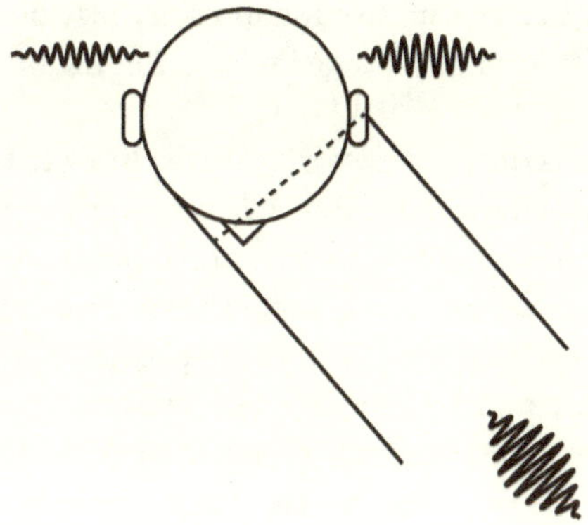

FIGURE 4.7. A sound from the left arrives earlier and louder at the subject's left ear than at the right ear, producing an interaural time difference (ITD) and an interaural intensity difference (IID) between the two monaural sounds.

position, this is a difficult problem. It is an example of an *inverse problem*. The pair of acoustical waves is produced from the sound source according to the laws of physics, which are linear but potentially complicated if we take into account diffraction. The information processing problem consists in inferring the source position from the two acoustical waves—that is, somehow inverting the laws of physics. *If* we further assume that this inference is performed by first calculating the interaural differences at each frequency, then this information processing problem becomes a computational problem.

But of course, there is another way to locate sounds: turn your head toward the louder ear until the sounds have equal intensity at the two ears. This is a feedback control solution. Behaviorally, the final head position is still invariant to the nature of the sound produced by the source. However, the solution does not involve solving an inverse problem. The final position is not computed—that is, it does not result from applying a series of calculations to the two input acoustical waves. In fact, this behavior does not solve an information-processing problem at all—the implementation of a mapping from input data to output data—because the input is not fully specified prior to the organism's actions.

This control system might be cast in an algorithmic way as follows: if the sound is louder at the right ear than at the left ear, then turn right; repeat. However, this is a cooking-recipe kind of algorithm, not an effective method to compute the position of the sound source. The problem of localizing the sound source gets solved not by a series of operations in the data domain, but by interaction with the environment. Therefore, there is a clear distinction between feedback control and computation. Of course, a control system can use computations. For example, sound intensity may be calculated by summing acoustic power over some time interval. But the behavior as a whole is not a computation.

Of course, animals do not necessarily use feedback to locate sounds, since they can locate transient sounds. But they certainly can when given the opportunity. Thus, feedback control is part of the behavioral repertoire of animals. Feedback control is one particular kind of interaction between organism and environment, which remains within the cybernetic tradition—there are other kinds, such as Ariadne's thread or trial-and-error processes. Its appeal is that control systems are dynamical systems with a teleological dimension: dynamical systems designed so as to achieve some goal, which is to ensure that some state variables remain within some bounds. Thus, the control view allows interpreting biological processes, which are typically modeled as dynamical systems, in teleological terms (what the processes try to achieve). This is why the control perspective on brain and behavior has become a popular alternative to computationalism, although it remains a heterodox view.

In psychology, William Powers developed a theory of behavior, Perceptual Control Theory, based on a hierarchical system of feedback loops (Powers, 1973). Behavior is interpreted in terms of control theory as the "control of perception," which undermines the classical idea of a "stimulus." In computational neuroscience, these considerations have motivated the emergence of *computational neuroethology*, a field that aims at developing coupled models of brain, body and environment, emphasizing the fact that behavior is not an output of the brain, but a result of its coupling with body and environment (Chiel and Beer, 1997)—the term "computational" is rather confusing since the models are simulated on a computer but are not based on a computationalist view of behavior.

But control is a rather narrow perspective, not just because agents can interact with their environment in other ways, but also because the distinction between agent and environment may not be so clear-cut in behavior.

Body in Behavior

The preceding examples are cases of an agent interacting with its environment, through sensors and actuators. This is the paradigm of control, where it is the agent that stabilizes the system, or some measured quantity. But such stabilization may also be performed directly by the body.

Consider a cupboard, standing firmly on its four feet. If the cupboard is pushed laterally so that it gets slightly tilted, then after this perturbation, it will come back to its original position (figure 4.8). How this works is that when it is tilted, the cupboard can only swivel about the two feet in contact with the

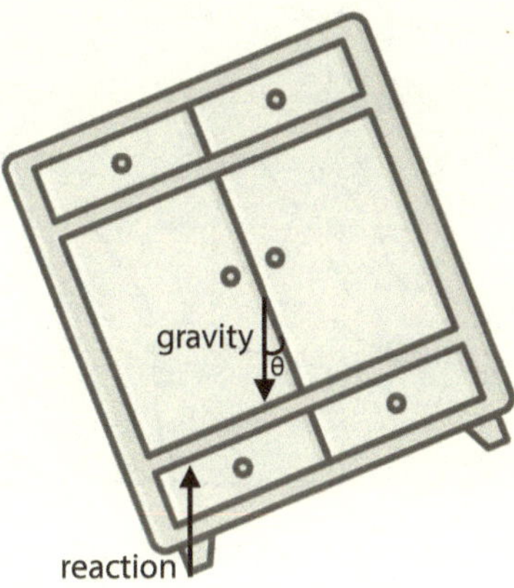

FIGURE 4.8. The cupboard returns to its standing position without measurement or actuation.

ground, and gravity exerts a vertical downward force at the center of mass. Thus, if the perturbation is small enough, then the torque of gravity realizes a negative feedback on the cupboard's angle: it brings it back to its initial state. In mathematical terms, the dynamics can be described by a typical feedback control equation: $d\theta/dt = f(\theta)$, where θ is the angle relative to the vertical, with $f(\theta) < 0$ when $\theta > 0$ and $f(\theta) > 0$ when $\theta < 0$.

However, this is not exactly a case of control since there is no controller. There is no meaningful separation between the controller and the controlled object (as in brain and body). There are also no sensors: the cupboard's angle is not measured. If we consider only the cupboard, there is in fact no way to map its structure to its angle, because the angle is a relational property of the cupboard-and-ground system (body and environment). This is to be contrasted with a thermostat: the thermostat has a temperature sensor—that is, a distinct physical part whose intrinsic properties vary with temperature in a lawful way; an actuator that switches the heater on or off; and some control mechanism connecting the two (which may or may not be a program).

The cupboard's useful responses to environmental perturbations are not computed, and they do not result from feedback control either. In neuroscience and cognitive science, this is related to the concept of *embodiment*, the idea that the body is fundamentally integrated with cognitive processes (Gallagher, 2006; Pfeifer and Bongard, 2006). This is most obvious in sensorimotor skills, such as reacting appropriately to mechanical perturbations, making a movement or exerting some force (Tytell et al., 2011). The cupboard could be a four-legged animal, for example. Surely, an animal has a vestibular system that is sensitive to the head's position, and it can exert forces by

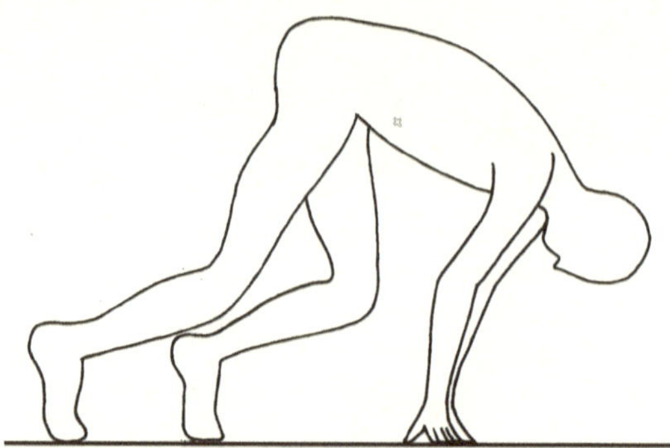

FIGURE 4.9. Anticipatory posture of a runner. Original figure by Le Mouel C. 2017. https://doi.org/10.3389/fncom.2017.00067 / CC BY https://creativecommons.org/licenses/by/4.0/.

contracting its muscles in response to proprioceptive signals. But the primary (and earlier) response to mechanical perturbations is a mechanical response of the musculoskeletal system. Accordingly, performance in sensorimotor tasks can be improved by better control, but it can also be improved by better structure (making a more stable cupboard). This can occur on the timescale of development or training (growth of muscles and bones), but it can also occur on the timescale of behavior, by adjusting the properties of the musculoskeletal system. For example, when standing in challenging balance conditions, humans co-contract their ankle muscles (Le Mouel and Brette, 2019). Humans also adjust their posture in anticipation of future movements (Le Mouel and Brette, 2017). An example is the posture of the runner in starting blocks, which allows the runner to use her own weight to produce torque for movement (figure 4.9). Postural adaptation is neither computation nor control, but anticipatory reconfiguration: a change in the relation between motor commands and the movements they induce.

These examples demonstrate that not everything useful and lawful is computational, and that these other ways of solving problems are an important part of the normal behavioral repertoire of organisms. We can now explain two important conceptual distinctions, which justify placing computation in a very different category from interaction or body properties. The first distinction is that dynamics is not computation, because computation is defined formally, independently of physical time. The second distinction is that interaction is not computation, because computation is *epistemically closed*—there is no unknown in a computation. These two distinctions cast serious doubts on the idea that the dynamical activity of interacting neurons is a case of distributed computation.

Why Dynamics Is Not Computation

Time in Computation

Computations are timeless. Sure enough, instructions must be carried out in a certain order, one after the other. But the same computation is carried out whether it takes an hour or a second. Considerations of time belong to implementation, and computation is by definition independent of implementation—this is one of the key tenets of the computational view of mind. Of course, one can see a computation as a dynamical system, with iterations representing "time." But this is the intrinsic time of a computation, an abstract notion meaning a mathematical structure isomorphic to the natural numbers. It is distinct from actual physical time. In computer science, complexity theory compares the efficiency of different algorithms by calculating the number of steps it takes for the algorithms to complete (Cormen et al., 2009). It is entirely irrelevant how much physical time each step actually takes, because complexity theory is only interested in how this time scales with the size of the problem. For example, a good sorting algorithm has complexity of order $n\log n$, which means formally that if the number of elements n is large enough, then the time it will take to sort them will be smaller than $\lambda n\log n$, for some number λ.

Two physical systems that implement the same computation might run in complete asynchrony: the computation would still be exactly identical. But as soon as the physical system is embedded in the world, the abstract computational notion of time cannot work anymore, because the input to the physical system depends on the actual time it takes for the system to run, not on the number of computational steps. The behavior of a physical system coupled to the world depends not only on the input–output mappings realized by its components, but also on their implementation.

An elementary example is a clock. We can certainly say that a clock implements a computation, namely counting. But this is not what makes it a clock. What makes it a clock is that each tick takes exactly one second, and this is not obtained by computation but from the physical properties of the device. For example, in a traditional clock (figure 4.10), it is the dynamics of the pendulum that sets the tick frequency.

A more sophisticated example is a feedback control system that would use some computation to process measurements, for example a proportional

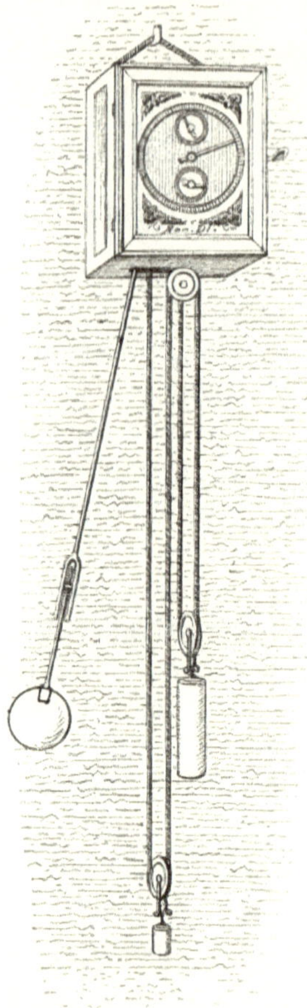

FIGURE 4.10. First pendulum clock, designed by Huygens in 1657 (Gerland and Traumüller, 1899).

controller: calculate the difference between the measurement and a target, and output a correction command proportional to that difference. This control system will not behave in the same way if the computation is quick or fast. The computation would be the same, but the behavior of the system would be different (stable or unstable, for example). For systems that interact with the world, implementation matters. In this sense, dynamics is not computation. This is one of the key points of the dynamicist view of cognition (van Gelder, 1995).

Time in Distributed Systems

The remark extends to distributed systems, whether they interact with the world or not. What a system does depends not only on the input–output mappings that its components realize, but also on their temporal embedding. In a classical computer, the operation of all components must be synchronized by a clock. Parallel programming relies on sophisticated systems to synchronize computations executed in different processors or threads. For example, the execution of one thread can be suspended until some other computation has finished. These systems ensure that the execution of the parallel program is independent of its implementation. But there are no such systems in the brain. Neurons do not suspend their activity while waiting for other neurons to finish whatever task they are supposed to do. The brain is not a "parallel computer" in any way near what an actual parallel computer is.

Seeing single neurons as implementing a timeless mapping, from a set of input numbers to an output number, is precisely what is done in artificial neural networks. As we have seen, the precursor of all artificial neural network models is the binary neuron model of McCulloch and Pitts (1943). In that model, the binary output is a function of binary inputs, and therefore can be seen as a logical operation. The key innovation in this work was not the incorporation of biological knowledge into a simple mathematical model. There were mathematical models of neural networks before, pioneered in the 1930s by Nicolas Rashevsky (who then became Rosen's mentor). Rashevsky developed mathematical models of neural networks featuring all the ingredients of the McCulloch-Pitts model: excitation, inhibition, threshold (Abraham, 2002; Rashevsky, 1936). The key innovation of McCulloch and Pitts was to discard time. Rashevsky's models were continuous-time dynamical systems. In such systems, a neuron influences other neurons for an indefinite amount of time, so that the activity of a neural network cannot be decomposed into a discrete sequence of operations. Instead, in the McCulloch-Pitts model, the state of a neuron is only a function of the state of other neurons at the previous step. This makes the neuron model implement a mapping in the binary domain—that is, a logical operation. Formal neuron models of later connectionist networks implement mappings in the domain of real numbers instead of binary numbers, but the temporal dimension is still absent.

Discarding time was a key step in linking neural models to computer science and logic, but it was a step away from biology. Biophysical models of neurons, whether those of Rashevsky or the more empirically informed models of Hodgkin and Huxley (1952), are typically modeled by differential equations, which are continuous-time dynamical systems. If the state of such a system is perturbed, the effect of the perturbation persists for an extended period of time. For example, when the neuron is excited or inhibited by another neuron, neurotransmitters are released in the synapse and bind on receptors on the postsynaptic neuron, which then triggers a current. Depending on the receptor, this synaptic current can last just a couple of milliseconds or hundreds of milliseconds. In addition, neurons are coupled with conduction delays that are also diverse, up to tens of milliseconds (Swadlow and Waxman, 2012). Therefore, the state of a neuron depends on the activity of other neurons on various extended timescales. This is not simply a mapping from neural activities at the previous time step to neural activity at the present time step. It follows that the dynamics of a biological neuron does not implement a mapping on real numbers—that is, an operation—and the dynamics of biological neural networks does not implement sequences of such operations—that is, a computation. We will revisit this issue when we discuss the notion of neural state in chapter 8.

Why Interaction Is Not Computation

Epistemic Closure

Our discussion of feedback control highlights another key feature of computation. Remember that a computation is an effective method in the domain of data (essentially, numbers), which means a sequence of simple steps that can be executed by a human with just pen and paper. This implies that the result of a computation only depends on its formal specification. Computation is *epistemically closed*: everything there is to know about the computation, such as its result on any given input or the number of steps it will take, is implicitly included in its formal specification. This is what it means that a computation is a formal procedure. In contrast, the result of an action, which is not formal but instead consists in interacting with the world, is not known in advance.

Compare the two ways of localizing a sound source. One way consists in computing the direction of the sound source from two acoustical waves, for example by calculating interaural time and intensity differences, then inferring the direction that is consistent with those cues. In this case, the mapping from acoustical waves to source direction is entirely specified by the computation. The other way consists in turning one's head toward the louder sound. In this case, the mapping from acoustical waves—say, the waves captured before the head starts moving—to sound direction is not specified by the description of the actions of the system and their conditions. It is only by executing these actions in the real world that one can map the acoustical waves to the source direction.

In the same way, the problem of finding the direction back to the nest can be solved by tightening a thread attached to the nest. When the thread is tightened, its direction indicates the direction to the nest. Thus, there is a procedure followed by the organism that leads to an estimate of nest direction. But the formal specification of the procedure ("pull on the thread") does not determine the outcome. Instead, the result depends on the realization of the procedure in the physical world. For example, when the thread is tightened, it might be blocked by a tree. In that case, the initial direction of the thread is toward the tree, not toward the nest. Thus, the result of the procedure to estimate nest direction depends on whether there is a tree in the way: this estimation is not a computation because the procedure is epistemically open. In contrast, a computation of nest direction from the time series of velocity measurements is epistemically closed: its result (whether correct or not) is independent of the actual configuration of the outside world; it depends only on the formal specification of the procedure and on the input. This remark holds even if the computation is stochastic: the outcome might be variable but is still independent of the outside world.

Computation as Anticipation

All operations of a computation are specified in advance, and their result is formally determined. In this sense, computation is fully anticipatory. In many cases, the result of the computation is supposed to match something about the real world, such as the location of a sound source. Therefore, it formally specifies a mapping between two measurable quantities in the world, which may be more or less accurate. In other words, a computation entails a certain kind of model.

In the case of sound localization, computation specifies the mapping between acoustical waves and source direction. Feedback control does not specify such a mapping. It remains anticipatory, since it is assumed that turning one's head toward the louder ear will result in reducing sound intensity differences between the two ears. But the anticipation is much more limited than in computation. Another way to put it is that computation incorporates much more knowledge about the world than control. For example, the computation of sound direction assumes that the head has a certain size, because the mapping from sound direction to interaural time differences depends on that size. Feedback control does not rely on such assumptions.

Therefore, more anticipation also means less robustness. Every bit of knowledge about the world that is implicitly assumed in a computation is a condition for the computation to fail—if the head grows, then the computation yields incorrect results. We will address these issues in more detail in chapter 7 when we discuss the notion of anticipation.

These remarks about epistemic closure actually apply more broadly to data processing, also called *information processing,* and not just computation. A system that processes data implements a formal mapping from input to output. Whether that mapping can be decomposed into elementary operations or not, the result of the processing only depends on the input data and on the abstract mapping, not on anything happening in the world.

Epistemic Closure in Connectionist Models

Epistemic closure is what allows modern artificial neural networks to adjust their weights by the backpropagation algorithm, or more generally what allows differentiable computing. The general principle of such systems is that the input is mapped to an output by a parameterized mapping ($\boldsymbol{y} = F(\boldsymbol{x}, \boldsymbol{p})$), and an error criterion E is defined on the output (figure 4.11). The parameters are then adjusted in a direction that decreases the error. One could try different modifications and choose the best one, but the key to the success of neural networks is to select this direction formally, without any trial. This is done typically by *calculating* the error change in every direction ($\partial E/\partial p_i$), based on the formal definition of both the error and the mapping. Parameters are then typically changed in the direction of the error gradient ($\Delta p_i \propto -\partial E/\partial p_i$). This trick is possible precisely because the system performs a computation, so that the effect of changes can be known in advance before those changes are actually realized, just by formal manipulation of the model.

For a closed loop system that includes the environment, gradient descent is not possible because the loop includes unknown elements—namely, the effect of the system's actions on the sensory feedback. For example, imagine a system that localizes sound sources by turning toward the louder ear. It has two sound sensors and two levers, but the spatial arrangement of the sensors and levers is unknown. For the system to work, the actions need to be connected to the correct lever, so that it turns toward the sound source and not away from it. We cannot calculate the error gradient, since the error is not formally defined: it depends on the arrangement of the real controlled system. The only option is trial and error. In robotics, a recent successful approach is precisely to use differentiable simulators of body and environment, so that gradient-based optimization techniques can be used (Newbury et al., 2024), but living organisms cannot rely on such strategies.

The issue of epistemic closure extends to distributed systems. Suppose that computation is distributed among nodes, each node doing some computation and exchanging with the other nodes. For a given node, the rest of the system

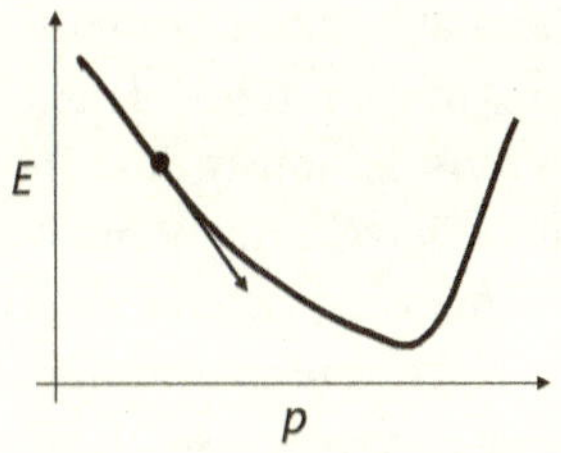

FIGURE 4.11. Gradient descent. The error E depends on a parameter p. To minimize error, p is moved in the opposite direction of the gradient (arrow), which requires prior knowledge of the gradient.

is the external world with which the node interacts. Consider again backpropagation in a formal neural network. The algorithm uses the chain rule of differentiation to distribute the calculation of the error gradient among the different nodes ("neurons") of the network. In practice, each neuron calculates some quantity that depends on the derivative of its activation function, and passes the result to neurons in the previous layer. Thus, messages are passed from neuron to neuron. For the calculation to work, the meaning of those messages must be formally specified, which means in practice that they must match a variable of the computation that the target neuron performs. In other words, in these exchanges, each neuron must assume that other neurons function in a certain prespecified way. If we turn to biology, this becomes problematic because cells do not have fixed properties that would be specified in advance. For example, individual neurons adapt, so that the relation between "inputs" and "output" changes over time.

The idea of a rigid meaning associated to neural signals is the concept of "neural representations" or "neural codes," which is central in neurocomputationalism. Here "meaning" simply refers to the computational state associated to the neural signals, such that neural processes can be interpreted in terms of computation. This is what it means that neurons "compute." We will discuss neural representations more fully in the next chapter. For now, we will show how it underlies the concept of "neural computation."

Neural Computation

Marr's Three Levels

As we have seen, not all behavior is computational. Some relies on interaction and dynamics. But the ant does not actually use a thread to find its nest. It seems to anticipate the direction of the nest based on its previous path. This is what motivates the claim that its brain "computes" that direction. But what does it mean exactly in terms of the activity of neurons? The concept of neural computation was perhaps most famously described by David Marr's "three-level" framework, which we have already mentioned (Marr, 1982). Marr claimed that information-processing systems must be analyzed at three levels, in order: the computational level, the algorithmic level, and the implementation level. He envisioned these levels as essentially independent. In sharp contrast with bottom-up approaches (chapter 3), this framework can be described as "top-down": it starts from a problem defined independently of the organism's biology, and then interprets the organism's physiology in that light. This is in line with the concept of machine.

The computational level describes the goal of the computation and the properties of the mapping from input to output. For example, in spatial navigation, the goal of the computation is to estimate the correct homing vector $\boldsymbol{x}$, which would lead the ant directly to the nest. The main property of this computation is that the homing vector does not depend on the outward path taken by the ant: the outcome of the computation must be path invariant. Then we specify the input. We consider that the velocity vector $\boldsymbol{v}$ is available to the ant, perhaps through the polarization of sunlight or a magnetic compass. Finally, the analysis of the problem led us to find that this mapping must be such that $\boldsymbol{x} = \int \boldsymbol{v}$. This is the "computational" level of analysis.

It should be clear by now that the term "computational" is rather unfortunate, since at this stage we have not discussed any method to implement the desired mapping, and not all successful methods are computations. It would be more accurate to call it the *problem level*: an analysis of the problem to be solved, and of its mathematical structure. A broader implicit assumption in this framework is that the problem is an information processing problem, that is, that of mapping some kind of data to some other kind of data. But as we have seen, not all behavior is of that kind; the person using a thread does not actually process data, even though it is still the case that $\boldsymbol{x}$ and $\boldsymbol{v}$ are related by the formula $\boldsymbol{x} = \int \boldsymbol{v}$.

The second level is the algorithmic level. It describes the computation that solves the information processing problem by manipulating the abstract variables that we have identified. In homing behavior, this would be the update: $\boldsymbol{x}(t+dt) = \boldsymbol{x}(t) + \boldsymbol{v}.dt$. In sound localization, this could be the calculation of the time difference between the two acoustical waves, followed by a table that maps that difference to sound direction. In other words, the algorithmic level is the formal description of the computation.

The third level is the implementation level, which interprets brain processes as the computation described formally in the second level. This interpretation typically takes the following form, as is described in various forms in the philosophical literature (Chalmers, 1996; Putnam, 1991; Shagrir, 2006) and earlier in Rosen's modeling relation (Rosen, 1985): a physical system is said to implement a computation when there is a mapping from physical states to computational states such that physical processes map to transitions in the formal model of computation.

For example, in the case of spatial navigation, the computational state is the homing vector $\boldsymbol{x}$, and the transition φ is the update over one time step: $\varphi(\boldsymbol{x}) = \boldsymbol{x} + \boldsymbol{v}.dt$, where $\boldsymbol{v}$ is the velocity vector, which is the input of the computation. The homing vector is calculated by iterating the transition. This algorithmic description is linked to brain processes by first postulating that there is a mapping f from the brain state $\boldsymbol{s}$ to the homing vector $\boldsymbol{x}$: $\boldsymbol{s} = f(\boldsymbol{x})$. We might say that the brain state "represents" $\boldsymbol{x}$ through the mapping f. The dynamics of the brain relates successive states: $\boldsymbol{s}(t+dt) = \Phi(\boldsymbol{s}(t))$. We then obtain the following diagram (strictly speaking, f should map $\boldsymbol{s}$ to $(\boldsymbol{x}, \boldsymbol{v})$, but $\boldsymbol{v}$ is omitted for readability):

$$\begin{array}{lcclr} \textit{Behavioral measurements} & \boldsymbol{x}(t) & \overset{\varphi}{\mapsto} & \boldsymbol{x}(t+dt) = \boldsymbol{x}(t) + \boldsymbol{v}(t)dt & \textit{Algorithm} \\ & f \uparrow & & f \uparrow & \textit{Representation} \\ \textit{Neural measurements} & s(t) & \overset{\Phi}{\mapsto} & s(t+dt) & \textit{Implementation} \end{array}$$

Cummins (1991) calls this the "Tower Bridge" model (in reference to London's Tower Bridge). What it means for the brain to implement the computation is that the diagram commutes: when the brain goes from one state to the next, the computational variables that the state represents go through one step of the computation. In mathematical form, we write: $\varphi \circ f = f \circ \Phi$. This diagram illustrates the different levels of analysis of an "information-processing" system according to Marr. Thus, the notion of representation is a key component of neural computation. But it should be stressed that "representation" is meant in a very narrow sense here: it is a purely formal relation between brain state and computational state, which is relative to a particular computational model. The computational state does not necessarily correspond to a mental state—in connectionism for example, it does not.

Let us look at the example of sound localization. A computation of sound direction might consist in calculating the interaural time difference (ITD) between the two monaural sounds $\boldsymbol{S}_L$ and $\boldsymbol{S}_R$, then mapping that ITD to the corresponding sound direction θ. The implementation of this computation might then look as follows:

$$\begin{array}{lccccc} \textit{Computation} & (\boldsymbol{S}_L, \boldsymbol{S}_R) & \overset{\varphi_1}{\mapsto} & \textit{ITD} & \overset{\varphi_2}{\mapsto} & \theta \\ & f_1 \uparrow & & f_2 \uparrow & & f_3 \uparrow \\ \textit{Brain} & \boldsymbol{s} & \overset{\Phi_1}{\mapsto} & \boldsymbol{s}' & \overset{\Phi_2}{\mapsto} & \boldsymbol{s}'' \end{array}$$

Here, the binaural sound $(\boldsymbol{S}_L, \boldsymbol{S}_R)$ is somehow "represented" in the brain state, in the sense that there is a mapping f_1 from brain state $\boldsymbol{s}$ to the sound. Then brain processes change the state $\boldsymbol{s}$ to a new state $\boldsymbol{s}'$ that can be mapped to the ITD of the sound, then to a state $\boldsymbol{s}''$ that can be mapped to the estimated sound direction, in such way that the diagram commutes ($\boldsymbol{s}''$ maps to a direction that equals the result of the computation). Typically, in the neuroscience literature, the states $\boldsymbol{s}$, $\boldsymbol{s}'$, and $\boldsymbol{s}''$ correspond to anatomically distinct brain regions. In sound localization, it is traditionally assumed that the auditory nerve ($\boldsymbol{s}$) represents the monaural acoustical waves, the medial superior olive ($\boldsymbol{s}'$) represents ITD, and the inferior colliculus ($\boldsymbol{s}''$) (or the auditory cortex) represents sound direction. In other words, the domain of the representation mappings (f_1, f_2, f_3) corresponds to distinct regions of the brain.

This concept of implementation based on a mapping model of representation (a "code") is the most central concept of neurocomputationalism, as it underlies computation, but also information (chapter 6) and prediction (chapter 7). Unfortunately, it is deeply flawed.

The Problem with Representation

The fundamental issue is that what we call a "representation" is an interpretation of neural activity by the observer, not something realized by the system. The issue has been noticed for a long time by philosophers of mind (for example, Searle [1990]; see Sprevak [2018] for a review), although some seem to believe that it might be fixable—it is not. For example, Hilary Putnam famously argued that virtually anything can be said to implement any computation, if we simply ask that there should be a mapping from physical states to computational states (Putnam, 1991). The basic idea of the argument is simply to assign each possible flow of the computation to a subset of trajectories of the underlying dynamical system, and to map physical states to the corresponding computational states. Although this particular argument can be questioned (Chalmers, 1996), it is easy to see that one cannot give a coherent account of neural computation by simply mapping neural states to computational states.

Suppose that we have found a brain area that represents the number of objects in a visual scene. The basis of this claim is that the activity of that area varies systematically with the number of objects, and not their size or color. Then a smarter neuroscientist arrives and claims to have found that the brain computes the square of the number of objects in another brain area, whose activity varies systematically with the squared number of objects. Therefore, the brain computes the square operation because the brain processes that relate the activity of the two brain areas map to the square operation in the domain of integers.

The problem, of course, is that we might just as well say that the second brain area represents the double of the number of objects, and so the brain now computes the doubling operation. Anything can be said, since it is the observer who comes up with the representation mapping. It has to be so, since neural signals do not belong to the domain of integers. But then, this means that if something computes $f(x)$, then it also computes $g(f(x))$, for any mapping g. Indeed, virtually anything computes anything. In fact, an information processing system that encodes the input as x, does some computing work, and outputs $y=f(x)$, already computes $f(x)$ *before* it even starts its work, since x might as well represent $f(x)$. For example, a camera can be said to compute the presence of Jennifer Aniston in the image, since there is a mapping from the camera image to the presence of Jennifer Aniston.

This is a problem to account for neural computation, but also to account for computation by any kind of physical system. For example, a machine that takes an input number specified by a knob and outputs a number specified by the level of a liquid might compute the next integer, or its double, or its square, depending on how we decide to read the liquid level.

It gets even worse: a pile of paper on a desk with a switch on the side may even be said to implement noncomputable functions. The switch is not connected to anything, but it can be in two states, on or off. This can be seen as an elementary automaton whose state consists of an input x that is written on the paper and a binary digit (0 or 1). Its transitions are simply that it goes from $(x, 0)$ to $(x, 1)$ and then stops—that is, $(x, 1)$ is a terminal state. The pile-of-paper automaton can be said to implement any function φ through the following commutative diagram:

$$\begin{array}{llcl} \textit{Abstract space} & x & \overset{\varphi}{\mapsto} & \varphi(x) \\ & f\uparrow & & f\uparrow \\ \textit{Pile of paper with switch} & (x,0) & \overset{\Phi}{\mapsto} & (x,1) \end{array}$$

where f is defined as follows:

$$f((x,s)) = \begin{cases} x & \text{if } s=0 \\ \varphi(x) & \text{if } s=1 \end{cases}$$

In particular, if x is a program together with its input, then we could choose the mapping φ that determines whether the program halts on that input. The pile of paper with switch could then be said to implement that mapping. Thus, one could write the program and its input on the pile of paper, push on the switch, and the pile-of-paper-with-switch would have computed whether the program halts on that input.

The problem is that φ is a famous example of a noncomputable mapping (this is known as the halting problem). Therefore, we have found an automaton that can implement something that no automaton can do. This paradox arises because we have authorized the representation mapping to be any kind of mapping, and in particular a noncomputable mapping. In other words, some of the work is done by the mapping, which is a formal construction of the observer, not something realized in the system.

We have encountered this issue in chapter 3: without constraints on the interface with the real world, virtually any connectome can be said to implement virtually anything, such as the identification of Jennifer Aniston in an image. The solution is *embodiment*: to consider not what the observer can do with neural activity, but how neural activity contributes to what the *organism* does.

But what about actual computers and machines? It seems that the implementation framework works with engineered devices. This is because, implicitly, we do not allow *any* mapping to be a representation, for otherwise we would use a pile of paper and a switch instead of computers, more powerful and less expensive. We implicitly assume that the engineer would choose "simple" representations, that is, representations adapted for the user, not something that requires some difficult cognitive work from the user. In the case of a squaring machine, the engineer would choose the same representation for the input and the output. All bits in a hard drive or a memory chip would be represented with the same convention. All of these make sense because machines are engineered, but none of these applies to brains, which are not engineered.

Many conceptual difficulties in brain theory arise from stubbornly trying to fit brains into the engineering mindset, instead of adopting the embodiment strategy—seeing action potentials as activity (as the name implies), not symbols. One of these engineering inspirations is the ubiquitous concept of neural "code" (Brette, 2019a). This terminology designates the representational mapping from neural activity to a computational variable, which is central in the concept of neural computation. It is also central in the idea that brains make "predictions" (chapter 7), as well as in the idea that the activity of a neuron can be "information" (chapter 6). We have started to see some of the conceptual issues with neural codes in the context of neural computation. We will now examine how neuroscientists have tried to address these issues. While some of these efforts have led to interesting findings, they have been essentially hopeless to provide a conceptual basis for computation, information or prediction in the brain.

5

Neural Codes

NEURAL REPRESENTATION plays a central role in computational theories of the brain, because it offers a bridge between brain processes and computational descriptions of cognition, through the concept of "implementation," as we have seen in the previous chapter. As philosopher of computationalism Jerry Fodor wrote bluntly: "there can be no computation without representation" (Fodor, 1981).

In the neuroscientific literature, a "neural representation" is typically something somewhere in the brain that stands for something else. For example, in the context of sound localization, a neural representation of the mouse's location in the cat's brain might be the firing rate of a neuron, or of a group of neurons, in some neural structure in the auditory brainstem. That firing rate covaries with the mouse's location, and in that sense, it is said to "represent" or "encode" the location. It is implied that this representation is then somehow "used" by the brain to direct its behavior, such as turning toward the mouse.

Neural representations are often called "neural codes," which is more explicit (Brette, 2019a). A code is a correspondence between two domains. A well-known example is Morse code, used to convey messages over telegraph lines: each letter corresponds to a particular sequence of dots and dashes, so that someone reading a sequence of dots and dashes may infer the original message in English (figure 5.1). In the same sense, spike trains of retinal neurons are said to encode or represent the image, or neurons in the hippocampus are said to encode the animal's position (Moser et al., 2008).

In many cases, this simply means that the activity of some neurons covaries with some property of the experimental configuration. Often, that property is varied while the context is fixed. For example, a neuroimaging study on language claims that "the meaning of language is represented in regions of the cerebral cortex collectively known as the 'semantic system'" (Huth et al., 2016). The basis of this claim is that certain areas of the brain tend to consume more

A	·–	N	–·
B	–···	O	–––
C	–·–·	P	·––·
D	–··	Q	––·–
E	·	R	·–·
F	··–·	S	···
G	––·	T	–
H	····	U	··–
I	··	V	···–
J	·–––	W	·––
K	–·–	X	–··–
L	·–··	Y	–·––
M	––	Z	––··

FIGURE 5.1. Morse code table.

energy than others when specific words are heard, which allows drawing a map of words over the cortex. Thus, there is a correspondence between something in the world (words) or in the mind (meaning of words) and something in the brain. Technically, this correspondence is generally statistical, in the sense of Shannon's communication theory (Shannon, 1948).

Neural activity has been described as "neural codes" for at least a century. In the 1920s, Edgar Adrian (recipient of the Nobel prize in 1932) examined the electrical responses of nerves to stimuli, such as touching the skin. He observed that the nerve would fire electrical impulses ("action potentials" or "spikes") at a frequency that increased with the strength of the stimulus (Adrian, 1928). He compared these series of spikes to Morse code. By the 1970s, entire books were already devoted to "sensory coding" (Somjen, 1972; Uttal, 1973), and a seminal review titled "Neural Coding" would illustrate its technical appendix with hieroglyphs (Bullock and Perkel, 1968).

There are several issues with neural codes, such as whether they can be considered as information for the organism (chapter 6) or whether it makes sense to consider them as brain states (chapter 8). In this chapter, we will focus on their role in the theory of neural computation. Neural codes are indeed used to substantiate what it means for the brain to "compute." For example, the retina encodes the image, the inferotemporal cortex (IT) encodes face identity, and so brain processes along the visual hierarchy (thalamus, visual cortex) implement the computation of face identity from the image. This is formalized by the following commuting diagram:

$$\begin{array}{llcl} \textit{Computational space} & \text{image} & \overset{\varphi}{\mapsto} & \text{face identity} \\ & f \uparrow & & g \uparrow \\ \textit{Brain} & \text{retina} & \overset{\Phi}{\mapsto} & \text{IT} \end{array}$$

But as we have seen in chapter 4, the problem with this framework is that the retina might just as well be said to encode face identity, through a more complicated encoding—namely, the composition of φ and f. The traditional proposition is to constrain "neural representations" to correspond to the states of anatomically separated brain elements. For example, it is trivial and uninteresting to say that the retina encodes face identity, but it is much less trivial to say that a single neuron in IT encodes the presence of Jerry Fodor—that is, that one can infer whether Jerry Fodor is in the visual field from the activity of just one neuron. But for this to be indeed nontrivial, the neuron's activity should not simply vary when a photo of Jerry Fodor is substituted for a photo of David Marr in the visual field, because a neuron of the retina would also pass the test (just pick one pixel that differs between the two photos). There must indeed be a *mapping* from the neuron's activity to the presence of Jerry Fodor, independently of the experimental context, and not just a contextual correlate. Unfortunately, this basic symbolic quality is usually not met by so-called neural codes (*Codes or Correlates?*). A proposed solution to this empirical problem is to introduce "neural population codes" or "neural manifolds," where a group of neurons jointly encodes a set of properties (*Population Codes*). But this amplifies the theoretical problem: now the neurons can be said to compute any joint function of all the properties, depending on the observer's imagination. Unfortunately, the conceptual issue with neural codes is fundamental, and undermines the entire neurocomputational edifice (*The Theoretical Problem with Neural Codes* and *The Fundamental Incompleteness of Parametric Codes*). In addition, cognitivists have been insisting for a long time that the kind of representations necessary to explain cognition must meet an additional criterion, namely, they must be *structured* representations, but neural codes are typically unstructured (*Structure in Neural Codes*).

In other words, neural codes fail both basic and advanced tests for representations, as needed for a theory of neural computation. To move forward, we will end this chapter by reflecting on the phenomena that representation is supposed to explain (*Why Representations?*).

Codes or Correlates?

The Tuning Curve

A key historical concept in neural coding is the *tuning curve*, which quantifies the responses of a neuron to parameterized stimuli. For example, in the late 1950s, David Hubel and Torsten Wiesel found that neurons in the occipital cortex of the cat fire spikes when edges of certain angles are moved across a specific region of the cat's visual field (Hubel and Wiesel, 1959). This observation can be quantified by measuring the neuron's firing rate as a function of bar angle. This relationship is the neuron's tuning curve for orientation (figure 5.2). As the firing rate varies systematically with bar orientation, it is then commonly claimed that the neuron *encodes* or *represents* orientation (Seriès et al., 2004).

Indeed, superficially, the firing rate seems to have the quality of a neural code for orientation: the firing rate can be mapped to the bar's orientation. But this relationship is contextual. If other properties of the bar are changed—for example, its contrast—then the neuron's firing rate also changes. Therefore, the firing rate does not actually stand for orientation, it only correlates with it. It follows that one cannot use the neuron's firing rate as a computational variable representing orientation in an algorithm.

Francis Crick (one of the discoverers of the structure of DNA, then turned neuroscientist) warned against the "fallacy of the overwise neuron" (Crick, 1979), by giving the example of a color-tuned neuron. Cones in the retina, for example, are broadly tuned to wavelength (Schnapf et al., 1987): when one varies the wavelength of light, the transduction current varies systematically in amplitude. Within the context of such an experiment, it is possible to deduce the wavelength by just observing the amplitude of the transduced current. Yet animals or humans with a single functional type of cone are color blind. Why is it that the experimenter can deduce the light's wavelength from the activity of a single type of cone but people with those cones cannot? The answer is simple: the deduction is only possible when one already knows that in this particular experiment, only the wavelength is varied while everything else is constant. For example, if light intensity were not fixed, it would be impossible to recover the wavelength from the current amplitude, because the amplitude also varies with intensity. In other words, inferring wavelength from the activity of a single type of cone is only possible for people who live in a world of pure monochromatic light sources of fixed intensity.

Obviously, the real world is more complex than a tuning curve experiment. When we marvel about the sound localization abilities of cats, we are not astonished by the fact that the cat's brain seems to be able to find entries in a

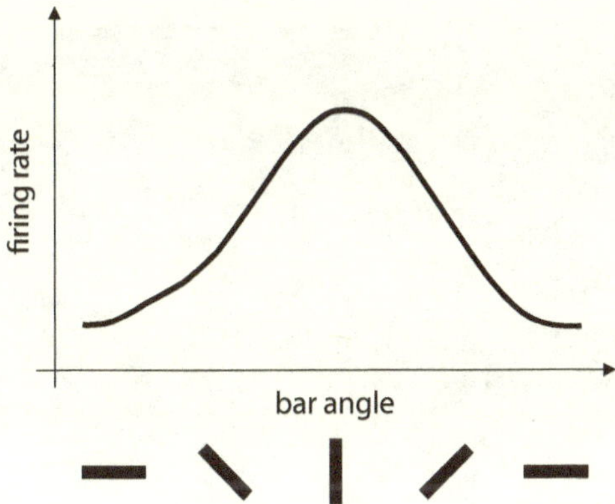

FIGURE 5.2. Tuning curve to the angle of a visually presented bar.

correspondence table (sound location versus firing rate). The prowess is rather that the cat's behavior seems to be undisturbed by a vast array of potential distractors, such as loudness, spectral properties, noise, reverberation, other sound sources, and so on. It is this invariance to nonrelevant features that is challenging from an engineering perspective. If a physiological signal varies with sound location, but also with loudness, timbre, and the cat's mood, then it cannot be used as a proxy for sound location in an algorithm that drives navigation behavior. Formally speaking, tuning curves do not fit the implementation formalism, because the relation between the neuron's activity and the formal variable is not a mapping.

Tuning curves quantify the sensitivity to a parameter, when everything else is fixed. Neurons of the primary visual cortex are sensitive to local visual orientation, but they do not "encode" orientation in any meaningful way. To make their firing rate a code in the computational sense, one should demonstrate instead the orthogonal property: that the firing rate is invariant to all the dimensions that are *not* encoded.

In some subfields of neuroscience, invariance to nonrelevant dimensions is, appropriately, a central point of ongoing discussions. For example, in the hippocampus of rats, there are neurons called "place cells" that fire whenever the rat is in a particular location (Moser et al., 2008, 2015; Rowland et al., 2016). The set of locations where the cell fires strongly is called the *place field*, analog to the tuning curve but in two dimensions. One of the key findings about place cells is that the place field is invariant to changes in the proximal environment, and it is maintained and updated with movement in total darkness. Thus, these are not simply visually sensitive cells.

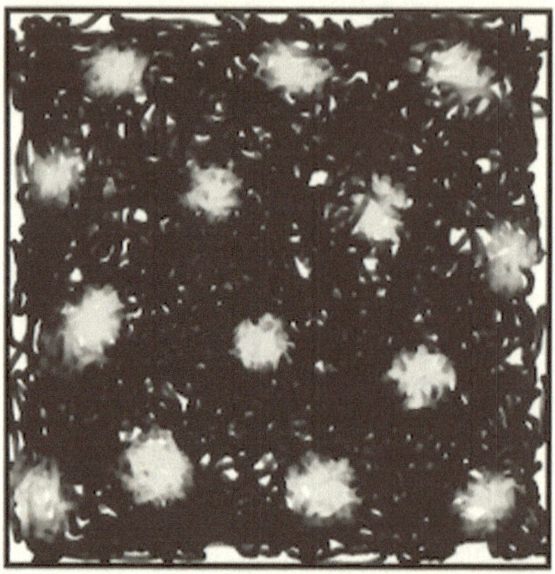

FIGURE 5.3. Spatial field of a grid cell of the hippocampus. The rat's trajectories are in black; white spots show regions where the recorded neuron spikes more. Original figure by Weber SN, Sprekeler H. 2018. https://doi.org/10.7554/eLife.34560 / CC BY https://creativecommons.org/licenses/by/4.0/.

There are also grid cells, which are neurons that fire at the nodes of a hexagonal grid (figure 5.3). These cells are active at the same positions independently of speed and direction of movement. Other types of cells, called head direction cells, are active at preferential orientations of the head relative to the environment, independently of the position in the environment. It is these invariance properties that have led neuroscientists to claim that the activity of these cells forms a representation of space. There are issues with this claim, which we will examine later, but at least this neural activity displays some degree of invariance: the cells fire when the situation has certain general properties, independently of the particular instantiations. This is a key prerequisite for the neurocomputational framework.

However, in other areas of neuroscience, the fallacy of the overwise neuron is surprisingly common.

Overwise Auditory Neurons

When a source on the right emits a sound, the acoustical wave arrives slightly earlier on the right than on the left (see figure 4.7). This timing difference is called interaural time difference (ITD). In mammals, many neurons in the auditory brainstem (especially the medial superior olive, MSO, and the inferior colliculus, IC) are sensitive to ITD (Joris et al., 1998): when a sound is played through earphones and the ITD is varied, the firing rate of those neurons changes. On the basis of such tuning curves, it has been claimed that it is possible to infer the relative position of sound sources from the activity of an individual neuron, with the same accuracy as humans, as measured psychophysically (Shackleton et al., 2003; Skottun, 1998).

What does it mean to compare the activity of a neuron with the perception of a person, without even considering the nervous system that the neuron belongs to? This relation is what Davida Teller called a "linking proposition," which is generally implicit (Teller, 1984). Here it is implied that the neuron's activity is a code for sound location. If it is indeed a code, then the finding is rather surprising, since the animal could rely on this single neuron's activity as a proxy for sound location, making the thousands of other ITD-sensitive neurons in the brain redundant. (There are about 15,000 neurons in the MSO of humans [Kulesza 2007].) This seems to be a waste of energy.

But this is not correct, for exactly the same reason that people with a single type of cone are color-blind. All neurons in the auditory brainstem are also sensitive to sound frequency within some range, and therefore to the spectrum of a complex sound, and most are sensitive to sound intensity. Therefore, these neurons are sensitive to ITD but they do not encode ITD. There is simply no way to infer anything about sound location from just the activity of a single neuron, just like it is impossible to infer the color of an object from the number of photons absorbed by a cone.

Similarly, one of the most influential theories of sound localization proposes that sound location is encoded in the relative average activity of two populations of neurons, one in each hemisphere (Grothe et al., 2010). This concept is called "slope coding." The empirical basis for this claim is essentially that when the ITD of a sound is varied, the relative activity of those two neural populations varies accordingly. This is the case in the auditory brainstem as well as in the auditory cortex. But this activity also varies with various aspects of sounds, and therefore it is not a code at all. It can be shown indeed that, because of this sensitivity to contextual variables, the sound's ITD cannot be accurately recovered from population averages unless the auditory world consists of identical pure tones (Goodman et al., 2013).

An astonishing feature of neural coding discussions is that they pretend to inform us about how brains drive behavior, but without ever considering the system responsible for that behavior, only the responses of an isolated neuron, or of an isolated brain area. If we did take some interest in that system, we would find for example that a unilateral lesion of the MSO or IC results in sound localization deficits in the contralateral field only (Jenkins and Masterton, 1982). This directly shows that it is not the relative activity of the two hemispheres that matters. In the same way, electrical stimulation of the cat's IC triggers an orienting response toward a particular contralateral direction, with stronger stimulations resulting in responses with a larger part of the body (one pinna, both pinnae, and eyes), but toward the same direction (Syka and Straschill, 1970). Again, the relative activity changes but the direction of the behavioral response does not.

Of course, it would indeed be very remarkable if we found neurons or neuron populations that really encode the sound's ITD—that is, which fire only as a function of the sound's ITD and nothing else. To show this, we would need to measure not ITD tuning curves, but responses to other kinds of changes. Such neurons have not been found.

Another major cue for sound location is the interaural intensity difference (IID), but here also, the situation is similarly confused. Many mammals, including humans, use IIDs to localize high-frequency sounds in the horizontal plane. The leading theory is that neurons in the lateral superior olive (LSO) encode IIDs by subtracting the intensity of the two monaural signals. To support this view, neuroscientists found that principal neurons of the LSO have monotonous tuning curves for IID. On the fringe of the leading theory, several authors pointed out that those neurons are also sensitive to ITDs (Joris and Yin, 1995), level, and spectrum (Tsai et al., 2010). This alone should cast doubts on the narrative that the activity of those neurons is a code for IID. It turned out recently that experimenters had actually been recording interneurons instead of the principal neurons projecting to other areas, which were missed because they fire only transiently (Franken et al., 2018). Benichoux and Tollin (2018) commented that this recent development "is a good example of how prior expectations can involuntarily mislead scientific endeavor."

Thus, all these so-called neural codes for ITD or IID do not meet the most basic criterion for a computational variable representing ITD or IID.

Context-Dependence of Neural Codes

The danger of tuning curves is that they tend to hide the fact that they are measurements done with a fixed experimental context. For example, we measure the firing rate of a neuron while the location of a sound is varied, but the sound itself is always the same pure tone of fixed frequency and intensity. When we look at the tuning curve, we can see that the firing rate is a function of sound location, and it gives the false impression that we have uncovered a general fact: the firing rate of this neuron is a function of sound location. But that is the case in the experiment by construction, not by discovery: the firing rate cannot be a function of anything else since everything else is fixed. In reality, the neuron's firing rate is not at all a function of sound location, it is a function of other stimulus properties, and very possibly of other things beyond the stimulus (e.g., mental state). It follows that the neuron's firing rate cannot be a computational variable standing for sound location.

As it turns out, the general case is that neural responses depend very much on context. This is well known in the spatial navigation system. In the same environment, place cells fire at reliable positions over extended periods of time, but minor alterations of the environment can result in large changes in the place field of hippocampal places cells, a phenomenon called "remapping" (Colgin et al., 2008). In fact, place cells may also respond to particular states in a nonspatial behavioral task, such as manipulating the frequency of a tone with a joystick (Aronov et al., 2017). Thus, if the firing pattern of place cells is a computational state, then it is one that is relative to a particular cognitive context. After all, this is often the case of computational variables in a program: the same variable *x* may be used in different algorithms within the same program, and stand for different things.

But for this to possibly work, there must be a fixed context, which the animal recognizes as such, such as an environment that the animal has already explored. This cannot work for what we call "context" in perceptual experiments, which is simply everything else that the experimenter considers irrelevant for *her* experiment. For example, the experimenter might be interested in the orientation of bar, but when the first bar is presented to the animal, the bar has an orientation, but also a width, length, and color. The experimenter has decided to keep those other dimensions fixed, but the animal does not know that in advance—let alone the recorded neuron. Thus, we find that the "neural code" is contextual, but what we call context is not the previous experience of the animal but the design of the experiment. The idea of contextual codes also leaves out the critical question of how the "context" is encoded, which we will examine later.

How common is this contextual dependence of neural responses? In fact, orientation-sensitive neurons in the primary visual cortex are sensitive not only to various other dimensions such as contrast but also to the surrounding spatial context (Bolz and Gilbert, 1986; Hubel and Wiesel, 1968). They are also modulated by locomotion (Pakan et al., 2018) and prior presentation of sounds (Chanauria et al., 2018). In the inferotemporal cortex, where neurons have been reported to encode the identity of specific faces, it has been shown that slight pixelwise image modifications can dramatically alter the neuron's response (Guo et al., 2022). In addition, the activity of visual cortical neurons also depends on what the animal is trying to do in the experiment (Gilbert and Li, 2013). For the same visual stimulus, the neuron's activity will vary with stimulus parameters in one particular way when the animal performs a given task, and in a completely different way when the animal does another task. Specifically, the activity of neurons appears to depend specifically on those object properties that are relevant to that task. It seems to make sense if the spikes produced by neurons are indeed framed as *activity* (what neurons do), but it does not fit so well with the idea that the number of spikes produced by a neuron is a computational variable standing for some particular property.

Let us see more concretely how neurocomputational theories of perception deal with context dependence with an example. An influential theory, the "Bayesian brain," holds that neurons in sensory areas of the brain represent sensory information through their tuning curves (Knill and Pouget, 2004). Specifically, the average number of spikes produced by a neuron i in response to a stimulus is a function of a parameter θ. This is what an experimenter measures and reports as the tuning curve $f_i(\theta)$. Once we know these tuning curves, it is then possible to guess the stimulus parameter θ from the number of spikes produced by the neurons, using Bayesian calculus, as explained in (Jazayeri and Movshon, 2006). In practice, this means that each spike produced by a neuron i increases the log likelihood that the parameter value was θ by $\log f_i(\theta)$ (figure 5.4). In other words, the tuning curve quantifies the likelihood of each parameter value given that the neuron spikes, which makes intuitive sense: if a neuron has a tuning curve peaking at θ and that neuron spikes a lot, then the stimulus was probably close to θ.

In this neurocomputational theory, the log likelihood of a particular stimulus parameter value can be computed by a weighted sum of the spikes of all neurons, where the weights are determined by the tuning curves. In the connectionist formalism, these weights correspond to synaptic strengths, which are presumably learned by experience. Thus, we have a neurocomputational theory of statistical inference, where the formalism of Bayesian calculus is implemented by the algebra of synaptic weights and spike counts.

All of this works beautifully as long the experimental setting is fixed. But as soon as the animal does something else, or the scene is different, the tuning curves change because they are context-dependent, and so the synaptic strengths do not correspond to the correct tuning curves anymore. It follows that the computation of the log likelihood is now false.

This failure occurs because the measured firing properties of neurons are not "codes" for the stimulus parameter, in a computational sense. The tuning

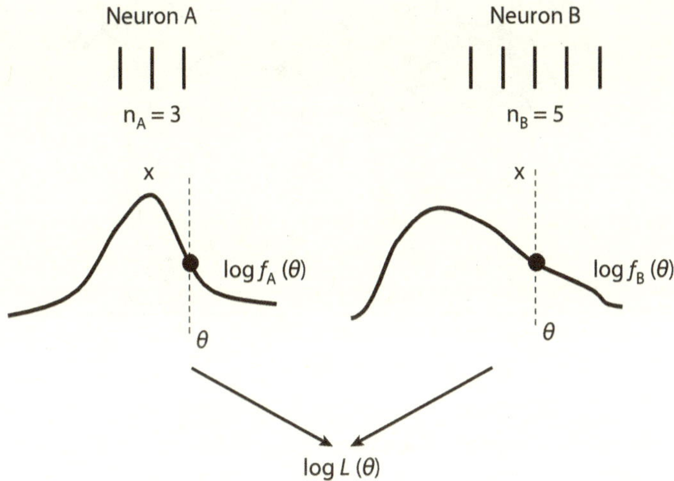

FIGURE 5.4. Bayesian inference of a stimulus parameter θ from neural activity (after Jazayeri and Movshon, 2006).

curve only measures the extent to which neural firing is sensitive to the parameter, in a given context. They cannot be used as the basis of a computational scheme.

The fact that neurons are typically sensitive to many dimensions of stimuli and tasks is now well established. For this reason, many neuroscientists now think that perceptual objects and properties are represented jointly by the activity of many neurons, rather than by individual neurons. In this case, what counts as a representation is a mapping from the state of a large neural population to some abstract, typically multidimensional, domain. Unfortunately, "population codes" also depend on context and time, and there is no satisfying way to group neurons into populations.

Population Codes

Hiding the Complexity

In textbooks, the properties of neurons are often presented as cartoons—for example, tuning curves look like bell-shaped Gaussian curves of equal width—and theories are based on those idealizations. But all neurophysiologists know the uncomfortable truth that transpires in more technical publications: surely, there are general trends in the activity of neurons in a given brain area, and these trends are the typical material for theory, but overall,

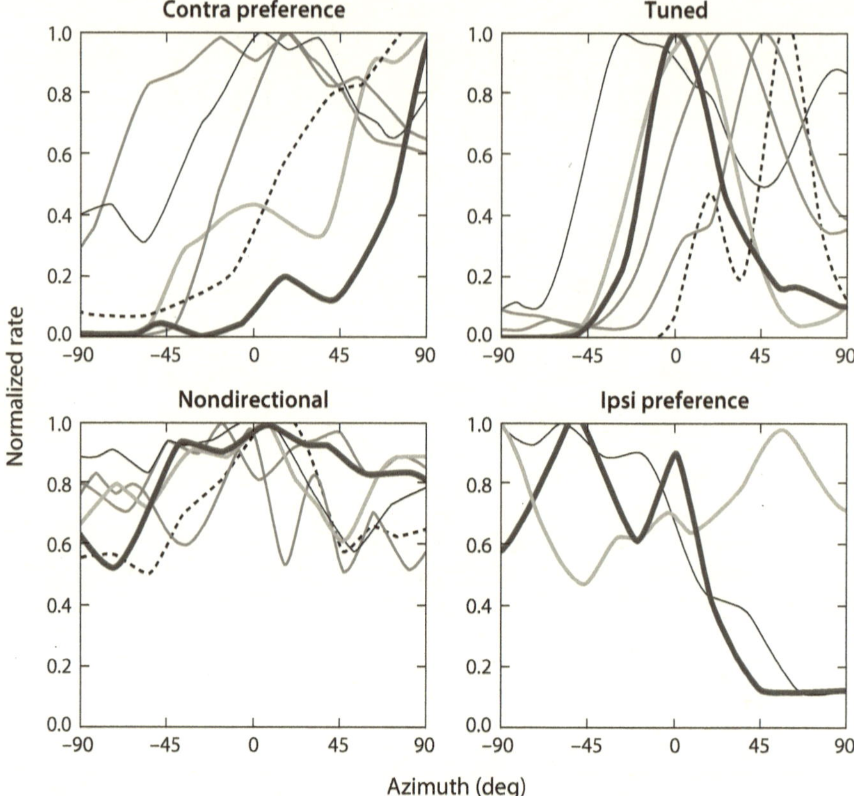

FIGURE 5.5. Tuning curves of neurons of the cat's inferior colliculus to the azimuth of spatialized sounds, grouped in four descriptive categories (adapted from Delgutte et al., 1999).

neural measurements are quite messy. It is not just that neurons are sensitive to many aspects of stimuli and tasks. In addition, each neuron reacts to stimuli in its own bizarre way. Tuning curves are generally not linear and often have multiple peaks (figure 5.5). Very often, the activity of different neurons covaries in certain ways, even when the stimulus is fixed.

The modern approach to this problem relies on the development of large-scale recording technologies, such as calcium imaging or multielectrode arrays. The animal is presented with some stimuli, or it performs a certain task, and the activity of many neurons is simultaneously recorded. This activity then forms a "population code" for whatever experimental parameters are being considered. Technically, in modern data analysis pipelines, this activity is represented as a high-dimensional vector of firing rates, which spans a vector space called a "neural manifold" (Ebitz and Hayden, 2021; Gallego et al., 2017;

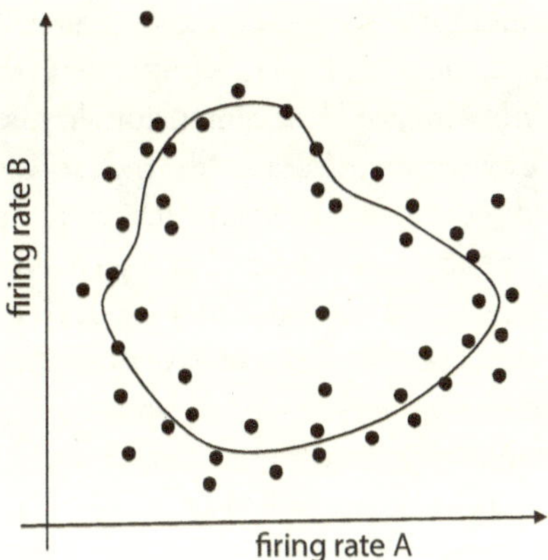

FIGURE 5.6. One-dimensional neural manifold (curve) summarizing the joint activity of two neurons.

Langdon et al., 2023). Of course, this is an idealization, since in mathematics a manifold is a smooth subspace, while measured neural responses consist of spikes and a finite number of observations.

Neural manifold analysis makes substantial progress over tuning curve analysis, because instead of measuring the correlation between one experimental parameter and neural activity, we try to describe all the activity of the population, whether it covaries with explicitly identified dimensions of the experiment or not. Typically, this amounts to identifying a lower-dimensional subspace (the manifold) such that most neural activity is close to the manifold (figure 5.6). One can then define a few coordinates on the manifold such that the activity of the entire population can be mapped at any given time to its coordinates (in figure 5.6, the position along the closed curve).

For example, the firing rate of neurons of the anterodorsal thalamic nucleus of mice during spatial navigation can be summarized by a one-dimensional neural manifold, with the topology of a circle or "ring" (Chaudhuri et al., 2019), similar to figure 5.6. It turns out that the position on that one-dimensional structure can be related to the direction of the mouse's head.

Note that the circular structure of the neural manifold during spatial navigation is not a very surprising finding, given the previous work on single neurons. Indeed, if the activity of neurons only depends on head direction θ, as characterized by the tuning curve $f_i(\theta)$, then obviously the joint neural activity can be summarized by the angular quantity θ. As θ varies during spatial navigation, the activity of all neurons draws a ring-like structure in high-dimensional space, which is the image of a circle by the joint mapping $\{f_i\}$. The ring topology is simply the topology of the input space.

Nevertheless, manifold analysis adds some systematicity to standard tuning curves analyses. Indeed, the remarkable finding is that, in contrast with tuning curve experiments, the observed low dimensionality is not due to the constrained nature of the experiment, because the mouse was freely moving, with many other variable dimensions than head's direction. Apparently, these neurons do not seem to care about anything other than head direction. Even more remarkable, the same ring topology was observed during REM sleep—that is, when there was no stimulus at all. It could have been the case that the activity of those cells also depended on some unmeasured internal factor, such that the neural manifold would have two dimensions or more. In fact, this is exactly what the authors found during non-REM sleep.

Thus, manifold analysis establishes the fact that during spatial navigation and REM sleep, the activity of those neurons appears to depend mostly on head direction, at least at a gross timescale of 50–100 ms (we will come back to this in chapter 8). However, we should not be led to believe that we have uncovered a particular way in which those neurons represent head direction, through a beautiful "ring structure" in high-dimensional space. As noted earlier, the fact that the neural manifold has a ring topology is a direct consequence of the fact that the neurons are mainly sensitive to head direction and direction is a variable with ring topology. Representing head direction by coordinates on a curved manifold hides but does not erase the weirdness of neural responses shown earlier, neither does it explain it. In contrast with models of this system, there are still neurons that fire at two different directions, others that do not seem to care about direction. Manifold analysis does not explain that fact, nor does it uncover any particular organization.

Nonetheless, we should not undervalue the remarkable finding that neural activity seems in this case to depend only on one particular variable of interest, head direction θ. Thus, there is indeed a mapping from neural activity to θ, as required by the computational implementation framework. However, we will see in chapter 8 that this is not sufficient to interpret the firing rate of those neurons as a computational state. But first, we ask whether neural populations have similar properties outside the navigation system.

Context-Dependence of Neural Activity

The navigation system might be one of the best fits to the computational framework, because many neurons seem to be sensitive mainly to certain specific abstract features (although the activity of place cells is environment-specific). What has manifold analysis revealed about other brain structures? First, it has confirmed the findings of single-neuron analyses: mixed selectivity is widespread, and responses are context-dependent. This is the case in particular in sensory cortex, motor cortex, and prefrontal cortex (involved in working memory). Neural manifolds often have low dimensionality, but the dimensions change depending on the task.

How significant is the fact that measures of neural activity have low dimensionality? If a task has *n* degrees of freedom and the activity of neurons is thought to command behavior, then we expect neural activity to vary roughly over *n* main dimensions. This is a consequence of the structure of the task, rather than of the particular way in which neural interactions achieve the task. It is known in particular that movements are typically produced by so-called motor synergies—that is, by the activation of specific groups of muscles. Animals have many muscles, so it would be possible in principle for an animal to produce the same task such as grasping an object in various ways. (This is known as *motor redundancy*; Bernstein [1967].) But instead, animals activate just a few groups of muscles, so that the contraction of all the muscles in the body can be described by a small number of dimensions. In this context, it is not completely unexpected that the activity of motor cortical neurons, believed to indirectly command the contraction of muscles, can be described by just a few dimensions (Gallego et al., 2017). However, motor synergies depend on the task—for example, walking, jumping, and swimming (d'Avella and Bizzi, 2005). This implies that there cannot be a fixed mapping from neural activity to the activation of motor synergies. How many motor synergies exist across all possible tasks? Here is a recent estimate: "the numbers of all task-specific and shared synergies combined may exceed the number of relevant muscles" (Bizzi and Cheung, 2013).

In the same way, in sensory tasks where input stimuli are varied along *n* dimensions, we would expect the activity of a large population of sensory neurons driven by those stimuli to also vary along *n* dimensions. This observation derives from the structure of the task rather than from the way the neurons deal with the stimuli. Quite logically, it is indeed found that the dimensionality of neural manifolds in sensory areas varies with the kind of stimulus (Jazayeri and Ostojic, 2021). It could not be otherwise, since dimensions are defined relative to a particular category of stimuli (e.g., the orientation of a grating, the size of an object, etc.), as we will discuss further in the next section. In the standard connectionist view, the structure of the neural manifold is determined by the connectivity structure of the network (Langdon et al., 2023), but this does not work if that structure changes every time the animal does something else.

Thus, in many cases, the relation between neural activity and task dimensions is (not surprisingly) task-dependent. In addition, it turns out that this relation also changes across the task or over longer periods.

Drift and Adaptation

This is the case for example in prefrontal cortical neurons during the course of a working memory task (Stokes, 2015), or with odor-evoked responses in mouse piriform cortex over days (Schoonover et al., 2021), a phenomenon called "representational drift" (Driscoll et al., 2022).

While the long-term instability of neural responses has only been uncovered recently, thanks to technical advances in long-term recording, it has been known for a century that the response of sensory neurons is also not stable over shorter periods. The most obvious fact is that neural responses to a constant stimulus take the form of a series of electrical impulses, "action potentials" or "spikes"—that is, a variable signal: there is no fixed relation between any stimulus property and the state of a neuron at a given time (this will be the subject of chapter 8). Neuroscientists typically circumvent this problem by reporting "firing rates"—that is, the number of spikes produced over some period of time, or over repeated trials.

But even with this trick, it is found that the firing rate changes over time for the same stimulus. For example, in the 1920s, Edgar Adrian examined the electrical responses of nerves to sensory stimuli, such as touching the skin (Adrian and Zotterman, 1926a,b). He observed that the nerve would fire spikes at a frequency that increased with stimulus strength, but the frequency would decrease over seconds. More generally, properties of sensory neurons adapt to input statistics, so that neural responses depend on the historical context preceding the stimulus, over several timescales (Barlow et al., 1957; Hosoya et al., 2005). Time-varying responses to constant stimuli have also been documented in detail in olfaction (Laurent, 2002; Mazor and Laurent, 2005).

These phenomena have been termed "representational drift" or "dynamical representations," but with the same logic, it could be claimed that the decimals of π represent the decimals of e, only with a representation that changes from digit to digit (figure 5.7). It becomes rather unclear what is "representational" at all about such time-varying correspondences.

Thus, the systematic analysis of the activity of neural populations has confirmed that "neural representations" are associations between experimental dimensions and neural activity that are stimulus-dependent, context-dependent, task-dependent, relevance-dependent, and finally, time-dependent across various timescales, from milliseconds to seconds to days. In fact, the "represented" dimensions also depend on the dimensionality reduction method (Langdon et al., 2023).

At this stage, the representational and coding terminology starts sounding like epicycles. In the geocentric model of astronomy, all planets were supposed to revolve around the Earth in circles, because circles are perfect geometrical

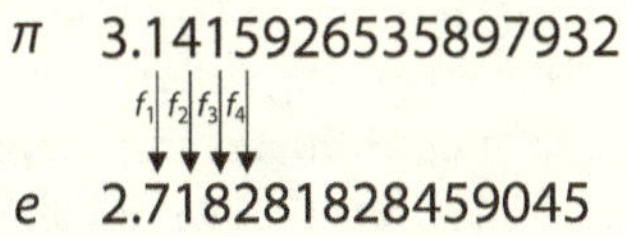

π 3.1415926535897932

f_1 f_2 f_3 f_4

e 2.718281828459045

FIGURE 5.7. Drifting representation of the decimals of π as those of e.

forms. But it certainly did not appear so, and therefore Ptolemy proposed that planets moved along small circles called epicycles, themselves orbiting around the Earth. Then increasingly more epicycles had to be added to account for empirical observations. In the same way, empirical research shows that neural activity cannot be mapped to symbols, unless we allow the mapping to change every time we make a new observation.

Thus, neural manifold analysis has provided techniques to summarize and visualize neural activity in various contexts, but it has *not* confirmed that this activity is a representation or a code in the computational sense. What neural population analysis has also *not* confirmed is that neural populations form computational units coarser than the individual neuron.

Neural Populations as Computational Units

Because of the early conceptual difficulties with the mixed selectivity and context-dependence of individual neurons, the neurocomputational framework has been shifted to coarser levels. In mean field approaches (a pioneer work being the Wilson-Cowan model [Wilson and Cowan, 1972]), the activity of neurons in a spatially delimited area is lumped together into a single variable. But systematic analyses of neural activity have shown that nearby neurons can have very diverse response properties. This is the case with place cells of rats, for example.

In neural manifold approaches, the population can be diverse but its activity can be summarized by an analysis of variance. For example, the activity of n neurons $(x_1, \ldots, x_n)$ might be summarized as a function of two variables y_1 and y_2: $(x_1, \ldots, x_n) \approx f(y_1, y_2)$. Here, y_1 and y_2 can be considered as coordinates on the neural manifold. Each coordinate is a function of the activity of the entire population, a particular projection. Thus, it has been proposed that the coordinates defined over the population are the units of computation or dynamics, rather than the individual neurons (Ebitz and Hayden, 2021; Saxena and Cunningham, 2019).

However, this is not correct for two reasons. The main reason is that, as reviewed earlier, neural manifolds are not stable entities. In particular, they depend on context and time. This implies that, depending on context and time, the coordinates will actually depend on the activity of individual neurons in different ways, either involving different neurons or with different weights. Even the number of coordinates is variable. Therefore, manifold analysis does not define coarse units that would allow us to ignore the finer level of individual neurons.

The second reason is that manifold coordinates are units of statistical description, not necessarily of dynamics or computation. To be units of dynamics, one would need to show not only that neural activity can be described by a few coordinates, but also that what drives the evolution of those coordinates is only their present state, and not the activity of the individual neurons. This might be the case, or not. For example, it could be that neural activity on a 100 ms scale is well described by just two dimensions, but that its evolution depends on activity occurring on a finer timescale. We will see in chapter 8 that this is a likely possibility.

Thus, whether based on individual neurons or populations, the "neural codes" regularly reported by neuroscientists do not meet the basic criterion of the implementation framework—namely, that they should be a mapping from neural activity to a formal symbol. What we will see now is that this framework is conceptually flawed.

The Theoretical Problem with Neural Codes

Neural Codes Are Formal Constructions of the Observer

Remember that to frame neural activity as computation, we interpret the activity of some neurons as a computational state, so that brain processes correspond to transitions in a computational model. This interpretation is what we call a neural code, it is a mapping from neural measurements to a formal symbolic space. But interpretations live in the observer's mind; they are not observed in the brain—contrary to what is sometimes claimed (Thomson and Piccinini, 2018). For a simple reason: in the brain, there are neurons, electricity, molecules; there are no bar orientations or face labels.

In chapter 4, I outlined a simple example of a number code. Suppose the firing rate of a neuron varies lawfully with the number of objects n in the visual system, and with nothing else. This neuron encodes n. This is the interpretation that the experimenter would normally make, since *she* decided to look for *how* neurons encode n. It is the experimenter who decides what is being encoded. But if instead the experimenter asked *what* neurons encode, she would face an infinity of alternative interpretations. For example, the firing rate of that same neuron also varies lawfully with $n+1$, therefore the neuron also encodes $n+1$. In fact, it encodes $f(n)$ for any mapping f. As we have seen in

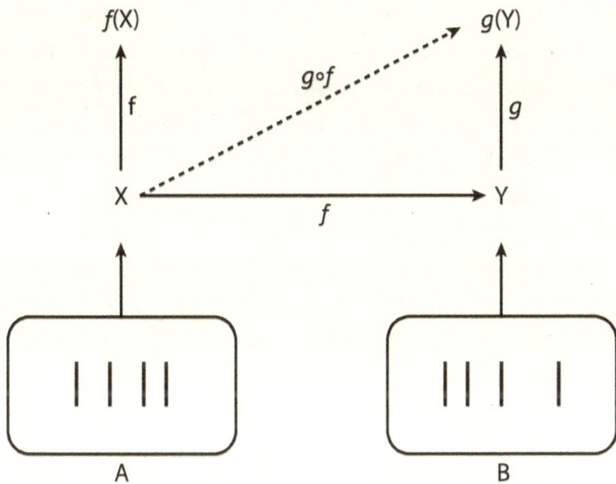

FIGURE 5.8. Brain region A encodes X and brain region B encodes $Y = f(X)$: the brain computes f. But region B also encodes $g(Y)$, and therefore the brain computes $g \circ f$ for any g. Region A also encodes $f(X)$, which is the result of the computation.

chapter 4, the mapping could very well be a noncomputable function, and the neuron would still encode it. The code lives in the observer's mind, not in the neuron. This cannot be otherwise, since the endpoint of a code is in a formal space, not in the brain.

This basic property of the implementation framework makes it impossible to account for computation. Suppose we have found a brain region A that encodes a property X, and that region projects to another brain region B that encodes a property Y, such that $Y = f(X)$ (figure 5.8). Then, following the interpretation framework, we conclude that the brain computes f. But since brain region B also encodes $g(Y)$ for any g, then the brain can also be said to compute the composition of the two mappings, $g \circ f$, for any g. For example, if f is invertible (such as $f(n) = n+1$), then brain region B also encodes X, and so the brain actually computes nothing at all. It also encodes $g(X)$ for any g, and therefore it computes any g, including a noncomputable mapping g. Conversely, brain region A also encodes $f(X)$, so it actually encodes the output of the computation. Thus, any information processing system can be said to compute its output in the input layer, without the need for the rest of the system. This makes statements such as "the brain represents" or "the brain computes" essentially empty.

Neuroscientists have been trying to address this issue by complementing observation with manipulation. Thus, to show that a region actually encodes something, they might perturb that region (say, by electrical or optogenetic perturbation, or lesion) and then observe whether the corresponding behavioral manifestation is affected. Unfortunately, this does not prove anything. For example, we could still say that the retina encodes the identity of Jerry Fodor versus David Marr, because if we stimulate the retina in such a way that it encodes David Marr, then the subject will report seeing David Marr. The issue is more fundamental than correlation versus causation.

Certainly, saying that the retina computes face identity because we can construct a mapping from retinal activity to face identity feels like cheating. All the computation actually occurs in the mapping that we have built. But this is exactly what neuroscientists do with neural codes in general, and it could not be otherwise. As we have seen, tuning curves of neurons are usually rather complicated, often nonmonotonous (see figure 5.5). The code is then equally complicated. In Bayesian accounts of brain computation, stimulus properties are inferred by calculating log likelihoods by pooling spike counts of many neurons in precise ways (see figure 5.4). When tuning curves of neurons are compared to psychometric measurements of behavior (such as sound localization), this is typically done through a formal construction called an "ideal observer," which combines neural responses in an optimal way to calculate the most likely stimulus, given the tuning curves and the known experimental design. Neural codes are *generically* complicated mappings built by the observer.

Perhaps then, we should restrict ourselves to biologically plausible decoding schemes, as is sometimes proposed. Unfortunately, this makes no sense: neurons do not output numbers or face labels, they produce spikes. There is no such thing as "biologically plausible" decoding. It is necessarily the observer who interprets those spikes as numbers and names of people.

This state of affairs also undermines other classical neurocomputational concepts, such as information and prediction. Indeed, a signal is said to constitute "information" about some property if one can map (decode) the signal to that property. For example, a neuron's firing rate is said to be informative about the number of objects n if one can map its firing rate to n. But if that is the case, then we can also map the firing rate to $n+1$. Thus, this firing rate is informative about the number of objects, but we cannot tell how many. How can we call that concept "information"? We will examine this issue specifically in the next chapter.

In the same way, the standard neurocomputational framework to account for anticipation is to say that the brain predicts something when some neurons encode the prediction. But as we have just seen, the same neurons may be said to predict the next integer $n+1$, or to make a wrong prediction ($n+2$), or to compute n^2, or to just repeat the input (n). The input neurons that encode n can also be said to encode $n+1$, and therefore to make a prediction. In a computation, the input predicts the output, so there can be prediction before any computation occurs. All of this shows that the concept of prediction based on encodings does not work. We will examine this further in chapter 7.

As we have seen, there is a solution to this problem, which is looking at how neural activity organized at the system's scale orchestrates behavior (the systemic/embodied approach), rather than trying to make neurons speak, which is simply not possible. This is admittedly difficult, but at least conceptually coherent.

This being said, to say that a neuron's activity encodes the number of objects is not an empty statement at all, and it can even inform us about how the system is organized. It would indeed be very remarkable if the neuron's activity depended only on that number and on nothing else. It is just that what the activity depends on can only be defined up to an arbitrary mapping. Thus, what a statement about coding actually says is a dimensionality reduction property: instead of depending on all possible aspects of a situation, the activity only depends on one particular dimension.

Dimensionality Reduction

When we say that a system computes face identity, we probably imagine something like this: a camera is connected to some electronic system, and an LED lights up whenever a face is captured by the camera. To claim that the LED encodes face presence is then not a trivial proposition at all: for example, the proposition would not be true if we just connected one of the camera's pixels to the LED. Here, the LED can be in just two states, and each of these two states can be assigned to one category, "face" or "nonface." There is a reduction of dimensionality from the space of all possible images (which have as many dimensions as pixels) to the space of possible states of the LED.

When we say that the computer encodes a program variable, what we mean is not just that we can define a mapping from the computer state to that variable. Otherwise, a program that just reads an image file would also be said to encode the identity of the person in that image. What is meant is that there is a mapping from one delimited place in memory, a small series of bits, to the program variable.

This is the idea of a "grandmother cell" in neuroscience: a neuron that would activate when you see your grandmother, under any angle of view, but not when your grandmother is not in view. It would mean that the system is organized in such a way that the state of a component depends only on a specific feature of the situation, which is not at all a trivial property. If that component controls another part of the system that determines behavior, then the behavior will only depend on whether the grandmother is in view.

Tuning curves are also an attempt to capture this idea: if it were the case that a neuron's activity varied systematically with sound location and nothing else, then a motor system driven by that neuron would produce behavior that only depends on sound location. This feature would then be a major contribution to sound localization behavior, since to make it an accurate sound localization system would then just be a matter of correctly associating the neuron's output to the corresponding motor command.

Unfortunately, as we have seen, neurons do not usually have this property. This is why neuroscientists have looked at increasingly sophisticated population codes, which simultaneously encode multiple dimensions in the activity of many neurons. But this move amplifies the conceptual problem with codes, since their only notable property is dimensionality reduction.

Population Codes Amplify the Conceptual Problem

Remember the example of a color-tuned neuron. The number of photons absorbed by a cone depends on the cone's absorption spectrum, but also on light intensity. For this reason, the activity of a color-tuned neuron cannot be a code for light wavelength. However, we could consider the activity of three cones of different type (S, M, and L), or even three different colored surfaces (figure 5.9). The number of photons N_i absorbed by a surface at a given wavelength λ is proportional to the product of light intensity I with the absorption coefficient at that wavelength a_i: $N_i \propto I.a_i(\lambda)$. Therefore, the ratio of absorbed photons between two different surfaces only depends on wavelength and not intensity: $N_i/N_j = a_i(\lambda)/a_j(\lambda)$. This way, we have constructed a quantity based on the responses of different cones, which is invariant to light intensity. This is a "population code" for light wavelength.

The problem is, again, that this code is an abstract mapping constructed by the observer, not something physically realized by the system. In the same sense, the retina can also be said to represent face identity: one can construct a mapping from the image projected on the retina to face identity—for example, the mapping implemented by a deep learning network trained for face recognition.

This makes it impossible, for example, to ask how "neural manifolds" might be the substrate of computations. Indeed, if a neural manifold encodes two dimensions x and y, then it also encodes $x+y$, or even a noncomputable function $f(x, y)$. This is just a choice of coordinates. The coding-implementation framework does not allow distinguishing between representation and computation.

In the absence of constraints on dimensionality, coding statements become essentially empty. For example, in a hierarchical model of face recognition, the

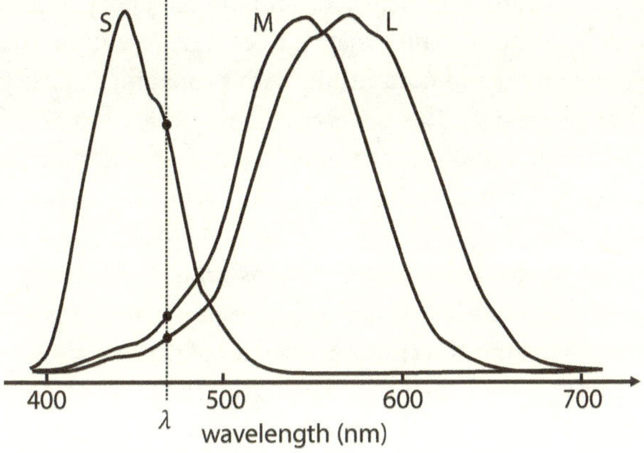

FIGURE 5.9. Absorption spectrum of cones. The relative absorption characterizes the light's wavelength λ.

input layer of the model is also a population code for faces, objects, and so on. This seems absurd, but this is exactly what the coding-computation framework allows.

We say that a system *has* a representation of something, or that it represents something. These linguistic uses cause much confusion (Zahnoun, 2020). Representation is not something that the system actually has or does, it is something that the observer of the system does. We say elliptically that "the system represents," but this is misleading: what it means precisely is that the observer interprets the system's state as something else, using a formal mapping she has constructed. Coding representations are never actually "in" the system.

We now come back to the context-dependence of neural codes. Could we save the concept of code by allowing the context to be encoded? Unfortunately, this is theoretically impossible.

The Fundamental Incompleteness of Parametric Codes

Encoding the Context

We have seen that "neural codes" are in fact typically context-dependent, whether tuning curves or neural manifolds. But there seems to be a straightforward solution to this context-dependence: to consider that contextual variables are encoded elsewhere. Thus, the complete encoding of a perceptual scene, or of the content of working memory, would be a mapping from an extended population of neurons to all the possible dimensions.

For example, a prominent neurocomputational theory of perception, the Bayesian brain hypothesis, claims that "the brain represents information probabilistically, by coding and computing with probability density functions" (Knill and Pouget, 2004). This presupposes that there is a set of predefined variables to which probability is attributed—examples of variables are the position of an object, the orientation of a visual grating. Individual neurons may be sensitive to many dimensions, but once all the neurons are taken into account, it is possible to retrieve each dimension independently from neural activity (a process called "inference"). For example, neurons in the primary visual cortex are sensitive to various properties of drifting visual gratings: orientation, contrast, speed, spatial wavelength. Thus, the orientation tuning curve changes if contrast or wavelength are changed, but one could define a mapping from the activity of all neurons to those four dimensions and thereby obtain a context-independent mapping.

The problem is that properties of an object, such as the orientation of a grating, only make sense once an object has been defined with those properties. If there is no visual grating, then there is no orientation of the grating. In the same way, the position of a sound source is only defined if there is a point sound source. If instead there is the sound of a river or of wind, then there is no such thing as the position of the sound source, and so the brain might not possibly be representing source position. Additionally, in real life (as opposed to controlled experiments), there are many potential objects apart from the one the experimenter is interested in, so it is not sufficient to encode properties: one has to encode properties *in conjunction* with what they refer to.

Thus, a perceptual scene cannot be encoded as a set of properties, because this leaves out what defines those properties. Cognitively, what defines properties corresponds to object formation and scene analysis, two fundamental aspects of perception that are not addressed by the concept of code (i.e., a mapping to object or task dimensions). This point challenges the idea that perception might be some sort of statistical inference.

Perception as Statistical Inference

A popular view of perception is that perception is a statistical inference problem. Sensory data are "noisy" or incomplete—for example, an object can be partially hidden—and the perceptual problem is to infer what is out there, given those data. The problem is formalized as statistical inference by considering that sensory data are produced by a certain statistical model that depends on hidden variables, which we want to estimate. Once expressed in this way, there is a known solution from probability theory, based on Bayes's formula. For example, one could empirically measure the distribution of interaural delays and intensities for a variety of sound sources at a given position. Then when a sound is captured at the two ears and we want to know where it comes from, we can calculate the probability that the source is at a given position given the acoustical cues, assuming a particular prior distribution of the positions (the default assumption).

But of course, statistical inference presupposes a particular model, which defines the variables. In this case, we have assumed that there is a single sound source, which has a well-defined position. If there are two sound sources, the inference formula will still give a result, but it will be meaningless, unrelated to the actual situation.

The general conceptual issue is that variables are necessarily defined within a particular model, which captures the structure of the scene (objects and their relations), but the difficult part of the perceptual problem is precisely to capture that structure. Donald Rumsfeld once captured this important distinction in a famous epistemological comment on foreign affairs (Rumsfeld, 2011): "there are known knowns; there are things we know we know. We also know there are known unknowns; that is to say we know there are some things we do not know. But there are also unknown unknowns—the ones we don't know we don't know." Statistical inference focuses on the known unknowns: we assume there are well-defined variables, which we only need to estimate. But this is a relatively simple problem. Thomas Bayes solved it in 1763. As Rumsfeld pointed out, the hard part is to deal with unknown unknowns.

As we will see in greater depth in the next chapter, building knowledge from observations is not generically a statistical inference problem. For example, Newton did not come up with his laws of mechanics by applying Bayes's formula to the trajectories of moving bodies. Applying statistical inference first requires building an appropriate model, in particular designing appropriate variables, such as the acceleration of the center of mass. In a word, statistical inference starts when all the difficult conceptual work has been done. The problem of perception is not that the world is noisy: the problem is that the world is complex, and this is not the same thing at all (Brette, 2018).

The same remarks apply to variations of inference theories of perception, in which variables are not defined a priori by the modeler but are instead unspecified "internal variables." For example, in predictive coding (Rao and Ballard, 1999) and related propositions such as the free energy principle (Clark, 2013; Friston, 2009; Nave, 2025), it is proposed that neurons encode the inferred value of internal variables of a generative model of the inputs—for example, the retinal image. Technically, this essentially means that the code is a parametric description of the image. These parameters are often abusively called the "hidden causes" of the input, because in statistical jargon, a cause is simply a variable in a statistical model (with no causation actually involved).

But a parametric description of an image is only possible if indeed the image varies along the dimensions specified by the parameters (e.g., position and size of an object).

If the "hidden causes" are supposed to be abstract features of the situation, such as the position and size of an object of interest, then these variables need to be defined, but there is no space in generative models for such definitions, only a fixed number of variables that are supposed to apply to all possible situations. Of course, it is possible to describe arbitrary images by a fixed number of variables, for example with the Fourier transform: any image can be written as a sum of periodic components, so the image can be equivalently described by the intensity of its pixels or by the coefficients of its Fourier components. But Fourier components are not abstract features that could be used to drive behavior.

There is no universal set of abstract features that could apply to all situations, because the very notion of an abstract feature is relative to a particular situation, which defines the feature (the orientation *of a bar*, the color *of a dress*). Therefore, perception is not a statistical inference process (at least not only), and there can be no general mapping from neural states to a set of variables, if those variables are supposed to determine behavior.

When Parametric Representations Seem to Work

Despite these remarks, we have seen that in the navigation system, the activity of some neurons can be described by a single dimension that corresponds to head direction (at least at a coarse temporal scale). This observation does not contradict our previous remarks: in contrast with the abstract features of perception, head direction is always defined. The rat always has a head, and only one, and it is always in some well-defined position. More broadly, whether it is empirically the case or not, it is at least coherent to conceive that there might be a mapping from the state of some neural population to some parametric description of the body configuration.

However, even in the navigation system, this is not exactly the case. Is it really possible that there might be a fixed mapping from the activity of some brain structure to the direction of an animal's head? This would require the animal to have some reliable magnetic sense. In rats, the preferred direction of head direction cells rotates when the visual landmarks are rotated (Taube et al., 1990). Thus, there is no fixed mapping from neural activity to head direction. Rather, the mapping is environment-dependent. In the same way, the firing patterns of grid cells rotate when cues on the walls are rotated (Hafting et al., 2005). They also expand or contract when the environment is scaled up or down (Barry et al., 2007; Stensola et al., 2012). In fact, within the same environment, grid patterns vary with the animal's travel history, in particular the time since it visited a boundary (Ginosar et al., 2023). Finally, the place field of place cells changes entirely when the environment is changed, a phenomenon called "remapping" (Colgin et al., 2008). Thus, all these representational mappings vary to some degree with the environment in which the rat navigates, and with the travel history of the animal.

In fact, place cells may also respond to particular states in a nonspatial behavioral task, such as manipulating the frequency of a tone with a joystick (Aronov et al., 2017). Thus, when we only observe the activity of a place cell, its spikes might signal the fact that the rat is at a particular place, or that it is hearing tones of a particular frequency. The experimenter knows, but only because she has extra information about the experimental context. Again, the activity of place cells, even the entire population of place cells, cannot constitute a complete parametric representation of behavioral tasks, and the required context (what the behavioral task is, what the "place fields" refer to) is not an additional set of parameters: it is the definition of parameters. In fact, when mice are trained on an auditory navigation task, the activity of hippocampal cells correlates better with anticipated actions than with sensory inputs (Zutshi et al., 2025). The observed spatial specificity of place cell firing appears to be a consequence of experimental design rather than evidence of spatial coding.

The limitation of parametric codes becomes even more acute as we consider cognitive processes involving memory.

Toward Cognition

Mainstream models of working memory propose that memory items are stored and represented in the persistent activity of neurons tuned to the relevant stimulus property (Constantinidis and Klingberg, 2016). This provides a way to store graded properties, like the position of a visual target or the pitch of a musical note. But what is the content of such a memory?

Suppose for example that a neural network in my brain stores the number 100. What have I memorized exactly? Was it the area of my apartment in square meters, or the height of my son in centimeters? Obviously, "100" is not a complete memory item. One also needs to specify what the number refers to. But there cannot be a specific network tuned to the area of my apartment, nor can there be a network tuned to my son's height. Therefore, working memory cannot be based on parametric representations.

This issue is related to a major dispute between classical cognitivists and connectionists: according to cognitivists, representations must be *structured*, whereas connectionist representations, the traditional neural codes, are not.

Structure in Neural Codes

Representing Relations

So far, we have seen that, empirically, neural codes fail the basic test of the computational implementation framework: they do not actually map neural activity to symbols. Yet, the representational mappings of the implementation framework carry a very elementary notion of representation, which philosophers call "receptor representation," some physiological state that is lawfully related to a mental state. The broad consensus among philosophers and cognitivists is that receptor representations are too crude to explain cognition (Ramsey, 2007). An additional criterion is that the representations must be *structured*. Formally, this means that the mapping from neural activity to the symbolic space must be a morphism (sometimes inaccurately called "isomorphism"), a mapping that maps objects as well as operations on objects to the representing space.

This connects to a major dispute within the computational view, between connectionists and cognitive scientists, which I mentioned in chapter 1. For cognitivists, representations are not just made of items but also of relations between items. Traditionally, this discussion focuses on higher cognitive processes, such as reasoning, speaking, or memorizing. For example, I remind myself to buy a present for Mrs. Smith's younger son. What kind of memory item might this be? Clearly, this is not just a value that could be represented by the firing rate of some neuron. This example features a number of abstract concepts: "buy," "present," "Mrs. Smith," "son," "younger." The brain structure in charge of representing this memory item must not only represent each of these concepts, but also the relations between them. For example, the concept "son" does not designate any particular person until it is applied to Mrs. Smith. In the same way, the proposition "I buy a present for Mrs. Smith's son" does not have the same meaning as "Mrs. Smith buys a present for my son."

In connectionism, a neural representation is typically a vector of activations of a population of neurons, which stands for a particular item, such as "Mrs. Smith." For example, some cortical neurons called "concept cells" respond specifically to pictures of a particular person such as Jennifer Aniston, as well as to their written names (Quiroga, 2012). But how do these concepts combine? For example, suppose there is a group of cells that represents Mrs. Smith by its firing, and another group of cells that represents the concept "son." How is the concept "Mrs. Smith's son" represented in terms of neural activation?

There are essentially two ideas: orthogonal representations, and combinatorial units. Orthogonal representation means, to a first approximation, that the representations do not overlap substantially. Thus, different neurons, perhaps in different places in the brain, represent Mrs. Smith and the concept "son." However, this only works for representing a conjunction ("Mrs. Smith" and "son"). More complex structures such as "I buy a present to Mrs. Smith's son" cannot be represented by a simple conjunction. In text search, this model is called "bag of words": a page of text is represented by an unordered set of words, and this can be compared to a query, also represented as a set of words (e.g., similarity is assessed by the number of words in common). This is sufficient to convey the general theme of a text, but not its precise meaning. In the same way, the conventional connectionist model where concepts are represented by orthogonal vectors is a "bag of neurons" model.

To represent more complex structures, another simple idea is the combinatorial neuron: the activity of a neuron, or of a group of neurons, represents a structured concept, such as "Mrs. Smith's son." But there are serious limitations to this idea. First, these combinatorial representations must preexist. Keeping in mind that a computational representation must be involved in some computation, this means that there must be brain processes identified to computations, which specifically deal with each of these combinatorial units, ready to be used on every occasion when a structured concept is formed. Second, a combinatorial unit does not represent its constituents, nor does it represent the relations. It represents the structured concept as a single symbol, without conveying the notion that it is somehow related to its constituents.

In cognitive science, the idea that representations must be structured has been popularized by Fodor's "language of thought" hypothesis (Fodor, 1980), according to which thinking occurs in some kind of mental language, something that is abstract and structured, and allows recursive composition (e.g., the father of the father of . . . Mrs. Smith). While this specific theory has been largely dismissed in neuroscience (or simply ignored), the problem of structure has triggered a debate in sensory neuroscience around the "binding problem."

The Binding Problem

While it seems clear that language requires some kind of structure, namely syntax, it is perhaps less immediately obvious that structure is important for perception: after all, an image is just an array of pixels. But consider the scene depicted in figure 5.10. A bald man wearing a turtleneck sweater is talking to a man wearing a tie and jacket, while a woman with a necklace is listening. How can we use a representation of this visual scene to answer questions such as: "What is the bald man wearing?"

Suppose there are neurons, or assemblies of neurons, that represent certain visual categories. For example, some neurons fire to signal "bald man," others

FIGURE 5.10. A visual scene with three people wearing different clothes.

for "woman," "necklace," "turtleneck sweater," "tie," and so on. We now have a conjunctive representation, with a list of visual categories detected in the scene. But from this list, we have no way to know whether the bald man wears a turtleneck sweater or a tie. This problem is known as the "superposition catastrophe" (Rosenblatt, 1962): when there are several objects in the scene, properties or parts need to be bound to the right objects, not just listed.

The classical concept of a cell assembly as the unit of representation is prone to the superposition catastrophe, because it is essentially unstructured ("bag of neurons"). This problem has motivated alternative representational theories, where properties belonging to the same object can be bound together. Specifically, the theory of "binding by synchrony" (Singer, 1999; von der Malsburg, 1981, 1995) proposes that the time of spikes, relative to an ongoing global brain oscillation, acts as a signature of objects. Thus, two neuron assemblies that fire at the same time signal properties of the same object. For example, the neurons representing "bald man" and "turtleneck sweater" would fire simultaneously, but in alternation with the neurons representing "woman" and "necklace."

The theoretical appeal of this synchrony-based theory is that it mingles representation with computation, as a neurocomputational theory should. Indeed, it is known that neurons are more likely to fire when it receives synchronous rather than asynchronous spikes (Abeles, 1982; Rossant et al., 2011), as we will see in more detail in chapter 8. Therefore, a downstream neuron receiving inputs from the "turtleneck sweater" neurons and the "bald man" neurons would fire, but a neuron receiving inputs from the "turtleneck sweater" neurons and "woman" neurons would not, even though all those neurons are active.

Although the theory has some appeal, the empirical evidence is disputed. I will not review this empirical work, but instead I will point out that the scope of the theory remains somewhat limited. Consider this other question about the visual scene: Who is the bald man talking to, the other man or the woman? This is a much more complicated question, because it only makes sense when we consider the entire scene, not the local features, and at the same time it is a question about how local aspects of the scene relate to each other. In other words, about the *configuration* of the scene. It is clear that labeling each feature with an object signature will not be sufficient.

One proposition to solve the binding problem within the standard connectionist framework is to consider that brain areas are not simply sets of feature representations but *maps*. That is, the firing of a neuron or group of neurons signals a feature at a certain position in the visual field. This is a form of combinatorial code. This proposed solution requires that space is then abstracted away by some mechanism, but in any case, we can see that maps are not sufficient to address questions about scene configuration, such as "Who is the bald man talking to?"

As a matter of fact, at the date of writing, state-of-the-art artificial neural networks have trouble dealing with structure, despite spectacular results on some tasks, such as object recognition. Convolutional neural networks are feedforward formal neural networks, with an architecture loosely inspired by the anatomy and physiology of the visual system of mammals. They have been trained to categorize images or recognize faces, with very high accuracy. These are standard connectionist networks, in which the activation of neurons in the last layer stands for specific labels. However, such networks perform very poorly on elementary relational tasks, such as deciding whether an image contains two identical objects (Kim et al., 2018). In the same way, at the time of writing, large language models have difficulties with compositional tasks (Dziri et al., 2023) and tasks that rely on abstraction (Lewis and Mitchell, 2024).

At this point, we might wonder why it seems so important to look for "neural representations" in the brain, when we find so little empirical evidence and the very concept appears plagued with fundamental flaws. After all, if there is one lesson in the success of modern artificial neural networks, it is that when we blindly optimize a network to perform a task, it is not clear at all what its neurons represent—except of course for the output neurons, which represent whatever the engineer has decided (e.g., object label).

Why Representations?

Representation-Hungry Behavior

Neural codes are a central concept in brain theory because they are an indispensable element of the computational implementation framework. This is why neuroscientists keep looking for them in the brain, even when all they find is contingent relations between experimental stimuli and experimental measurements—contingent on stimulus category, context, task, relevance, and time. As we have seen, the concept also has deep theoretical issues. This is why a number of thinkers have proposed to get rid of representation entirely. Roboticist Rodney Brooks famously wrote that "the world is its own best model" (Brooks, 1990), as he pioneered a new approach to robotics, called "behavior-based robotics," which emphasizes interaction with the world rather than mapping and planning. Others have proposed nonrepresentational theories of the brain that focus on feedback control (Powers, 1973; van Gelder, 1995).

These theories have been heavily criticized for being too crude. That is, much sophisticated behavior cannot possibly be based on feedback. This kind of behavior has been (abusively) called "representation-hungry" (Clark and Toribio, 1994).

Consider the ant's homing behavior. Sure enough, path integration can be performed by tightening a thread, rather than by calculating the integral of the velocity vector. But this is not what the ant does. Indeed, if the ant is moved to a different place, the homing direction is unchanged: the ant returns as if its nest had been displaced in the same way (Wehner, 2020). Therefore, the homing direction is not determined by a feature of the local environment, but by the integrated velocity vector. This vector cannot be measured at any given moment in the environment. Rather, it depends on a precise and sophisticated way on the past behavioral history of the animal. It depends on the animal's trajectory, but it is invariant to many aspects of those trajectories: whether the animal halts for a while, whether it circles around a tree before resuming its course, and so on. It is also invariant to whether the animal is hungry, or cold, or smells food. It is this invariance to nonrelevant aspects that is highlighted when it is claimed that such behavior is "representation-hungry": the behavior only depends on some abstract feature that is not directly measurable. Since the brain determines behavior, the brain must somehow "represent" that feature.

Sound localization is another example of a "representation-hungry" behavior. The cat hears a mouse scratching the grass, then runs toward the source and catches the mouse. This is a prowess because, at a given position, the mouse can produce a variety of sounds, and yet these various sounds result in the same adapted behavior: running toward the source of the sound. We have seen in chapter 4 that there is a simple feedback-based solution, which consists in moving toward the louder ear. But the cat does not do that (at least not only). Instead, it might hear a mouse scratching the grass for a fraction of second, then move toward the mouse's expected location. In many psychophysical experiments where sound localization performance is measured, the auditory stimulus is a click, a very short sound. In such cases, the cat's pinnae and eyes turn accurately toward the sound source, after the sound is over (Populin and Yin, 1998). Thus, sound localization by the cat does not rely on interaction (or at least not necessarily so). Instead, it appears to be a sophisticated anticipatory behavior: the cat moves its eyes and pinnae toward where the sound source is expected to be, and this movement can occur after the stimulus is gone.

Why would these behaviors require a neural representation? In the ant's homing behavior, the homing direction is determined by the specific actions of the ant (unlike when pulling the thread). Therefore, if we assume that the state of the brain determines which action will be taken, then there must be a mapping from that state $\boldsymbol{s}$ to the homing direction $\boldsymbol{x}$, which is the integrated velocity vector. This mapping is a neural code for $\boldsymbol{x}$. In the cat's sound localization behavior, the movement of the cat toward the sound source is determined by the source's position, while the sound stimulus itself might have ended before the movement starts. It follows that, at the time when the sound ends, there must be a mapping from the brain state $\boldsymbol{s}$ to the source direction $\boldsymbol{\theta}$, a neural code for $\boldsymbol{\theta}$.

Thus, at first sight, it does seem that certain kinds of behavior are hungry for representations. But unfortunately, such statements are essentially empty. Surely, if you recognize someone's face and act accordingly, there must be a mapping from your brain's state to the face's identity. And there is one, which maps the retina's image to the face's identity. But what insight is that into how the brain helps you recognize faces? Exactly none, because it just a way of rephrasing the behavioral observation. If our account of neural representation or code is so liberal that it can be any formal mapping from the state of the entire brain to any kind of behavioral observation, then it imposes no constraint at all on the brain's organization. Indeed, in this broad sense, the ant's brain represents its spatial position, but it also represents its state of hunger on a scale from 1 to 10 added to the angle of the homing direction in radians. In this same sense, a planet can be said to represent its future position, since there is a mapping from present to future position.

To say that such behaviors are "representation-hungry" is entirely abusive, as others have noted (Zahnoun, 2021). The representation in question is not realized in the system: it is merely a descriptive tool constructed by the observer. Therefore, it is not something the organism feeds on. That the cat can localize the mouse does not imply that there is something somewhere in its brain that specifically stores the mouse's location, in the manner of a computational variable.

What Does Representation Try to Address?

Although the phrase "representation-hungry" is wrong, because it conflates problem and solution, it does point to certain kinds of behavior in need of explanation. The phenomenon in question is *anticipation*. As we have seen in chapter 2, anticipation is a major feature of behavior, and of life more generally. This is the feature that is lacking in feedback theories of cognition. The fact that the ant's behavior depends on other things than the ant's immediate surroundings is not particularly remarkable. What is remarkable is rather that the ant correctly anticipates that by walking in a certain direction, it will arrive at its nest. In the same way, the cat anticipates on the basis of what it heard, that if it runs in a certain direction, it will probably catch the mouse.

Behavior is made of actions: activities that the organism engages in so as to achieve something. This is why actions depend on abstract properties that extend both spatially and temporally, rather than just on proximal stimuli. Obviously, it is not the fact that behavior depends on many things that is particularly notable: it is that fact that behavior (often) *works*, that it is anticipatory. The fact that behavior can depend on absent or abstract features both derive from anticipation. Anticipation is a difficult subject that we will address specifically in chapter 7.

Thus, the philosopher Mark Bickhard argues that organisms do represent, but in the sense that they have anticipatory activities driven by certain situations, not in the sense of encodings (Erdin and Bickhard, 2018). Whether the representation terminology remains useful or confusing is an open question, given that representations are systematically identified to encodings in much of the literature.

But how does the cat know where to find the mouse? This brings us to another major concept of brain theory, *information*. Classically, information goes hand in hand with the concept of code. A neuron's activity is considered information about the position of the mouse if that activity can be mapped to the position of the mouse. But if that is the case, then the neuron's activity can also be mapped to the *wrong* position of the mouse—for example, the symmetrical position with respect to the cat's orientation. Therefore, the neuron's activity is also information about the wrong position of the mouse. How can such a concept of information work? (Spoiler: it does not.)

6

Information in the Brain

INFORMATION IS EVERYWHERE in the neuroscientific literature. For David Marr (1982), information is knowledge, and knowing the world is the primary function of perception:

> If we are capable of knowing what is where in the world, our brains must somehow be capable of representing this information.

It is natural that information is a central concept in neuroscience and cognitive science, since cognition is the acquisition and manipulation of knowledge. Information is also fundamental in consciousness studies. In integrated information theory, a basic "axiom" is that "consciousness is informative" (Oizumi et al., 2014): to be conscious of something is not only to have a certain experience, but also to *know* that there is something in particular out there (or in there). According to global neuronal workspace theory (a computational cognitivist theory of consciousness), "a conscious process corresponds to information that enters a large-scale reverberant network" (Mashour et al., 2020). In sensory neuroscience, the retina is thought to convey "visual information from the eye to the brain" (Meister and Berry, 1999).

A striking characteristic of the latter quotes is that information is conceived as something that exists out there in the world, which the brain extracts, and which can move around in neural networks. Indeed, the phrase "information processing," which supposedly characterizes what the brain does, implies that information is what gets processed: it is the input to brain processes. It is indeed common to read that information is literally everywhere in the world, not just in the brain or in the neuroscientific literature (Chalmers, 1997; Koch, 2012).

This perspective on natural phenomena is what philosophers call *substance metaphysics*. It consists in explaining phenomena by the presence of some substance with a special quality. For example, in the seventeenth century, burning used to be explained by the release of a fire-like element called phlogiston. Wood contains phlogiston, which is absorbed by the air during combustion, and later captured by growing trees. In biology, vitalism (in its simplest form) used to explain life as the result of a particular substance infusing organisms. In the eighteenth century, preformationism held that the embryo contains a preformed miniature version of the animal, which simply unfolds during development. The major conceptual weakness is that the postulated substance already has the properties that need explaining: something can burn because it has a fire-like substance; an organism is alive because it has a life-bringing substance; the embryo develops into a grown adult because that adult is already there from the beginning. Outside biology, the course of science has progressively replaced substance metaphysics by *process metaphysics*, as explained by Bickhard (2009, 2024)—we will come back to this subject when we discuss the concept of "neural states" in chapter 8.

Thus, in neuroscience, information is typically conceptualized as a sort of epistemic phlogiston: animals know because they have "information" in their brains, which enters through the eyes and moves up to higher cognitive centers. This is a fundamental misconception based on a reification of a cognitive process (Zahnoun, 2020). That is, what is actually a process (getting informed, understanding) is framed as the acquisition of some epistemic phlogiston that we call "information."

Note that, even though it is now obsolete, phlogiston theory was a serious theory that explained a number of empirical findings about combustible bodies. After all, there is nothing logically incoherent in the idea that matter has a particular kind of constituent that allows it to burn. In contrast, we will see that the idea that neural signals intrinsically possess information about something else is incoherent.

First, I will try to clarify what is normally meant by "information" in the context of cognition, and the properties it is supposed to have (*What Is Information?*). Then I will examine the mainstream conception of information, the one that is implied by neural codes, namely *information by reference*: something (a neuron's firing rate) constitutes information by reference to something else (the orientation of a bar) (*Information by Reference*). As we will see, this view is irremediably flawed and leads to various paradoxes, such as the fact that a cognitive process can only degrade information, and that the most efficient way to encode information is when it is undistinguishable from noise. A variation is to conceive *information as difference*: something is informative not because it correlates with some external property, but because it could have been something else (*Information as Difference*). Unfortunately, this view is equally flawed. The computational way out of this problem is to consider it more pragmatically: something is information by virtue of its role in a computation, which may result from evolution (*Pragmatic Information*). We will see that this does not work either. I will then argue that information can only make sense within an embodied perspective, in analogy with science: knowledge built from relations between observations, and from the effect of actions (*Embodied Information*). This provides a much richer notion of information than a simple correspondence table (*Properties of Embodied Information*). I will end with a discussion of what information might mean in terms of biological processes (*How Information Informs*).

What Is Information?

Information in the Common Sense

In the colloquial sense, a piece of information could be for example the sentence: "Donald Trump is the new president." This is a proposition that could be true or false. Perhaps it could be "fake news." Information can be more or less accurate, it can be confirmed or denied. It is this common sense of information that supports what we think of as cognition. When Marr writes that the brain must represent information about the world, information is meant explicitly as *knowledge*. To explain the sound localization behavior of cats, one might say that cats extract information about the mouse's location, and what is meant is that the cat knows where the mouse is: of course, the cat could be wrong.

Obviously, propositions are not things one can find in the world: they are mental constructions about states of affairs, concepts that live in the mind of cognitive agents. Yet, we also say that there is information in a newspaper, or that spikes carry information. What does it mean?

Information in Things

How can there be mental constructions in a newspaper? There cannot, obviously. Rather, we might be informed by reading the headline "Donald Trump Is the New President" in a newspaper. The information is not to be found literally in the ink. There is no intrinsic information about Donald Trump in the shape of the letters either. The title might very well be written in Chinese or Braille, it would make no difference in terms of information. What happens is rather that you get informed by reading the article, and this involves the letters in the article, but also linguistic conventions to associate letters to meaning, and of course, the reader: a cognitive agent who knows those conventions, and for whom "president" means something. For an ant, for example, the same title is rather a scented track.

Each of these three elements is necessary for the newspaper to have any information: the physical symbols that mediate information, what the symbols are supposed to mean, and who the information is for. But only the mediating substrate (the ink) is actually in the newspaper.

Linguistically, talking about "information" as if it were a thing is a case of *hypostatization*, as Zahnoun has pointed out for the related concept of representation (Zahnoun, 2020): we use the grammatical category of nouns for something that is not substantive, like an action or a property, as when we say: "I am going for a walk." There are no such things as walks in the world. In the same way, we say that we "acquire information": it simply means that we inform ourselves, not that there actually is some kind of epistemic phlogiston that we pick up in the world and put in our brain. Surely, you need letters to read, just like you need a ground to walk on. But the ground is not a walk.

Unfortunately, this linguistic peculiarity causes much confusion. Consider for example the common phrase "the organism extracts information" to designate perception. We can make some sense about the claim that a reader extracts information from a press article: the writer knew something and now the reader knows the same thing, and the transfer of knowledge was mediated by the press article. But the world does not know anything, so it makes little sense to depict perception as taking information from the world.

This is more than an unfortunate manner of speaking. Linguists have pointed out for a long time that language shapes our thought, in particular metaphors (Lakoff and Johnson, 1980). In this case, conceptualizing information as a thing hides a fundamental feature of perception: it is a constructive process, not a matter of extracting and filtering concepts from the environment. Consider for example a nose. A nose can be big or small, pointy, hairy, and so on, but what kind of physical object is a nose? A nose has no clear molecular delimitation, and it even includes holes. The delimitation is one made by our cognitive systems. More generally, objects are cognitive concepts shaped by our interactions with the environment. Therefore, contrary to Marr's claim, to perceive the world is not a matter of finding out what is out there, of inferring a preexisting objective description hidden by our limited sensory apparatus or degraded by "noise." Rather, perceptual objects are cognitive constructions of subjects interacting with the world. This fundamental point disappears in the mainstream view of information: information by reference.

Information by Reference

Coded Information

By far the most popular notion of information in neuroscience is the concept of neural codes (Brette, 2019a), which is related to the notion of representation: the "neural code of the retina" (Meister and Berry, 1999); the "neural code of pain" (Kucyi and Davis, 2017); the brain "codes our thoughts" (Dehaene, 2014); understanding sensory perception is "cracking the neural code" (Panzeri et al., 2017). Neurons are thought to "encode information," and also decode it or extract it. For example, about neural responses to sensory stimuli: "A stimulus activates a population of neurons in various areas of the brain. To guide behavior, the brain must correctly decode this population response and extract the sensory information as reliably as possible" (Jazayeri and Movshon, 2006). One can apparently dig into neural signals and find information there, written in some coded form.

It is reasonably clear what is meant by neural encoding, as we have seen in chapter 5: some measurable property of neural activity (e.g., the firing rate) covaries with a property of something in the world (like sound intensity), or in the mind (like pain). But what is meant by "the brain decodes"? The brain does not transform spikes into sounds, or into symbols. Thus, this is a metaphorical use, which might mean that the neural activity triggers an appropriate mental experience, or some behavior, or that the signal stands for some symbol in a computational view of brain function. For example, in his book on sensory coding, Somjen (1972) writes: "Information that has been coded must at some point be decoded also; one suspects, then, that somewhere within the nervous system there is another interface . . . where 'code' becomes 'image.'" The Cartesian flavor of the coding/decoding framework is evident.

Neural coding is a communication metaphor that has been used for many decades in neuroscience. For example, in a seminal review titled "Neural Coding," Bullock and Perkel (1968) depicted "the nervous system [as] a communication machine." This sense of information has been formalized by Shannon (1948), in a statistical context—note that the founding paper is titled "A Mathematical Theory of *Communication* [emphasis mine]." In this sense, there is information at the end point of a communication channel to the extent that one can recover the original message from the encoded message with good accuracy.

Because this mathematical concept has been given the name "information" and the theory is called "information theory," it may give the impression that information actually *is* what information theory defines, despite the warnings issued by Shannon himself (as well as by many others). For example, we read in the neuroscience literature that "the abstract definition of information is well motivated, unique, and most certainly relevant to the brain" (Simoncelli, 2003). This is certainly false, as distinct notions exist even within the mathematical domain, such as Kolmogorov's algorithmic information (also named algorithmic complexity). But most importantly, one should keep in mind that names given to mathematical concepts are only convenient place holders, with some more or less vague resemblance to the usual meaning of the words. A map in mathematics is not actually a map: it associates things with others, but you cannot use a mathematical map to navigate. In the same way, the information of information theory does not designate information in the common sense, the one that is relevant to cognition and behavior, because, as we have seen, that information has truth values and is tied to cognitive systems.

Nonetheless, information in the sense of Shannon does have some recognizable features of information in the common sense, which is what makes the theory useful in appropriate contexts—namely, communication contexts. For example, there is information in a Morse encoded message about the original message, as I can decode it and understand the original message. The great asset of information theory is that it captures a quantitative notion of information. For example, if instead the message were encoded with dots for vowels and dashes for consonants, there would still be some information in the encoded message, but clearly much less, and information theory can quantify it.

But as we have seen, the dots and dashes allow me to understand the original message only because I know the encoding, and because I would be able to understand the original message in the first place. There is obviously no information literally *in* the dots and dashes: the same dots and dashes may refer to specific letters, or to vowels and consonants, depending on the convention. Information in codes is information *by reference*: the symbols of the encoded message stand for something else, and I am informed by these symbols to the extent that I know what they stand for and that I can be informed by the original message they refer to. Information by reference presupposes a cognitive agent with preexisting knowledge, neither of which are in the things that supposedly "contain" information.

The error is to conceptualize information as what can be *potentially* recovered from a signal by a knowledgeable observer. This recovery process is called *inference.*

Perception as Inference

It is commonplace to frame perception as a process by which the organism *infers* what is in the world from the sensory signals it gets, as Helmholtz already argued in the nineteenth century. Our eyes capture an incomplete projection of the real scene, which is noisy and deformed. The image on the retina is even upside down. If the spike trains produced by retinal ganglion cells are considered as information about the visual world, then it seems that the brain must infer ("decode") the visual world from those electrical signals.

This raises the issue of "the view from inside the box" (Clark, 2013). The external observer can look at a tree and can measure the spike trains that visual neurons produce when the organism sees a tree (figure 6.1). Then, from the

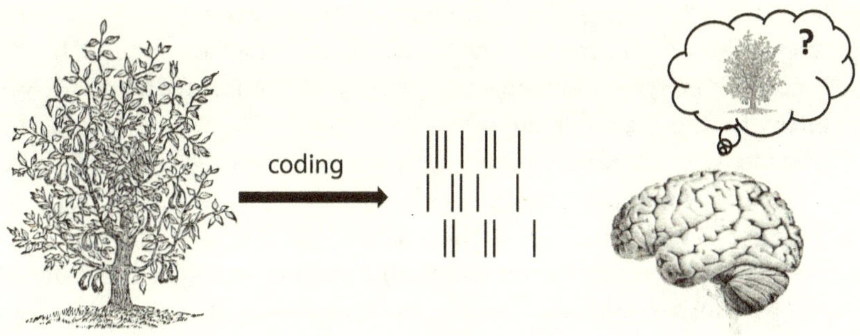

FIGURE 6.1. The view from inside the box: how does the brain infer that a pattern of spike trains corresponds to a tree?

spike trains, she might be able to infer that the organism was looking at a tree. But the neurons that receive the spike trains never have independent access to the tree. How could the brain, then, infer that the spike trains refer to a tree? John Eccles, a respected neurophysiologist who got the Nobel prize for his work on synapses, expressed the problem as follows:

> In response to sensory stimulation, I experience a private perceptual world which must be regarded, neurophysiologically, as an interpretation of specific events in my brain. Hence I am confronted by the problem: how can these diverse cerebral patterns of activity give me valid pictures of the external world? (Eccles, 1965)

This is related to the symbol grounding problem in cognitive science (Harnad, 1990): how do spikes, the symbols of the neural code, make sense for the organism? How does the brain know that a particular pattern of spikes refers to a tree and not to an elephant?

The solution that is most often proposed is *inference*. Inference uses deduction and prior knowledge to draw conclusions about what *should* be there. For example, you might see two arms sticking out of a tree, and because you know that arms are normally attached to a person, you deduce that a person is actually behind the tree. However, it is not true that inferring is the same thing as perceiving: in a sense, you might "see" the person but still, you do not see the face even though you *know* that there is a face behind the tree. Knowing and seeing are different phenomena.

But this is not the key point with respect to the present discussion. The key point is that in order to infer, one needs prior knowledge. In this example, prior knowledge is constituted by prior visual experiences of persons who are not hidden by trees. This works for occluded objects, but it does not work for objects that are always hidden. Therefore, inference cannot possibly solve the problem posed by information by reference: if indeed spikes produced by sensory neurons are symbols for objects in the outside world, then there is no way to infer what the symbols refer to.

The idea of inference as the primary access to knowledge is logically incoherent: inference presupposes that knowledge has been previously acquired. Eccles was fully aware of this issue, which is why he was, logically, a dualist. For him, cortical columns form units (dendrons) linked with mental units (psychons), which can interact. Most neuroscientists do not embrace dualism, but for those who claim that spikes carry information, that makes them incoherent.

As we have seen when discussing neural codes (chapter 5), a variation is that perception might be a matter of *statistical* inference—that is, of inferring the most likely values of a number of statistical parameters. This is even more wrong, because parameters are characteristics are perceptual objects, which then presuppose perceptual processes.

Information by reference is a homuncular concept that presupposes an omniscient observer. This has a number of paradoxical consequences. We will discuss three of them: (1) cognitive processes degrade information; (2) the best neural code is indistinguishable from noise; and (3) adaptive codes can carry information without communicating it.

Cognitive Processes Degrade Information

Consider the example of Morse code again. The original message is turned into dots and dashes, and this transformation is invertible: one can accurately infer the letters from the dots and dashes. Thus, there is just as much information in the encoded message as in the original message. But if we decided to encode vowels as a dot and consonants as a dash, then there would be less information in the encoded message. What can never happen is that there is more information in the encoded message than in the original message. This is known as the *data processing inequality*: processing can only reduce information, never increase it.

Thus, perception as "information processing" is a process by which information about the world gets degraded. This is what happens with information by reference: the reference to the original message becomes less and less reliable as it is transformed. It would seem that the best way to know the world is to do nothing with the sensory inputs: in that way, all the information remains there, intact. Just like phlogiston, information by reference is not created: it already exists out there in the world, so the organism can only move it around and dissipate it.

Efficient Coding

What if the sensory inputs need to be communicated to some higher brain center for further processing? For example, the visual cortex is on the occipital region of the brain, on the opposite side of the skull relative to the eyes. Electrical signals from the retina are sent through the optic nerve to the thalamus and then to the visual cortex, but the optic nerve has a limited bandwidth. How should the spike trains of retinal ganglion cells encode the image so as to make maximal use of that bandwidth? Framed in this way, this is a compression problem that naturally fits the tools and concepts of information theory.

Efficient coding theory postulates that neurons encode stimuli such as images in such a way as to transmit the maximum amount of information (Barlow, 1961; Olshausen and Field, 2004). This works by reducing redundancy, as in data compression. For example, in the Lempel-Ziv-Welch (LZW) algorithm, repeated patterns are assigned to shortcut symbols. In the same way, spikes are postulated to encode common patterns. This might sound like a reasonable principle, but the LZW algorithm is useful only because the table of correspondence between shortcut symbols and data patterns is also communicated, in addition to the compressed data. Without the table, one cannot decode the compressed data. Unfortunately, neurons do not communicate tables together with the encoded messages.

One might object that the table does not really matter for the brain: the world is just presented to the brain in a different format, and it is not particularly useful to recover the original format. But it is easy to see that this cannot work, when the concept is pushed to its limits. Consider the ideal compression scheme, the most efficient code, which would jointly encode all sensory signals from visual, auditory, or tactile modalities, all proprioceptive signals and efferent copies (copies of motor commands). With that ideal code, there would be no repeated patterns at all in the messages. Every time we find a lawful relation between signals, we replace it with some shortcut symbol, so there cannot be any lawful relation in the encoded messages. Even statistical relations must disappear. Like in the Huffman compression scheme, we encode more common patterns with shorter codes, so that all symbols and patterns of symbols are equally likely. In other words, the ideal code produces messages that are undistinguishable from random noise.

Thus, the most efficient code, the one that maximizes information by reference per symbol (per spike), is the one that removes any possible way to make sense of the world. This unfortunate consequence occurs because only messages are communicated, not encoding tables. Efficient coding consists in putting all information in the common sense (observable regularities) in the encoding table, leaving only unidentified place holders for the observable regularities. The problem becomes particularly salient when we consider time-varying codes.

Efficient Adaptive Codes

Consider for example neural adaptation, the fact that the firing rate of many sensory neurons tends to decrease progressively when a constant stimulus is applied. This phenomenon was observed already in the 1920s by Adrian (Adrian and Zotterman, 1926a,b). In other cases, lightning conditions may change on a slow timescale, and neurons in the retina "adapt," in the sense that they continue to be responsive to fast intensity fluctuations. A classical interpretation of this phenomenon in terms of information theory is that the input–output function of neurons adjusts itself so as to maximize the information content of its output (Laughlin, 1981; Wark et al., 2007).

Figure 6.2 shows the example of a signal that has a slow component. In a short temporal window, the fast variations are obscured by the slow variation. The idea of an adaptive code is to map the signal to values in such a way that it maximizes Shannon information within that temporal observation window—that is, in such a way that it uses the bandwidth uniformly. In signal processing, this is simply called *filtering*, and as the terminology implies, it involves removing part of the signal, namely the low-frequency content.

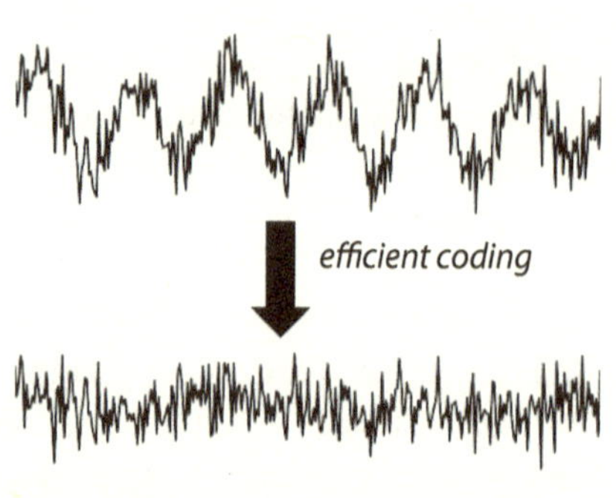

FIGURE 6.2. Efficient coding of a signal with a slow component.

The problem with interpreting adaptation as efficient coding is that only the output of the code is communicated, not the adjustments of the code. The information in the spikes is by reference to some input signal, but the reference is a moving target: how is the receiver supposed to recover the original signal? By considering "adaptive codes," we forget that the information in a code always involves a triad: the source domain, the target domain, and the mapping. Only the target values are transmitted, but to recover the source values, we need the mapping, which is not transmitted. The interpretation of adaptation as efficient coding then misinterprets low-pass filtering as a case of efficient coding, and misses the fact that filtering irremediably removes the slow varying signal.

This is related to the concept of "representational drift" that we discussed in chapter 5. If we allow the neural code to be time-dependent, then we make the concept of code entirely empty. For example, figure 6.3 shows an adaptive code of the decimals of π. Every digit is encoded with a zero, but according to a time-varying code. If this represents the activity of a neuron, with 1 for a spike and 0 for no spike (over some time interval Δt), then we have a neuron that succeeds in transmitting information about these digits in the most possible efficient way, without spiking at all.

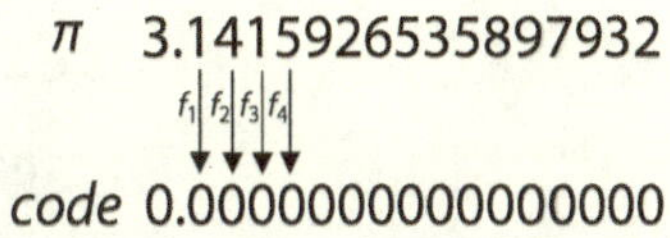

π 3.1415926535897932

f_1 f_2 f_3 f_4

code 0.0000000000000000

FIGURE 6.3. A very efficient adaptive code. The code changes on every digit, such that digits are always encoded to zero.

Oddly enough, neurons do not seem to do that. The issue with efficient coding is not "efficient": it is "coding." Surely, there is an evolutionary case to be made that living organisms are efficient at whatever is important for their survival. But why would they be efficient at coding? As we have seen, a communication system that is ideal in terms of encoding information by reference is one that provides unintelligible signals to its receiver, because all the useful information (in the common sense) is buried in an uncommunicated reference table. That cannot be an appropriate normative principle for living organisms.

We now turn to a variation of information by reference: information as difference.

Information as Difference

We have seen that information by reference raises the inference problem: how can you infer what is out there from indirect signals if all you ever have access to are those signals? Information as difference attempts to solve this problem by considering that the reference is not in the external world, but in a world of possibilities.

Consider an extraterrestrial animal whose sensory system consists of a single light-sensitive neuron, with a binary state S (figure 6.4). This neuron is the only access of the animal to the visual world. When the light is on, the neuron is in state A; when the light is off, in state B. I might want to call state A "light" and state B "dark," but these are arbitrary names: the animal does not know about light or dark, except through its state, which can just be A or B. Nonetheless, when I observe A, I am informed because it could have been B. This is something that can be said from the black box. If light and dark are equally likely, then observing S reduces my uncertainty about the state S by 1 bit, and this is true even without making reference to light.

This is the view of the proponents of Integrated Information Theory: it is stipulated that "consciousness is informative" in the sense that "an experience of pure darkness is what it is by differing [. . .] from [. . .] other possible experiences" (Oizumi et al., 2014). Note the difference with the concept of

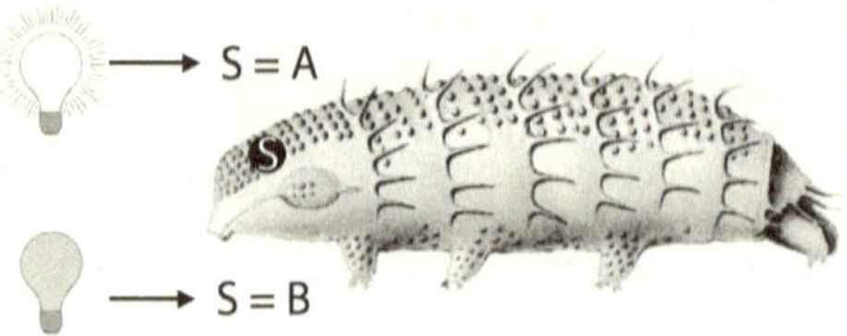

FIGURE 6.4. An extraterrestrial animal with one light-sensitive neuron (tardigrade drawn by Ferdinand Richters in 1900).

information that we have discussed so far: a physiological signal is not informative because it covaries with something in the world, such as light intensity, but simply because it varies. Observing a state is informative in that the state could have been something else. In this way, there is no need to ground the state value on some external state of affair.

However, this is still information by reference—namely, to a set of hypothetical values of the state, possibly with associated probabilities. There is some progress relative to the previous view of information, since here the reference could be to some knowledge acquired from the observation of the state. But that knowledge is still not *in* the state itself, it is in a cognitive agent watching the state, dependent on a history of observation (Brette, 2022b). Information is neither out there in the world nor within the organism's states.

Consider the claim that the amount of information carried by a state is the uncertainty reduction when its value is observed. For example, before you observe S, you know that S could be A or B with equal probability, and these possibilities will be reduced to a single possibility if you observe S. The reduction of uncertainty is 1 bit. But who is certain or uncertain, and who observes? Obviously, not the state itself.

If we forget this, it is easy to get rather paradoxical results, because possibilities do not exist out there in the world: possibilities are imagined by a cognitive agent. Suppose for example that our animal lives all its life in a box, then the state of its neuron will always be B, and therefore there would be no reduction of uncertainty. The state would not be informative. In another scenario, the animal, which we assume is exactly the same system, is sometimes moved out of the box, maybe half of the time. So now the state is informative: observing the state gives 1 bit of information.

Now compare these two scenarios. At the beginning of each scenario, the animal is in the box. Thus, the situation is exactly the same, both for the box and the system, in the two scenarios. In the first scenario, the animal is not informed, but in the second case, it is—that is, the animal knows it is in the dark. It is not coherent. The reason is that we have attributed prior knowledge to the animal when *we* have this knowledge, as the ones who conceive the scenarios and therefore the possibilities that the state refers to.

Consider a third scenario: the animal is put in a box that may be opened at some point, but maybe not, with some probability. Thus, the animal could live all its life in the box, just like in the first scenario, but in this case, it would be informed because the situation *could* have been different. She would know that she lives in the dark. Now the state of information depends not just on future events, but on *potential* events, even if those never actually occur. We now have two exactly identical animals that live in exactly the same world, but only one of the two animals is informed of the state of darkness. This is possible because one of the worlds *could* have been different, although it was not.

This view is mistaken. The confusion arises because information is defined by reference to a set of possibilities, but it is attributed to the state itself. There is no information in a state. If information is defined by reference to something external, then that thing is not in the state. If information is defined by reference to a set of possibilities, then those possibilities are not in the state either. If information is uncertainty reduction, then someone has to be certain or uncertain in the first place. In other words, information by reference, whether external or hypothetical, presupposes the phenomenon it is meant to explain, namely knowledge by a cognitive agent.

Reconsider the earlier quote: "an experience of pure darkness is what it is by differing [. . .] from [. . .] other possible experiences." The claim is superficially appealing: we appreciate the particular quality of darkness in comparison with different experiences we have had before, such as lightness. But this works if we already had previous experiences. In other words, it presupposes that we are able to experience and to remember. One cannot define consciousness by presupposing both consciousness and memory. Thus, the fact that Integrated Information Theory is a panpsychist theory of consciousness is not an insight of the theory, but an assumption: the theory presupposes that there is a cognitive agent behind every state configuration, without explaining either cognition or agency.

Both information by reference and information as difference define information as a potentiality: what an observer can infer from a signal. These are homuncular views, which take as prerequisite what they are supposed to explain, since inference requires prior knowledge (of external things, or of alternative possibilities). One cannot explain knowledge by assuming knowledge.

Computationalism offers another way to deal with information—namely, to essentially ignore the issue.

Pragmatic Information

Information in a Program

In a program, it does not really matter what variables mean. In many cases, programmers rather struggle to find meaningful names to variables, and would often settle for *i*, *dummy*, or some swear word. The only thing that matters is that the program produces the right output for any given input. Thus, if a sound localization program correctly outputs the direction of the sound source, then it is said to know what the direction is. This is essentially the view behind Alan Turing's "imitation game": if a human person discussing with a machine cannot tell whether it is indeed a machine or another human, then the machine is said to think (Turing, 1950). On this basis, large language models that can answer questions are sometimes said to "understand" the questions. This is line with behaviorism in its most basic form: what is in the skull does not matter, all that matters is how it looks like for an external observer.

This can be described as a pragmatic view of information: a computational variable is information about its consequences. For example, a robot calculates the direction of the sound source, say an angle in degrees, and then points to the sound source. The output of the calculation is then considered informative about sound direction. Let us examine this view.

First, consider a planet orbiting around the sun. Given an initial position, it will be at a particular position in six months. Surely, we could say that the initial position is information about the future position, but this is information for the astronomer, not for the planet. The planet does not know its future position, even though it will go there. Clearly, we have a confusion between the observer's perspective and the subject's perspective. To do something that has some effect is not to know those effects. In the same way, to output a number that will then be used to point in some direction, or worse, that will be interpreted by someone else as a direction, is not to know that direction.

Pragmatic information appears again as a variation of information by reference, this time by reference to its consequences. But by definition, consequences have not happened yet, and so they cannot determine the current situation.

Let us examine a concrete example. Imagine a sound localization program (figure 6.5). It takes as input the acoustic signals captured by two microphones, then it calculates a cross-correlogram and finds the time lag Δ with maximal correlation. Then Δ is transformed by a formula into an angle θ. This is what is typically called "information processing."

When there is a single sound source in a quiet environment, the estimated angle is correct—assuming the formula is right. Aren't we then entitled to

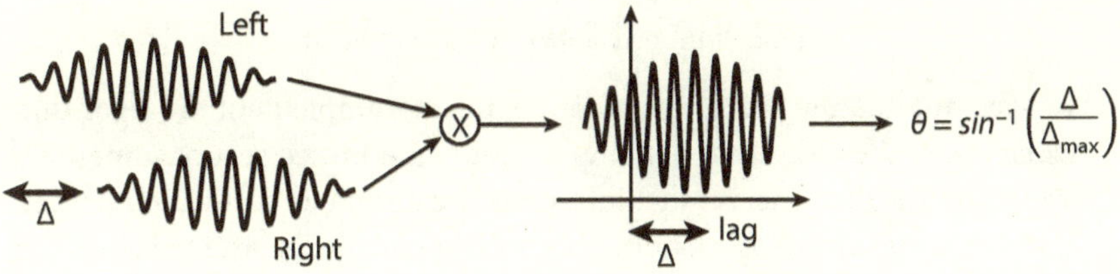

FIGURE 6.5. A sound localization algorithm.

claim that the program knows the source direction, and that the output variable θ holds that information?

But the program output is a value that the observer can interpret in terms of angles, a concept that the observer, not the program, knows. It is the observer that can tell whether the output is correct or not, not the program. Therefore, the program output is information for the observer, not for the program. Suppose for example that the sound localization program is altered—say, a bug is introduced, or some electronic component in the computer malfunctions. We now get a different mapping from acoustical waves to θ. Now the program is in error, but this fact is transparent to the program: all it does is output some value. It is the external observer that considers the output as an error, and the system as dysfunctional. The error is not "system-detectable" (Bickhard, 2009), and therefore the program variable is not information for the system itself.

This is important in the context of biological organisms because, as we have seen in chapter 2, organisms do not have components with fixed specifications. The organism develops and its structure is constantly renewed; the structure drifts and adapts. A meaningful concept of information for organisms must be robust to such indeterminacy.

When we talk of programs as "information processing," it should be clear that we are actually talking about information for the observer, because at no point is there information for the system that runs the program. This is implied by the very term "information processing": information is already in the input, and the useful part is extracted from there. As we have seen, information in this sense can only decrease with further processing, so it cannot describe the process of an organism being informed.

A key issue is that in a program, the epistemic norm is external—it is the programmer who can decide whether the output is correct or not, and who fixes the bugs when the program does not run as expected (the programmer expects, not the machine). But what if the programmer is evolution?

Evolution as a Source of Information

We have pointed out that if the components that implement the algorithm change, then the algorithm will implement something else, not something wrong, for it is the observer that interprets the output as correct or relevant. However, there is a biological norm that does not involve an external observer: Darwinian selection.

For example, imagine a fictional organism that runs the sound localization algorithm, which we will call the computational iguana. The output of the algorithm is connected to a motor apparatus, such that the organism turns its head toward the specified angle and swallows whatever object is in front of it. The iguana's survival depends on whether it can feed, and therefore on whether the sound localization algorithm is correct. In that case, the algorithm is what it is because it has been selected by evolution. In this sense, the algorithm can be said to incorporate information about the world.

However, this is misleading because the kind of feedback that the iguana gets when it is wrong is that it dies. But a dead organism cannot be informed. Therefore, evolutionary theory does not describe how organisms get informed.

Consider Huntington disease. This is a neurodegenerative pathology due to the mutation of a particular gene. Motor and cognitive symptoms start around the age of forty, due to the progressive loss of neurons in the striatum. Now consider a twenty-year-old person who carries the mutation. Those symptoms will occur later because of the genes she carries now. But the person does not know it simply by virtue of having the altered gene. She would still need to inform herself, even though the condition is constitutive of her own structure. There is indeed such a thing as unknowingly carrying a mutation.

Let us go back to the computational iguana. It has evolved an algorithm to compute the angle of sounds, and then to face the sounds. The iguana swallows at random times. This way, it can survive. Can we say that the iguana knows where the frog is? To see the issue more clearly, consider a different species of computational iguana, which computes an arbitrary combination of the two sound waves, and then vaguely points its tail in some direction, without even turning its head. But it turns out that the iguana lives in an environment where frogs move to face iguanas. The iguana, who also swallows at random times, manages to survive. In fact, there appears to be no difference at all from the iguana's viewpoint between the two species: both compute something from the sounds and manage to eat. Neither can be said to have information about the location of sounds.

Evolution is a source of change, not a source of information. Yet, we often talk of "genetic information": what does it mean?

Genetic Information

The genome is often said to hold information about the organism's morphology and physiology. It is in this sense that instincts are often said to be "hard-wired." But textbooks describe the usual result of the development of individuals of a given species, not its diversity. Individuals of the same species often look similar, but in details there is considerable diversity, and occasionally they can be spectacularly different. Mark Blumberg (2010) describes many examples of "monsters," individuals with an unusual body yet still functional living beings. A famous example that we mentioned earlier is a goat born without forelegs, a developmental defect—not a genetic change. Stunningly, this goat developed an upright walking gait. Not only did it learn to walk, but its body also adapted: its spine curved, its muscles strengthened. The same mechanisms at play during normal development allow the unusual young animal to develop into a functional adult.

It is often claimed, in particular within evolutionary psychology, that the theory of evolution implies that important behavior should be hard-wired. But that is an implication of Lamarckism rather than Darwinism. With Lamarckism, adaptation within the lifetime of the organism gets inherited: the goat progressively learns to walk, and the learned solutions are passed on to the next generation. But with Darwinism, useful phenotypes are merely conserved and reproduced. This does not require that the phenotype gets "hard-wired": the process that results in a functional adult is selected, whether that process is rigid or adaptive. If all goats were born without forelegs, then goats would be a bipedal species, without walking upright being hard-wired.

But there is a deeper point. In Darwinism, the possibility of evolution requires that the cognitive system is *not* rigid, that it can adapt to a different body. In other words, it requires that information is *not* inherited. Indeed, if a mutation leads to a change in the body, then for that change to be functional, the nervous system must be able to adapt to it. The alternative is that there is a simultaneous mutation resulting in a change in the nervous system that happens to exactly match the change in the body.

Consider for example the sound localization algorithm discussed earlier. The acoustic waves are mapped to an interaural delay Δ, which is then mapped to a motor command of the head. Suppose now that a mutation moves the ears toward the back of the head, so that they are now closer. The same sound direction will now result in a smaller interaural delay. But then the mapping to the motor command must be adapted so that the same motor command can be triggered by a smaller delay. The anatomical change must be mirrored by a matching change in the control of behavior.

If we believe that body and nervous system are genetically hard-wired, then a genetic change in either body or nervous system structure is detrimental and will therefore not be selected, except in the unlikely event of simultaneous matching mutations. The more reasonable alternative is that body and nervous system coadapt during the lifetime of the organism, in particular in the course of development. Of course, it is entirely possible that structural changes in the nervous system that favor and strengthen the new behavior then get selected—a phenomenon called the "Baldwin effect" in evolutionary theory (Baldwin, 1896; Kull, 2014). But for an anatomical change to be viable in the first place, there must be some degree of flexibility in sensorimotor control: physiological signals cannot be symbols with fixed meaning.

Thus, if we allow for mutations and developmental variability, or simply for normal development, then there cannot be genetically hard-wired information about sensors (which sensors exist is not even fully genetically determined).

Evolutionary psychologists like to deride the idea of a "blank slate" (Pinker, 2003), where all behavior would get written through experience, and they propose instead that the slate is already filled at birth with genetic information. But what is ridiculous is not that the slate is blank: it is that there is a slate at all, which stores information to be read by the organism. What kind of "information" do genes store anyway? They do *not* store a map of body, as we have seen in chapter 2, let alone cognitive algorithms. A gene is not a program. To a first approximation, it specifies the primary structure of a protein. In animals, the genome also influences development, but not by specifying a map of the adult as in the obsolete doctrine of preformationism. The genome is a constraint on the development and physiology of the organism. Consider for example the blind spot in the eye. There is a part of the retina where there are no photoreceptors. If we put the tip of a pen at that specific position (while closing the other eye), it disappears from sight. But usually, we are not aware of the blind spot, it is as if the missing part of the visual field were "filled in." Everyone has a blind spot, roughly in the same place, so could we say that the genome stores that information, the position of the blind spot? That would be highly misleading. The reason why there is a blind spot is because it corresponds to the start of the optic nerve. It is a result of geometrical constraints on development: the nerve has to go somewhere. The position of the blind spot is not stored anywhere: it results from the development of the organism with both genetic and physical constraints.

All the notions of information that we have reviewed commit the same error: they consider that the information is to be found *in* a neural signal, *in* a computational variable, or *in* a gene, and at the same time be about something else. Information is conceived as a sort of epistemic phlogiston inside things, which can be extracted, filtered, and moved within the system. But if information is a relation between a state and something in the world (information by reference), then information is not *in* the state. If information is a relation between a state and some alternative possibilities (information as difference), then information is not *in* the state either. If there is a relation between a gene and the later phenotype of an adult, then that relation is not *in* the gene. If information is defined as a relation, then it is not in any of the related entities.

Therefore, we now move on to a relational perspective, where information is tied to the organism's body and actions.

Embodied Information

Information from Inside the Box

Let us go back to our little animal with a state that correlates with light and dark (see figure 6.4), but now consider that the animal might be much more complex: it might have eyes, it might be able to move, and so on. The light-sensitive neuron would just be one of its neurons, responding to global illuminance. If we only consider the state of this neuron, all we can see is that sometimes S = A and sometimes S = B. We might want to call A "light" and B "dark" but these are arbitrary names that we cannot relate to actual light and dark if all we ever observe is the state S.

However, for the organism, light and dark are not at all interchangeable situations, like two values of a bit: in light you can see, perhaps you can navigate using visual landmarks, and in dark you cannot. Here A is informative not because it is different from B, but because it indicates particular potentialities for the organism. The cognitive agent notices that whenever S = A, it can do certain things. On that basis, the state becomes informative because it indicates potentialities for the organism, and not just for an observer. This does not require independent knowledge and observation of light. The information also does not depend on whether the state could be B rather than A: the information depends on the *present* situation indicated by A, and not on alternative situations that might be envisaged by an external observer.

This notion of information is relational: here, information is relative to what you can do, or to other signals you can observe. The information would not be the same if the animal had only the one light-sensitive neuron and no eye, for example. Information in this sense is necessarily embodied, because what the information is about is defined as a function of the organism's sensors and possible actions. This point appears in several prominent embodied theories of perception: what you can know depends on what you can do.

Information in Embodied Theories of Perception

In 1905, the mathematician Henri Poincaré reflected about the nature of space and geometry and argued that to localize an object does not require to have a preexisting conception of Euclidean space (Poincaré, 1968). Instead, he proposed that to localize an object is simply to imagine the movements necessary to reach that object, that is, the muscular sensations that accompany those movements ("it only means that we represent ourselves the movements necessary to reach that object; [. . .] the muscular sensations that accompany them and that have no geometrical character" [translation mine]). What this means is that the mathematical structure of Euclidean geometry is already present in the relation between motor commands (or proprioceptive signals) and sensory signals. For example, if I make one step forward, my visual perspective changes. There is another movement that I can make to recover the initial visual perspective. This and similar properties endow sensorimotor space with a group structure (movements form a group action on sensory inputs). Then a notion of distance can be added with the time it takes to go from one point to another, or the number of steps it takes. In the end, we obtain the structure of Euclidean space based on relations between movements and sensory signals.

There are two key elements in this embodied view of information. One is that information is based on relations between sensory signals (Poincaré considered visual or tactile signals and proprioceptive signals), or between sensory signals and motor commands. There is no information in the value of an individual signal: it is to be found in relations between signals. The second key element is that such relations are considered informative because they can form the basis of an anticipation. That is, in a certain situation, the visual signals allow me to anticipate that a certain movement *would* produce a certain change in visual signals. In this way, we recover the main property of information in the common sense, the one that is important for behavior: it is something that can be correct or not. I anticipate that a movement will have a certain effect, but it might not, and I can check by actually performing the movement. Thus, the embodied view of information connects information with anticipation. In particular, Mark Bickhard has developed an alternative view of representation that is meant to address the failure of encodingism, namely that "representation is constituted in pragmatic future oriented indications of potentialities for interaction in complex interactive agents" (Bickhard, 2009). We will discuss anticipation specifically in chapter 7.

The embodied view of information is the basis of alternative theories of perception, in particular James Gibson's ecological theory of perception (Gibson, 1979). Two important Gibsonian concepts are invariant structure and affordances. Gibson rejected the traditional concept of information in

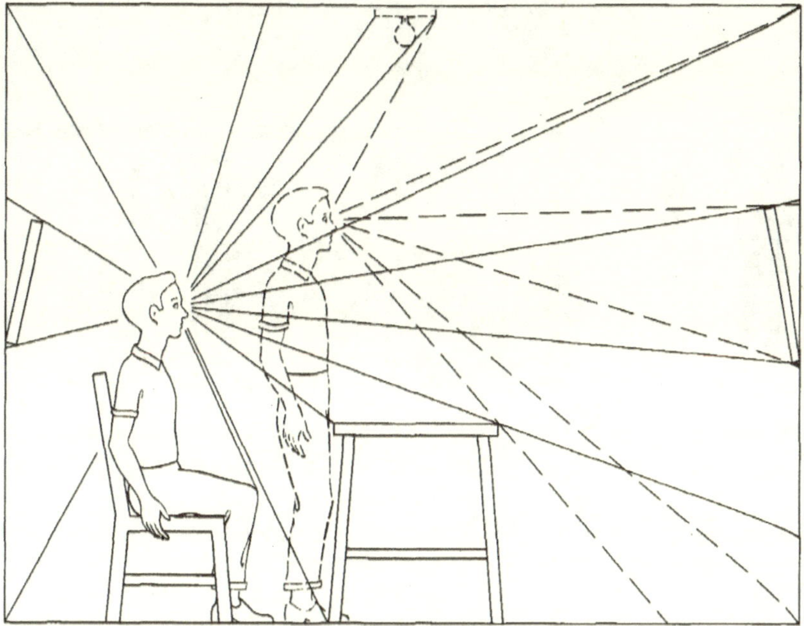

FIGURE 6.6. Information as relations that remain unchanged when the subject moves (Gibson, 1979). Original figure by Todd JT. 2020. https://doi.org/10.1177/2041669520952097 / CC BY https://creativecommons.org/licenses/by/4.0/.

cognitive science, namely that the world must be inferred from indirect signals seen as "codes" for the objective world. Instead, he insisted that there is information in the *sensory flow,* but not in the sense of information by reference. Rather, this information is constituted by the laws that the sensory signals follow. For example, visual signals change in certain lawful ways when the head and body move, because of the laws of perspective (figure 6.6). These laws that characterize a scene are called the *invariant structure*: it is the set of relations that remain unchanged through a sensory flow. See again how this relational notion of information has the property that it can be correct or not: maybe the laws hold, or maybe they do not, and maybe they hold for a limited time (when the configuration of the scene changes, for example).

Gibson argued that the alleged necessity of perceptual inference was largely exaggerated as a result of reductionist methodologies in psychology. Indeed, if information is to be found in relations, then isolating one component makes information disappear entirely. For example, consider the problem of determining the size of an object. The apparent size of the object on the retina gives an ambiguous indication because it depends on distance: the object will look

FIGURE 6.7. Visual ambiguity resulting from an unrealistic lack of context.

smaller if it is farther away. However, in an ecological setting, the object will appear on a background that is subject to the same laws of perspective (figure 6.7). As a result, the *relative* size of the object compared to the neighboring background pattern will not depend on distance. This makes it possible to compare the size of objects at different distances. The world is not "noisy," as the statistical inference view would insist: it is complex (Brette, 2018)—that is, it obeys a complex set of laws, and these laws are informative rather than a source of noise. Thus, Gibson famously stated: "Ask not what's inside your head but what your head is inside of."

Gibson considered that this kind of information is not inferred, by reference to some preexisting knowledge, since it is present as invariant structure in the sensory flow. This is why he talked about the "pick-up of information" by the perceptual system. This terminology certainly caused some confusion, as it is reminiscent of similar terminology about information by reference, such as "extraction of information." Thus, classical representationalists like David Marr thought that Gibson saw perception as some kind of magic process. This is ironic given that, as we have seen, conceptualizing information by reference to something else, as Marr embraced it, ultimately leads to a Cartesian dualist position where meaning has to be magically assigned to symbols. The key difference between Gibsonian "pick-up" and computationalist "extraction" is that Gibsonian information is relational while classical representationalism places information within signals themselves.

The second important Gibsonian concept is *affordances*. An affordance is a possibility for action that a sensory scene indicates. For example, a chair affords sitting. For Gibson, the higher-order perceptual information is constituted by

FIGURE 6.8. Affordances of a potty for different animals.

affordances. Again, this is a relational notion, this time between the organism and the environment. The affordances of an object are not properties of the object itself, but of the pair constituted by the organism and object. For example, consider figure 6.8, which could appear in a children's book. It shows a potty. What appears as a potty for a human baby, however, may not appear so for another animal. For example, for a mouse, this might be a slide. For a frog, it could be a bathtub. For a dog, it could be a bowl, where the dog could drink.

Traditionally, in artificial intelligence, categories such as chairs are defined as arbitrary sets of images to which we assign a common name. With the concept of affordances, a perceptual category is not an entirely arbitrary concept. It is constituted by interactional properties specified by the visual scene (or more precisely, the sensorimotor flow). As the potty example illustrates, these perceptual categories are not exactly categories about the environment, but about the *Umwelt*—that is, the environment seen as possibilities for meaningful interaction. Affordances depend not only on the world, but also on the kind of organism that perceives them.

Like invariant structure, affordances can be correct or not. For example, the frog can wait for rain and try whether the potty is actually suitable to take a bath. This stems from the fact that affordances are anticipatory: they refer to the anticipated effect of actions that the organism might take, and the anticipation can turn out to be more or less accurate. Thus, affordances also have the desired property of information in the common sense.

A related embodied theory is the sensorimotor theory of perception (Noë, 2004; O'Regan, 2011; O'Regan and Noë, 2001). Gibson described the sensorimotor laws of the visual world, called "ecological optics," in terms of the structure of the visual field (or "optic array"). But light is received by a discrete set of sensors, each one receiving an optical signal with a specific relation to the visual world. These sensors do not form a projection screen (for one, they are a discrete set). For example, in the eye there is a blind spot with no photoreceptors, there are inhomogeneities in spatial and color sampling, geometric aberrations, and so on. To explain how the organism might deal with this messy interface with the visual world, O'Regan and Noë (2001) propose the following experiment of thought:

> Imagine a team of engineers operating a remote-controlled underwater vessel exploring the remains of the Titanic, and imagine a villainous aquatic monster that has interfered with the control cable by mixing up the connections to and from the underwater cameras, sonar equipment, robot arms, actuators, and sensors. What appears on the many screens, lights, and dials, no longer makes any sense, and the actuators no longer have their usual functions. What can the engineers do to save the situation? By observing the structure of the changes on the control panel that occur when they press various buttons and levers, the engineers should be able to deduce which buttons control which kind of motion of the vehicle, and which lights correspond to information deriving from the sensors mounted outside the vessel, which indicators correspond to sensors on the vessel's tentacles, and so on.

Here, O'Regan and Noë propose a solution to the problem of knowledge from inside the black box. How relevant is the analogy for biological organisms? As we have seen, anatomy is often stereotypical and appears well organized, but this is just the result of the usual course of development, not the reading of a genetic map of the organism. Sensors and actuators are not organized in a rigid way, and unpredictable genetic or developmental events can disrupt this organization. Behavior must be robust to a certain lack of specification of the organism's anatomy: this is a prerequisite for Darwinian evolution. Thus, the goat born without forelimbs could walk. The thought experiment of the mixed-up cables captures this property of development in a somewhat exaggerated way.

In the sensorimotor theory of perception, information is to be found in *sensorimotor contingencies*, which are relations between motor commands and sensory signals. This is related to Gibson's invariant structure, but the signals are seen as physiological (rather than, say, optical and kinematic) and the emphasis is on the relation between actions and their effects on the sensory signals.

These embodied theories of perception have in common that information stems from a combination of observation and interaction with a previously unknown environment. This relates to the way scientists build knowledge, based on instruments and experiments.

The Analogy of Science

Information by reference is inspired by the way we get informed by reading a book or a newspaper: the symbols refer to a particular situation, real or imagined, which we can understand. It is a case of communication, where someone informs someone else by communicating her knowledge. But except in that specifically human sociocultural context, the way animals inform themselves by observing and interacting with the environment is not a case of communication. Reading a book is not a good analogy.

The embodied view of information emphasizes laws in the sensory signals and the effect of actions on sensory signals. This evokes another way in which we get to know the world: science. Scientific knowledge comes from observing and acting in the world: making measurements and experimenting. Experimenting consists in intervening in the world and observing the measurable consequences. From experiments and measurements, laws are constructed. The scientific view is that there is no other access to the world than through these means—in contrast, the religious view is that information comes from the Book. In the same way, an organism has sensors and effectors, and knowledge can be built in terms of their relations (see figure 6.9).

I will now discuss some properties of embodied information by examining an example in detail.

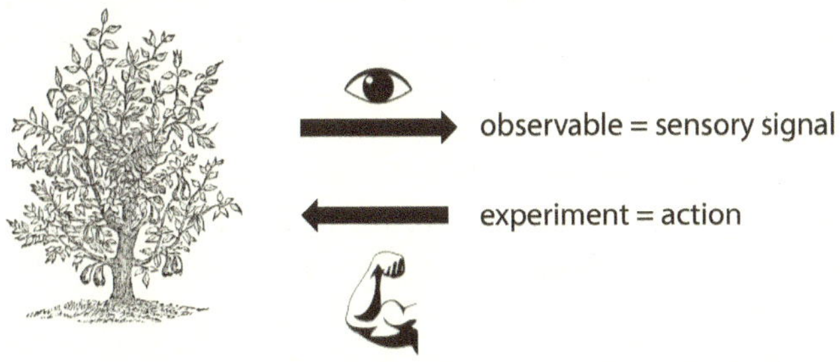

FIGURE 6.9. Information from relations between observables and the effect of actions.

Properties of Embodied Information

Embodied Information in a Martian Iguana

The set of laws in sensory signals and their relations to the organism's actions is constrained by the interface with the environment, which is specific of the organism. In analogy to physics, I previously called these laws "subjective physics" (Brette, 2016), but this should be understood as the laws constrained by the organism–environment interface, and not the cognitive constructions that are actually built by the subject. Indeed, it must be recalled that information presupposes a cognitive agent (someone who gets informed), and embodied information is not different in this respect—for example, the villainous monster thought experiment is explicitly homuncular. We will address this issue in the last section of this chapter. Here, I will ask what kind of laws the cognitive agent can construct, given a set of sensors and actuators, without prior knowledge about the external world.

To discuss embodied information in more detail, I will take the example of a fictional agent. (See Brette [2016] for a more detailed discussion.) Many accounts of perception are focused on vision, but we tend to have false intuitions about how we see. For example, we tend to think that the retina captures images like a camera does, which we then analyze. But as we have discussed, the light pattern on the retinal photoreceptors is messy and dynamic, unlike our visual perception. Therefore, I will choose the auditory modality, where these misleading intuitions are not as strong.

Consider then a Martian iguana (Brette, 2019a; figure 6.10A). We do not know the physiology of that alien animal. It might even be a robot. All we know is it has two ears, one on each side. It lives on the ground, still, in an environment where there is occasionally a Martian frog that produces sounds. The frog is usually still, but sometimes it jumps to another position.

When the frog produces a sound, two sound waves S_1 and S_2 arrive at the iguana's two ears (figure 6.10B). These two waves are related because they originate from a single sound. The physics of sound tells us that this relation, to a first approximation, takes the following form: $S_1(t) = S_2(t-\Delta)$, for some delay Δ (we neglect sound diffraction). This relation holds true as long as the sound lasts: this is what Gibson calls "invariant structure," a particular law in the sensory flow. This relation is information in the sense that it expresses a constraint on the sensory signals that occurs at a given moment, and which did not before. The information is about the two terms of the relation, not something external or something possible.

At this point, there is no indication of space in this relation. The iguana cannot possibly know where the frog is from just the observation of this

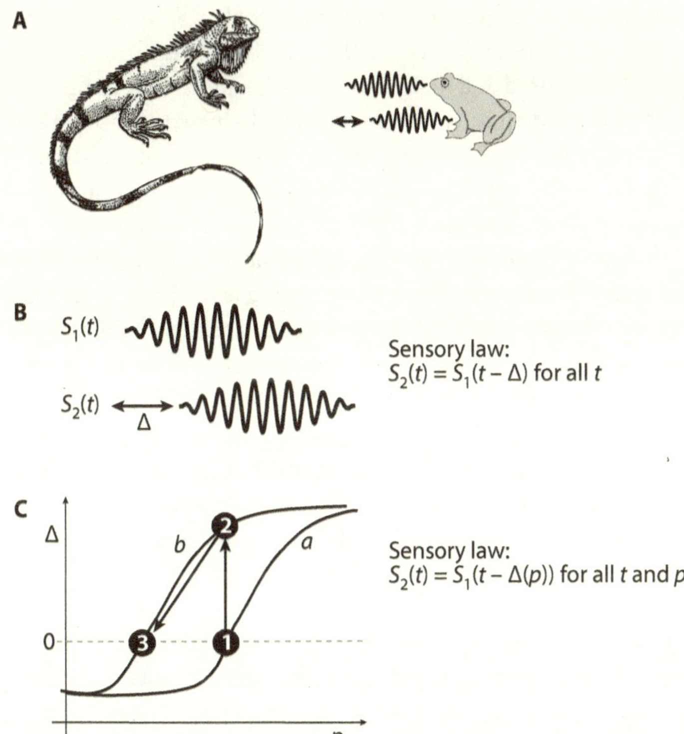

FIGURE 6.10. Embodied information in a Martian iguana. A: The Martian iguana hears Martian frogs. B: Its two ears capture acoustic signals with delays related to the frog's position. C: Relations between interaural delays and proprioceptive signal of the neck for two frog positions *a* and *b*. First, the frog is at position *a*, facing the iguana (1); then it jumps to position *b* (2); then the iguana makes a compensatory movement to face the frog again (3).

relation. Indeed, suppose for example that the two acoustic sensors are on the same side of the head. In that case, the relation $S_1(t) = S_2(t{-}\Delta)$ would hold specifically for periodic sounds of frequency $1/\Delta$, independently of the source position. The relation would then be specific of melodic pitch, not space.

However, if we consider that the iguana can turn its head, then the frog can notice that this relation is disrupted when the head is turned; in this sense, it is a spatial relation. But this does not tell the frog *where* the source is. Suppose now that the iguana has a proprioceptive sensor that provides a signal p related in some lawful way to the head's direction (figure 6.10C). Now if the frog produces sounds, the iguana can observe the relation $S_1(t) = S_2(t{-}\Delta)$, but in addition there is another relation between the delay Δ and p when the head is moved. Thus, a higher-order relation holds: $S_1(t) = S_2(t-\Delta(p))$ for all times and for all head positions (curve *a*). This relation is specific of the frog's position. It can be explored and registered by the iguana as long as the frog is still. This gives us an elementary notion of space, as discussed by Poincaré. Suppose that the frog is initially in front, so that the two acoustic signals match ($S_1(t) = S_2(t)$; point 1 in figure 6.10C). Suddenly, the frog jumps to another position and the iguana registers a different sensory relation (point 2). Where is the frog? What we know is that there exists a compensatory movement that turns the head so that the two acoustic signals match again—that is, such that the frog faces the iguana again (point 3). Thus, for each position of the frog, there is a particular position of the iguana's head (more precisely, a particular value of the proprioceptive signal) such that the two acoustical signals match. This defines the position of the frog in terms of the iguana's sensorimotor structure.

This is close to the concept of sensorimotor contingency (O'Regan and Noë, 2001), except the spatial relation is not a mapping from proprioceptive to sensory signals but a higher-order relation on the sensorimotor structure—for some value of p, a relation holds (matching for a particular delay Δ); for another value of p, another relation holds.

The notion of space defined by this higher-order relation is very elementary because it has no associated metrics. It is only defined in terms of compensatory movements. Suppose for example that the frog is in front of the iguana, then jumps to 45 degrees on the right, then to 90 degrees on the right. Is there any way to know from just the acoustic signals and the proprioceptive signal that the frog made the same angular displacement twice? Unless p is linearly related to the angle θ, this is not possible.

To address this issue, we need to consider not just proprioceptive signals, but also actions (figure 6.11). Consider then that the iguana can turn its head by issuing rotation commands. For example, the iguana might be a robot with a head mounted on a step motor. It can then decide to rotate its head by a certain number of motor steps. The iguana's cognitive system has access to a copy of these commands—what neuroscientists call the "efferent copy." That is, it has access to some transient signal a that is lawfully related to the angle increase $\Delta\theta$. When the iguana turns its head by an amount a, the proprioceptive signal changes from p to p'. Thus, an action corresponds to a mapping from p to p'. It can then be seen that the composition of actions endows the space of actions with the mathematical structure of a commutative group: for each action, there exists an inverse action; two actions have the same result independently of their order; and so on. Similar remarks can be made on how a rotation command changes the interaural delay Δ. In this way, we have constructed an algebraic structure on the proprioceptive space, which allows the

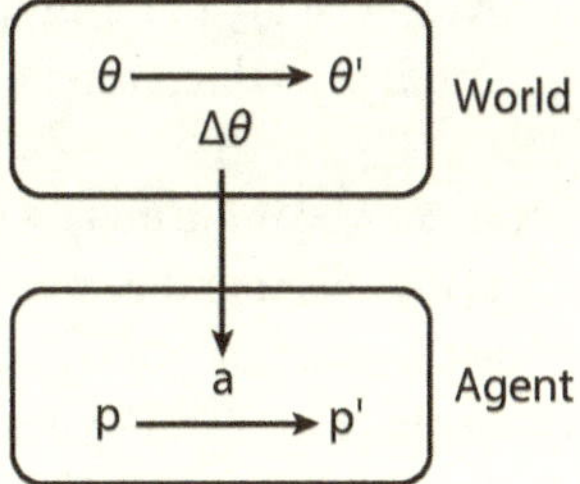

FIGURE 6.11. Actions transfer the commutative group structure of angles to the proprioceptive space.

iguana to identify identical movements of the frog. This is essentially Poincaré's insight.

Suppose for example that the frog turns by 45 degrees, so that the relation between p and Δ changes. This change can be canceled by a particular compensatory action of the iguana (e.g., a number of motor steps). If instead the frog turns by 90 degrees, then the new relation between p and Δ can be compensated by the iguana executing twice the same action, even though p might depend nonlinearly on the frog's angle. This is what it means that the frog's angle is twice as large.

Note that this would not work if the motor commands specified a particular target head angle (e.g., by appropriate contraction of antagonistic muscles), rather than a particular *change* in head angle (i.e., rotation). In other words, it works only because we have described a case of locomotion: the body is in the same configuration at the start and at the end of the command, while the relation between body and environment changes.

Thus, in the embodied view of information, information depends on the kind of action that the organism can make. Let us see a few ways in which action is involved in information.

Action in Embodied Information

It is commonplace in neuroscience to say that perception is for action, and perhaps to grant that what is perceived might be biased toward what is useful for action. What is less obvious is that perception itself might depend on action, or on the structure of the action set. But that is essentially Poincaré's insight: it is the possibility of locomotion that allows us to grasp sensory signals in terms of space. Action is also central in ecological and sensorimotor theories of perception, but there are various ways in which action can be involved, which we list (nonexhaustively) here:

1. Action might not be involved at all. For example, a musical note has a particular structure, periodicity, which can be noticed independently of any action.
2. The possibility of action may be indicated by the sensory structure, without having a role in establishing that structure. This is essentially the concept of *affordance*: a chair affords sitting.
3. Action may disrupt sensory structure. For example, the binaural structure of the acoustical waves captured by the ears when a source emits a sound is disrupted by head movements. This fact gives a spatial character to the auditory structure.
4. Action may cause the sensory flow, within which structure is observed. For example, when a bird flies, the optic flow has a particular structure radiating from the flight direction. Movement is necessary for that sensory structure to exist, but the structure is sensory rather than sensorimotor. This is closest to Gibson's concept of *invariant structure*.
5. Action may be involved through proprioception. For example, in the iguana's example, there is a relation between the proprioceptive signal p, correlated to head position, and the interaural delay Δ, correlated to the source direction. This is a variation of the previous case (4), where the signal is not related to the world but to the body. This is closest to the concept of *sensorimotor contingency* (O'Regan and Noë, 2001).
6. Action may be an event causing a specific change in sensory structure—here we focus on the property of the event, rather than on its result on sensory signals. In this case, an action is a certain operation acting on structure space. For example, a rotation of the head might change the binaural relation between acoustic signals in a way that depends on the chosen rotation. This is closest to Poincaré's argument about the origin of our understanding of space. It is a very rich notion, as it endows sensorimotor space with an algebraic structure.

This list emphasizes the various nuances that figure in embodied theories of perception: invariant structure, affordances, sensorimotor contingencies, Poincaré's algebraic sensorimotor structure.

We can now revisit the question of perceptual inference.

Inference with Embodied Information

We have pointed out that the argument that perception must be based on inference because things in the world are only indirectly accessed through sensory signals is incoherent. Inference requires knowledge, and if the world is considered inaccessible then it cannot be known and inference is impossible.

However, if information is based on relations between sensory signals, rather than between sensory signals and inaccessible external things, then the premises of the argument for inference vanish. Noticing that two signals are delayed versions of each other does not require inference. It is in this sense that Gibson considered that perception is direct.

However, this claim must be properly qualified. Suppose that the frog is on the right of the iguana and emits a short sound, say a tongue click. The two acoustical waves at the iguana's ears are delayed copies of each other, with a particular delay Δ. However, since this is a transient sound, the frog cannot explore the relation between delay and head movement, through its proprioceptive signal *p*. What we can say is that, based on prior exploration of such relations, the observed delay Δ indicates that the delay *should* change in certain specific ways with *p if* the head were moved and another sound were produced. This is an anticipation based on prior knowledge, which we might call inference. Indeed, this is implicitly acknowledged in the sensorimotor theory of perception: "Vision is a mode of exploration of the world that is mediated by knowledge, on the part of the perceiver, of what we call sensorimotor contingencies." (O'Regan and Noë, 2001): prior knowledge is used to explore. Thus, there is a role for some kind of inference in the embodied view of information, precisely because information is tied to anticipation. But the key point is that inference is logically secondary: first one must be able to grasp and build the objects that might later be inferred.

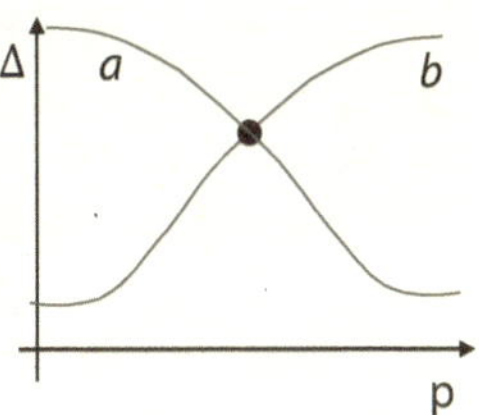

FIGURE 6.12. In some cases, the same sensory signal is compatible with several sensorimotor relations (*a* and *b*).

It is also the case that sensory signals can be ambiguous. For example, sounds produced at symetrical positions relative to the interaural line (i.e., same angular position in the front and in the back) will produce the same interaural delay Δ (figure 6.12). This could lead to misestimations of sound direction, which are called "front-back confusions" in auditory psychophysics. In relational terms, the delay Δ is consistent with two sensorimotor relations between proprioception and delay (*a* and *b* in the figure), and therefore it is ambiguous. But the very concept of ambiguity presupposes that there exist well-defined alternatives, which preexist to the ambiguous "stimulus." In the duck–rabbit illusion (figure 6.13), one might see a duck or a rabbit in the same image, but that is possible only because one has seen unambiguous ducks and rabbits before. Ambiguity and illusions cannot be a foundational fact of perception.

Thus, perhaps surprisingly, it is precisely when perception is viewed as an inverse problem, as David Marr did, that inference cannot be properly defined.

FIGURE 6.13. The duck–rabbit illusion.

Suppose that seeing a tree correlates with a specific pattern of spike trains in the early visual system. The perceptual system might then interpret that pattern as a tree. But what if it were actually a chair? There is no way to know that the inference is wrong if all information is by reference to something inaccessible. This is what Mark Bickhard calls the problem of "system-detectable error" (Bickhard, 2009): the encoding view of information and representation does not allow for the possibility of system-detectable error. In contrast, the anticipation of relations between sensory or sensorimotor signals does allow for system-detectable errors.

The idea that information is based on laws that govern sensorimotor signals, in analogy with science, evokes the popular idea that the brain constructs an "internal model" of the world. A prominent theoretical framework based on this idea is *predictive coding theory* or *predictive processing theory*, which proposes that the brain constructs a generative model of the world. Unfortunately, this is not the right kind of model.

Predictive Coding and Generative Models

Predictive processing theory is a popular theory meant to address the problem of information by reference (Clark, 2013; Friston, 2009; Rao and Ballard, 1999). It claims that the brain uses internal models to predict the sensory signals, and that it is then the variables of these models that represent perceptual facts. The classical paper of Rao and Ballard (1999) describes a hierarchical neural model of visual processing with feedforward and top-down connections. The top-down connections are tuned so as to cancel as much as possible the input of the model. This cancellation is called "predicting" the input—we will see in chapter 7 that this terminology is misleading. Thus, these connections implement a mapping from the activity of higher-order neurons to input—that is, an encoding of the input. This mapping is called a *generative model* of the input.

Generative models are very particular kinds of models, since they are essentially encodings. Models of physics and subjective physics are typically very different; they take the form of laws that the observable quantities follow. For example, the law of ideal gases relates different observable quantities (pressure *P*, volume *V*, number of moles *n*, and temperature *T*) by stipulating a certain relation between them: $PV = nRT$. This is not a generative model at all, yet it can be used to make predictions. For example, if you increase temperature by 10 percent while keeping the volume constant, you expect pressure to increase by 10 percent. The model describes abstract relations between different aspects of reality; it is not meant to account for, say, the position and momentum of all particles in the gas, as a generative model would be designed for.

According to predictive processing theory, "to successfully represent the world in perception [. . .] depends crucially upon cancelling out sensory prediction error" (Clark, 2013). But to deal with gases, it is neither necessary nor useful to predict the position and momentum of gas particles (let alone that this is physically impossible). The same holds in binaural hearing: to know where the sound comes from, it is useful to know that the two monaural inputs S_L and S_R are such that $S_L(t) = S_R(t-\Delta)$, but it is not necessary to predict the time-dependent pressure of both acoustic waves. Thus, it is false that perception necessarily depends on canceling out sensory prediction error.

Perceptually guided behavior depends on abstract properties of the sensory environment, but a generative model is meant to generate a particular input, not an abstraction. The activation of the higher-order layer of a generative model, which represents the input, cannot at the same time represent an abstraction, because as the example of gases illustrates, one cannot go from abstract (pressure, etc.) to particular (a particular set of positions and momenta). One can only generate a particular example of the abstraction. Thus, the following claim of predictive processing theory cannot be true:

> For example, the meaning of a phrase (encoded in semantic areas) predicts words (encoded in lexical areas), which predicts letters (encoded in ventral occipital areas), which predict oriented lines and edges (encoded in visual areas). (Adams et al., 2013)

Consider the phrase "two persons of the same height." One can generate an example image consistent with that phrase (figure 6.14), but one cannot predict what the image actually is in terms of "lines and edges."

Thus, generative models are encodings of the input, not laws followed by sensorimotor signals. They are not models in the same sense that scientific

FIGURE 6.14. Two persons of the same height.

laws form a model of the world: they are compression schemes. Despite the terminology, "predictive coding" is not the right framework for anticipatory behavior, as we will see in more detail in chapter 7. This relates to the fact that generative neural network models, such as large language models trained to predict the next word, have great difficulties dealing with abstractions beyond their training base (Dziri et al., 2023; Lewis and Mitchell, 2024). For this reason, Yann LeCun argues that generative models are intrinsically limited, because a generative model is not a world model (LeCun, 2022).

We have introduced the concept of embodied information to address the problem of prior knowledge, which is a prerequisite of information by reference. We examined what can be known from inside the black box, which must be based on sensory or sensorimotor relations, and the effect of actions. But of course, the black box perspective remains homuncular: we still need a cognitive agent able to notice those relations and act accordingly. In other words, how does information inform, in terms of the organism's processes?

How Information Informs

Beyond Epistemic Phlogiston

The mainstream view of information sees information as some kind of epistemic phlogiston, some stuff that exists in things by virtue of their being in a certain state. As we have pointed out, this is a commitment to substance metaphysics (Bickhard, 2009, 2024). In the context of information, substance metaphysics is not coherent. There cannot be information in some signal just because that signal correlates with something else: by assumption, the phenomenon is about the relation between the two things, not just one of the two things, so it cannot be in just one of them. In the same way, there cannot be information in some signal just because that signal could have been something else: those possibilities are not contained within the signals; they are enumerated by someone.

Embodied information improves on this view by describing information as relations between signals that are considered accessible by the system, in analogy with the laws of science. Such relations can be true or false, or more or less accurate, and this can potentially be verified by the system.

Nonetheless, to go beyond substance metaphysics, we still need to address the problem of reification. For example, Gibson claims that the organism "picks up information," but information is not a thing that exists out there; rather, it is a relation noticed by someone. More accurately, we mean that the organism gets informed. But what does it mean in terms of the organism's activity?

Looking back at embodied theories of perception, we can see two ways to relate information with the organism's activity: information as a constraint on the coupling between organism and environment, as in Gibson's "invariant structure"; and information as a basis for anticipatory behavior, as in Gibson's "affordances."

Information as Constraints

Gibson's concept of "invariant structure" exemplifies the concept of information as constraint—although he did not express it in this way. Invariant structure is a set of laws that are satisfied by sensory or sensorimotor signals, at a given moment. For example, when one moves, the laws of perspective combined with the constancy of the environment result in a particular structure in the visual signals (Gibson called these signals the "optical array"). This means in particular that the optical signals captured by the retina belong to a manifold with lower dimensionality than the number of photoreceptors. Thus, these sensorimotor laws express particular constraints on the coupling between the organism and the environment, and these constraints can then have a causal influence on the system (as we have seen in chapter 3).

The idea of constraints as causes plays an important role in modern theoretical biology (Deacon, 2012; Juarrero, 2002, 2023; Kauffman et al., 2008; Montévil and Mossio, 2015). Thermodynamically, constraints are what turns heat into work: igniting powder releases heat, but when air molecules are constrained to move in one dimension as in a cannon, mechanical work is produced. A related concept in physics is resonance. For example, the open holes of a flute constrain the vibration of air such that a periodic sound with a certain wavelength is produced. More generally, a dynamical system driven by a periodic input will often settle in a periodic regime, with a mode that depends on both the input and the structure of that system (this is called "periodic forcing"). In such cases, the dimensionality of the system's activity is reduced in a particular way.

This is why Gibson conceived the relation between invariant structure and the perceptual system as a case of resonance: "A percept is related to a stimulus invariant by the resonance of a perceptual system" (Gibson, 1983); "the visual system [. . .] tunes in on the invariant structure of the ambient optic array that underlies the changing perspective structure caused by his movements" (Gibson, 1979). Of course, this remains rather vague. Gibson was a psychologist who did not seem to have much interest in physiology.

However, this idea can be found in neurophysiology in discussions of brain oscillations and neural synchrony. We have seen in chapter 5 that the lack of structure in "neural codes" motivated the proposition that the timing of spikes could be a signature of perceptual objects, such that only neurons dealing with the same object would interact together—"binding by synchrony" (Singer, 1999; von der Malsburg, 1999). For example, different visual parts of a moving object would display similar variations in light intensity, or in optic flow, and this could drive synchronous activity in neurons responding to those different parts. In the original proposition, input similarity is superimposed on an endogenous brain oscillation, resulting in firing at similar phases of the oscillation—a well-known general nonlinear phenomenon called "phase locking" (Brette, 2004). Since neurons are much more sensitive to synchronous rather than asynchronous inputs (Rossant et al., 2011), this would in turn engage neural circuits specific to those combinations of neurons that are driven by the same object. Brain oscillations figure prominently in neurophysiological explanations of cognitive phenomena (Buzsáki, 2011, 2019), although this is of course disputed by classical computationalists.

In this example, the presence of an object with coherent motion introduces constraints on the activity of visually responsive neurons, resulting in particular lower-dimensional modes of activity. There are other examples in the neuroscience literature where sensory relations result in neural synchrony. One is the Jeffress model of sound localization (Jeffress, 1948; figure 6.15). Binaural neurons receive inputs from monaural neurons on the two sides, with different conduction delays. When input spikes arrive simultaneously, the neuron spikes. This would rarely occur if the two ears captured unrelated acoustic signals. But when something produces a sound, the acoustic signals at the two ears are related in a particular way: $S_L(t) = S_R(t-d)$, where the delay d varies with source position. When this delay matches the difference in conduction delays to a binaural neuron ($\delta_R - \delta_L$), that neuron receives

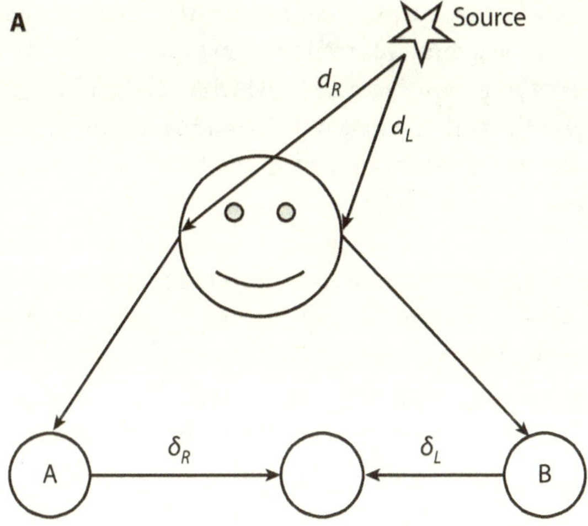

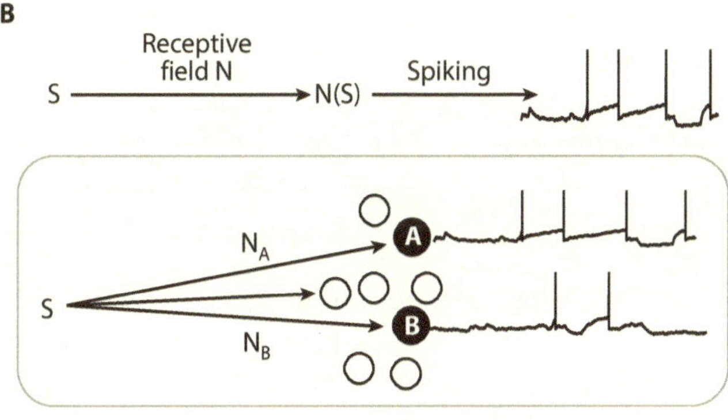

FIGURE 6.15. Neural synchrony induced by sensory relations. A: The Jeffress model of sound localization. B: Synchrony receptive fields. Original figure by Brette R. 2012. https://doi.org/10.1371/journal.pcbi.1002561 / CC BY https://creativecommons.org/licenses/by/4.0/.

synchronous spikes and fires. Thus, the neuron fires specifically when the sensory signals are constrained in a particular way. In Gibson's terminology, this neural circuit is "tuned" to a particular invariant structure.

I have generalized this interpretation of Jeffress model with the concept of "synchrony receptive field" (Brette, 2012), which describes how a constraint on sensory signals may result in a specific pattern of neural synchrony (figure 6.15B). The synchrony receptive field of a pair of neurons is the set of sensory signals that elicit synchronous responses in those neurons. In a simplified model, the time-varying sensory signal S (which can be multidimensional) is filtered in some way through the neuron's (classical) receptive field, giving N(S), and then transduced to a precisely timed spike train. Neurons A and B are then synchronous when $N_A(S) = N_B(S)$—that is, when the sensory signal is constrained in a certain way, or has a certain "invariant structure." This framework has been applied to models of pitch perception (Laudanski et al., 2014) and to sound localization in realistic environments (Benichoux et al., 2015; Goodman and Brette, 2010).

This line of research has been relatively unexplored, compared to classical computationalist modeling, but it gives a glimpse of how perceptual information may be conceived as constraints acting on the nervous system.

Anticipation

A key point about the idea of information as constraint is that it is a dynamical concept. Indeed, to say that sensory signals are constrained in a certain way requires those signals to change. For example, it is trivial and uninformative to say that the numbers $X = 5$ and $Y = 3$ are related by the identity $X = Y + 2$. But if we observe that $X(t) = Y(t) + 2$ for all t, then we can say that these two signals are constrained, such that the pair (X, Y) belongs to a one-dimensional space, instead of a two-dimensional one.

The very concept of invariant structure presupposes that we find something that does not change within a flow. The constraint is realized at a certain moment, but it exerts its effects on processes. The observed relations are then properties of an "extended present," like the optic flow: it is about changes occurring at a given moment.

Affordances also express relations, between sensory signals and actions. Seeing a chair informs us of the possibility of sitting—the chair affords sitting. We might call this a constraint (on the kind of actions that we can do and their effects), but it is not a dynamical concept anymore. What will happen in the future cannot directly exert a constraint on neural activity in the same way as sensory signals do. Rather, seeing the chair allows us to *anticipate*. We will now examine this concept.

7
Anticipation

OUR ANALYSIS of what information might mean for an organism led us to anticipation. To see a chair is not just to attach a label to a particular pattern of pixels: it is to anticipate a possibility of interaction with the environment—namely, sitting. This is the concept of *affordance* in ecological psychology (Gibson, 1979). Anticipation is also a core concept of the sensorimotor theory of perception. Poincaré remarked that "to localize an object simply means to represent to oneself the movements that would be necessary to reach it" (Poincaré, 1968). The modern sensorimotor theory stipulates that "vision is a mode of exploration of the world that is mediated by knowledge, on the part of the perceiver, of what we call sensorimotor contingencies" (O'Regan and Noë, 2001). To use knowledge for action (exploration) is to anticipate the effects of action. Bickhard's interactivism (Bickhard, 2009) is explicitly a theory of anticipation: "representation is constituted in pragmatic future oriented indications of potentialities for interaction in complex interactive agents" (Bickhard, 2015).

In fact, all theories of mind and life revolve, explicitly or not, around anticipation. The classical concept of representation is motivated by the fact that some behaviors, called "representation-hungry," can unroll in the absence of the stimulus (chapter 5). Representation is the proposed solution, not the phenomenon to be explained, namely anticipation. Another alleged characteristic of "representation-hungry" behavior is that it depends on abstract features of the situation, rather than proximal stimuli. But what are those abstract features? Certainly not any arbitrary set of situations that we can logically construct, since that would be everything and therefore could not characterize any kind of behavior. The ant's behavior does not exactly depend on the nest's position, since it does not get back to its nest if it is displaced. More accurately, the ant's behavior depends on where the nest can be *expected* to be, based on previous navigation history. Again, the phenomenon to be explained is that the ant anticipates the relative position of the nest. It is relative to a goal that we associate to the ant's behavior, namely going back home. In the same way, when the cat turns its gaze to the squeaking mouse, it anticipates where the mouse should be seen given the acoustic signals at the cat's ears. The abstract features that "representation-hungry" behavior depends on are those that are informative of potential interactions or of future events, in relation to the organism's goals.

The computationalist objection against anti-representationalist theories is generally that they are based on feedback. This is the case for example of the dynamical theory of cognition (van Gelder, 1995, 1998) and of behavior-based robotics (Brooks, 1986, 1991). The objection is justified, because behavior is generically anticipatory: action is directed toward goals, often indirectly linked to the maintenance or the reproduction of the organism.

Thus, what computationalism tries to explain is anticipation. Its proposed solution is that the brain calculates what needs to be done in order to reach a certain goal. Since computation occurs entirely in a formal domain (as opposed to behavior based on feedback), computation is in a sense maximally anticipatory. But the reason why computation superficially seems to work is not that behavior is based on computation—there is little direct evidence of that, as we have seen. The reason is simply that computation is a kind of behavior, and behavior is anticipatory, goal-directed. Obviously, computation has this feature too, where the goal is the correct result of the computation. Computation is a kind of behavior, but the converse, that behavior is a kind of computation, does not logically follow. As we will see in this chapter, computation does not explain anticipation, since goals are externally attributed: it assumes anticipation.

Anticipation is not just the central concept of brain theory. It is also central to all theories of life. Much discussion in philosophy of biology revolved around the problem of explaining the apparent goal-directedness of biological processes. Many biologists were not at ease with the idea that biological processes might have goals, and yet there were clearly anticipatory phenomena to be explained—for example, plants seek light, animals behave, and so on. This is why Pittendrich (1958) coined the term "teleonomy" to distinguish it from teleology while trying to convey essentially the same idea, but without an implicit maker pulling the strings. Mayr (1961) tried to provide a more precise definition, which Monod later adopted (Monod, 1970), by proposing to attribute teleonomy to "systems operating on the basis of a program of coded information," thereby turning the maker into an engineer. Again, the phenomenon to be explained is anticipation, while information and programs are proposed solutions.

The theories of life we have reviewed in chapter 2 are also about anticipation. For example, the autopoietic organization produces materials and constraints that satisfy its future needs. This is why enactivists see mind and life as co-extensive (Thompson, 2010): whether animal or unicellular, a living system acts so as to anticipate its future needs, based on the sensed environmental configuration. Theoretical biologist Robert Rosen (1985) also saw anticipation in the circular organization of causes in living processes (an effect being also a cause).

Thus, anticipation is the central concept of cognition and life. And yet, it is not so straightforward to explain what anticipation is, let alone how organisms anticipate. Therefore, much of this chapter is devoted to explaining not how organisms anticipate, but what anticipation is. Literally, to anticipate is to "take" in advance. Thus, at first sight, it seems that to anticipate is to behave as a function of future events. But of course, anticipation is not prescience. The future cannot influence the present. If the anticipated event turns out to not happen, this has no retroactive effect on the prior behavior. Anticipation is behavior determined not by future events, but somehow by an *expectation* of some future event. But "expectation" is a mental term that does not seem appropriate for many anticipatory phenomena. Therefore, the working definition that I will converge to in this chapter is that anticipation is the *exploitation of regularities*.

To get a clearer idea of anticipation, I will start by making an overview of anticipatory phenomena, through a number of examples (*An Overview of Anticipatory Phenomena*). The main theoretical account of anticipation is that the brain encodes predictions, but this framework fails because, as we have seen, the coding view of information is incoherent (*Action Based on Prediction*). A variation on this theme is predictive coding theory, also called predictive processing, which aspires to explain both perception and action under a unified predictive principle, but that theory trivializes both the notion of prediction and the concept of action (*Action to Meet Prediction*). Anticipation is not just future-oriented but also goal-directed. Therefore, we will need to clarify what it means for a biological process to be goal-directed (*Goals in Anticipation*). Ultimately, goals are grounded in the precariousness of the living organization, which define the primary level of anticipation (keeping the living processes running). From there, more sophisticated forms of anticipation can be built, which can be framed not as progress but as specialization: anticipatory processes that depend on certain regularities (*The Ladder of Anticipation*). These processes implicitly incorporate knowledge of regularities, and this knowledge must come from somewhere. Computation incorporates the programmer's knowledge, but in biological systems, anticipation must emerge (*Learning*).

An Overview of Anticipatory Phenomena

Anticipation in Human Behavior

There are many kinds of anticipatory phenomena in nature. Some involve predictions that we explicitly make. For example, you might predict the next number following a sequence of numbers (1, 2, 3, 4, . . .). This very special case is a numerical prediction, and it serves as a template for the main theoretical frameworks of anticipation, as we will see in the next section.

But quite often, the prediction is not numerical at all. For example, in a game of hide and seek, the searching person might see a leg sticking out of a tree. She would then predict that, if she goes around the tree, she should see the person. In this case, the (implicit) prediction is not a particular visual pattern, but perhaps a particular category of patterns. In addition, the prediction might be realized later, at an unspecified time. It might also not be realized, if she is not interested in playing. The prediction is conditional on an action that may or may not be taken.

Another example in human behavior is the use of an umbrella. I notice there are clouds, I predict that it might rain, and I anticipate that event by taking an umbrella. Here again, the prediction is not a particular pattern of sensory inputs. In fact, several predictions might be made: the visual downward motion of drops, their sounds, the wet sensation on my head. We do not predict the individual timing and position of each drop. Some predictions are essentially unrelated to the action (taking the umbrella), but the wet sensation is: it is because I want to avoid getting wet that I take my umbrella. Here, successful anticipation consists in making sure that the prediction is *not* realized. Thus, the predicted sensory event (I might get wet) is abstract, not precisely timed, and if I do anticipate, then it will not happen. It might also not happen anyway (in case it does not rain).

This should already make us skeptical that anticipation can be reduced to prediction.

Anticipation in Perception

Anticipation is ubiquitous in perceptually guided behavior. Consider for example a cat listening to its environment. At some point a mouse scratches in the grass and the cat turns its head toward the source of the sound. The cat anticipates that turning its head in a certain way should make the mouse appear in its visual field. By this, I do not mean that the cat imagines a certain scenario, especially as this behavior is a very quick reflex. Simply, the reflex is anticipatory, in that it appears to be directed toward a goal, and the anticipation could fail—maybe the mouse is hidden, or at a different place. It is on this basis that we say that the cat perceives the sound's direction. This corresponds to Poincaré's view on the perception of space: "to localize an object simply means to represent to oneself the movements that would be necessary to reach it" (Poincaré, 1968).

We can associate a prediction to this anticipatory behavior: that the mouse will be in sight if the cat makes a certain movement (note that *we* make the prediction, the cat does not speak). Again, this prediction is not a particular visual pattern, but rather some abstract category of visual scenes. In the auditory domain, the prediction could be that the acoustic waves at the two ears match if the cat does a certain movement. Here, the prediction is that a certain sensory relation should be satisfied (Gibson's invariant structure), not a numerical prediction. It is also not about what *will* happen, like the Oracle's predictions, but about what *could* happen, if the cat made a certain action. The cat may decide to do another action, such as fleeing from the source of sound: this behavior can also be anticipatory, although the cat goes in the opposite direction. This shows that, in anticipation, we can distinguish two distinct components, an epistemic component that we tend to call "prediction," and a goal, such that behavior is the selection of an action that *should* reach the goal.

When we see a chair, we anticipate that we can sit on the object. This is the concept of affordance (Gibson, 1979): we perceive possibilities of interactions. This anticipatory view of perception is also a central feature of interactivism (Bickhard, 2009). Here again, the associated predictions are not about what will happen, but about what would happen if we did certain things, or even about opportunities (what actions we can do). This makes sense: what is the interest of predicting what will happen to us if there is nothing that we can do about it? Thus, to anticipate is not to know the future, but rather to exploit certain regularities so as to produce certain effects.

Motor Anticipation

In the motor domain, a well-known case of anticipation is when one is asked to tap along a regular rhythm. Usually, people do this by tapping slightly before the next beat (Aschersleben, 2002). Thus, they anticipate the next beat. In any case, given the traveling time of action potentials from the brain to the muscles, even tapping at exactly the right time requires anticipation. People may also continue the rhythm when the beat stops. Here, the anticipatory behavior is a form of continuation.

A very different case of motor anticipation is postural adaptation. For example, when standing in challenging balance conditions, humans co-contract their ankle muscles (Le Mouel et al., 2019). Humans also adjust their posture in anticipation of future movements (Le Mouel and Brette, 2017). An example is the posture of the runner in starting blocks, which allows the runner to use her own weight to produce torque for movement (see figure 4.9). Anticipation does not consist in performing the action in advance, but in reconfiguring the body so as to improve future performance, at an indeterminate time.

Anticipation in Physiology

Anticipation is seen not only in animal behavior, but also in the biological processes of all cells. For example, all eukaryotic cells have mechanisms to repair the membrane in case it is ruptured (Brunet and Arendt, 2016). When there is a hole in the membrane, calcium enters the cell and activates different repair processes: contraction of the membrane around the hole, exocytosis of vesicles (figure 7.1, left). In addition, modern cells have mechanosensitive channels in the membrane. When there is a slight touch on the membrane, before the membrane is ruptured, the channel opens briefly and lets calcium in (figure 7.1, right). This in turn triggers the repair processes before the membrane is actually ruptured—if it is ruptured. This phenomenon is described as "anticipating damage." Here we have a case of anticipation, but the prediction is clearly made by us: given the mechanical stress on the membrane, we predict that membrane damage might occur. In the cell, there is no prediction to be found that could be distinguished from anticipatory action: all there is is a process that leads to membrane repair in advance of the rupture event.

Of course, it would be possible to identify a signal in the cell and call it a "prediction" of that rupture event. For example, we might say that the calcium concentration predicts membrane rupture, but this only means that elevated calcium concentration is often followed by membrane rupture. In the case of the cell with no mechanoreceptor, the membrane stress predicts membrane rupture in the exact same sense, and yet that cell does not anticipate. Thus, there are cases of anticipation that rely on prediction, as when imagining a future event, and (many) others that do not.

Another important example of anticipation in physiology is the circadian rhythm. Circadian rhythms are cyclic variations in behavior and physiology, according to the day/night alternance. Around noon, your stomach might start gurgling. This happens because the digestive system gets active, the stomach fills up with enzymes and acids. As a result, digestion can start as soon as food is ingested. The digestive system appears to have anticipated your consumption of food based on the time of the day.

In fact, circadian rhythms are found across all domains of life, including bacteria. For example, cyanobacteria extract energy from light. During the night, the chromosome is more compact, which reduces the expression of all genes, and then it opens up before dawn so that the cell is ready to exploit the energy of the sun (Johnson et al., 2008). This behavior does not merely react to light variations, since there is no light before dawn. Rather, it anticipates dawn. Again, we might want to say that the chromosome's compaction *predicts* dawn—that is, dawn will normally follow compaction. But we could equally say that this event predicts noon, the next sunset, and so on, so why do we say

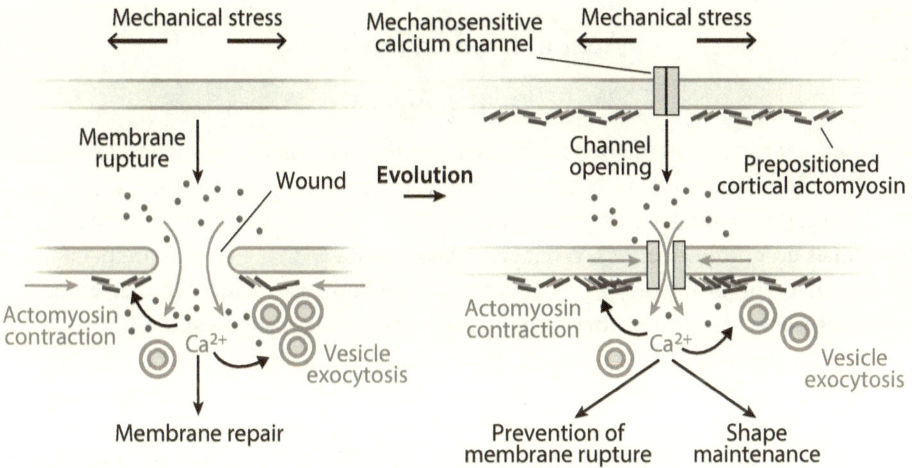

FIGURE 7.1. Anticipatory damage repair in cellular membranes. Original figure by Brunet T, Arendt D. 2016. https://doi.org/10.1098/rstb.2015.0043 / CC BY https://creativecommons.org/licenses/by/4.0/.

that the cell anticipates dawn specifically, and not the next sunset? It is because compaction *prepares* the cell for important physiological events occurring at dawn, namely photosynthesis.

Our final example is a classic anticipatory phenomenon studied in neuroscience: classical conditioning. While studying the salivation of dogs in response to food, Pavlov found out that the dog would start salivating before food is presented, when hearing the footsteps of the assistant bringing the food. Here, the footsteps are initially a neutral stimulus: it is only in the specific context of the experiment that it is associated with food. Thus, initially, the dog would not salivate, but after a few experiments where the footsteps are consistently associated with the future presentation of food, the dog would start salivating upon hearing the footsteps. The dog anticipates the food. We call it anticipation not just because there is a physiological event (salivation) that precedes food—that was already the case of auditory nerve's response to footsteps—but because saliva is produced so as to facilitate future food digestion.

As we have seen, there is a vast diversity of anticipatory phenomena in biology. How do organisms anticipate? The conventional answer is that biological systems *encode* predictions, and then act as a function of those encoded variables. Our quick tour of anticipation should already alert us that anticipation cannot be reduced to prediction. As we will see, the coding framework fails anyway to provide a coherent account of prediction, just like it fails to provide a coherent account of computation and information.

Action Based on Prediction

Predictive Models

A popular account of brain function is that the brain makes predictions about the world, more precisely about the value of sensory signals, and acts accordingly. But to predict is to say in advance, and neurons do not speak—most animals do not speak either. What can it possibly mean that a system predicts?

In theoretical biology, one of the most well-known accounts of anticipation is Rosen's: "An anticipatory system is a natural system that contains an internal predictive model of itself and of its environment, which allows it to change state at an instant in accord with the model's predictions pertaining to a later instant" (Rosen, 1985). Here, prediction is mediated by a *model*: the system has a model of the world, which for Rosen means essentially that there is a mapping from world states to system states (the *encoding*), and most importantly a mapping from system states to world states (the *decoding*). This idea that a part of the system somehow mirrors the world is the main theoretical framework for prediction by biological systems. It is central in predictive coding theory, which we will discuss shortly, but it was already well developed in the cybernetics movement. Thus, the famous "good regulator theorem" of Conant and Ashby (1970) claims that any "simple" optimal controller of a natural system must be a model of that system, that is, there must be a mapping from the state of the modeled system to the state of the regulator.

In what sense is a model "predictive"? As indicated by the quote earlier, for Rosen, a model is predictive if the decoding mapping is from current states of the system to *future* states of the world. Thus, in the example of predicting the next integer in a sequence, there would be a mapping from the activity of some neuron to the next integer (figure 7.2). Formally, there is a brain state S that depends on the past input $X_1, \ldots, X_n$ (1, 2, 3, 4 in the example). Neurons produce electrical events, not integers, so to cast their activity as prediction, we need to interpret a neuron's activity as an integer. Thus, we say that the brain state *encodes* an integer, which means that there is a mapping P from S to an integer $P(S)$, which is the prediction. Neurons encode predictions in a certain electrical format, and we decode it to the actual prediction with the mapping P.

The problem, as we have seen in previous chapters, is that P is a formal mapping. It is a construction of the observer, not something that exists in the brain, but it can do a large part of the predicting work in this framework. For example, the brain of a person who does not predict the next integer but just repeats the last one could also be said to predict the next integer. Indeed, take a neuron whose state S correlates with the last spoken number, that is, such that there is a mapping M from S to the last number $M(S)$. Then there is also

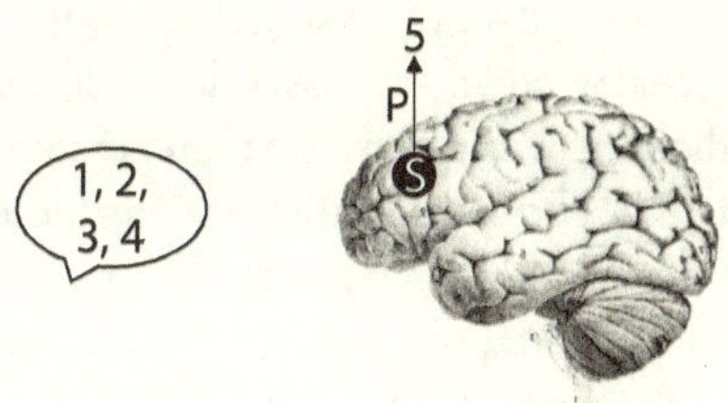

FIGURE 7.2. The brain predicting the next integer.

a mapping from S to next integer: $P(S) = M(S)+1$, which by definition means that the neuron predicts the next integer.

This problem is unavoidable because, since neurons do not speak, their predictions cannot be realized in the brain. They need to be interpreted. This is the central weakness of disembodied frameworks: they rely on the observer's interpretation. Thus, a state S predicts X if the observer can map S to X. But then, a state S also predicts $f(X)$, for any mapping f. For example, a state that encodes n predicts $n+1$ and memorizes $n-1$. The position of the planets in the solar system predicts their future positions. The input to a program predicts its output. Furthermore, a state can simultaneously make correct and wrong predictions of the same event, since we can map the state to different values. This theoretical framework is simply incoherent.

However, we have seen in chapter 5 that coding statements become nontrivial when coupled with dimensionality reduction. This leads to a more interesting framework called *predictive information.*

Predictive Information

Consider now a small variation of the task: the person hears two integers and then must predict the sum. For example, someone might say a series of integers such that each integer is the sum of the two previous ones, and the person predicts the next integer (this is called the Fibonacci sequence: $x_{n+1} = x_n + x_{n-1}$; figure 7.3). Suppose that the state S of a neuron can be mapped to the right prediction $P(S)$. As before, we could then also say that S can be mapped to the wrong prediction—for example, $P(S)+1$ (14,

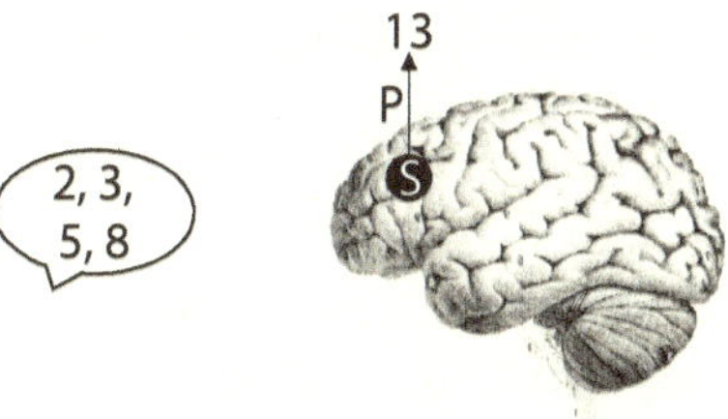

FIGURE 7.3. The brain predicting the Fibonacci sequence.

in the illustration). However, it still says something interesting about the neuron. Indeed, a neuron that would merely encode the current integer would not also encode the prediction, because the next integer depends on the last two integers. Thus, a state that correctly predicts the next integer must be a function of the sum of the last two integers, which is not a trivial property. In the framework of Shannon information, we say that the neuron's state has *predictive information* about the next integer (Bialek and Tishby, 1999). The prediction is still not realized by the system, but here we can say that the system does some of the work: the initial problem was to predict the next integer from the entire history of previous integers, and it is now to predict the next integer from a single value. It is clearly a simpler problem, that of building a correspondence table.

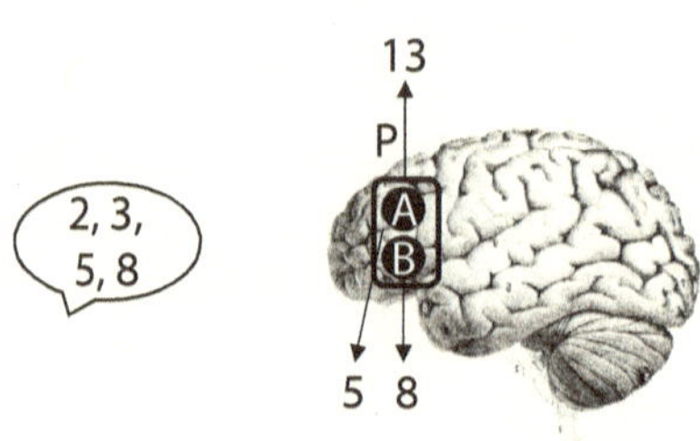

FIGURE 7.4. A pair of neurons that memorize the last two integers can be said to predict the next one in the Fibonacci sequence.

However, there is an important subtlety with the notion of "state." We have seen in chapter 5 that what we call neural state might be the activity of a neuron, or of a group of neurons, or sometimes a complex function of the activity of many neurons, as in neural manifold analysis. But with such a liberal definition of state, the notion of predictive information becomes trivial again. Suppose for example that there is a neuron that encodes the last integer and another one that encodes the previous one (8 and 5, in figure 7.4), but no neuron that encodes the next integer. No neuron has predictive information about the next integer. But if we consider the total activity of the pair of neurons (a "population code"), then it encodes the next integer. Memorizing becomes predicting. We have seen this problem before, when we noticed that the retina, or a camera, can be said to compute face identity, since we can construct a mapping from its activity to face identity.

Thus, in the encoding framework, the observer is introduced in two ways in the phenomena that we want to explain: (1) in the interpretation of the system's states as symbols (e.g., integers); and (2) in the description of the system in terms of states. This problem can be reduced by introducing constraints: by reducing the dimensionality of the mapping, and by allowing only certain observables to qualify as states, for example with anatomical constraints (e.g., properties of single neurons). For example, predictive information has been applied to the retina (Palmer et al., 2015), which makes sense because the number of axons leaving the retina (the optic nerve) is two orders of magnitude smaller than the number of photoreceptors—the optic nerve is a "bottleneck"—so that there is a strong anatomical constraint. Thus, we may ask what part of the visual field is discarded. Note however that, although it might initially seem that only predictive information matters for behavior, this is actually not the case, because a visual signal might carry no information about the future visual signal while carrying information about something else. For example, lightning (light) predicts thunder (sound).

In any case, the reliance of the encoding framework on the observer can be mitigated but never totally avoided. Indeed, as a disembodied framework, it necessarily relies on interpretation. Thus, the framework still cannot account for predicting the next integer, for example. This makes it impossible to frame anticipation as action based on prediction.

Action Based on Encoded Predictions

In encoding models of prediction, a neuron is said to predict some observable if there is a mapping from its state *S* to that observable. Then an action based on prediction is an action that depends on *S*. As we have seen in chapter 6, the problem is that the state *S* itself does not hold any information: the information—that is, the actual prediction—relies on the mapping, which is a construction of the observer.

For example, the current position of a planet predicts its future position. The motion of a planet depends on its current position; therefore the planet acts as a function of a prediction and can then be said to anticipate. The same could be said of any deterministic system. In the same way, any system that reacts to its sensory input can be said to anticipate, as soon as the observer can make a prediction based on the sensory input. Again, the problem is that the prediction is not done by the system but by the observer. This cannot work.

A popular theory called *predictive coding theory* or *predictive processing* tries to get rid of the observer's interpretation with two tricks. First, neurons predict quantities within the system (inputs to other neurons), and second, action is framed as trying to meet the prediction—this is called "active inference." Unfortunately, we will see that this theory has major shortcomings.

Action to Meet Prediction

Predictive Coding

In predictive coding theory, neurons encode the input to other neurons (Clark, 2013; Rao and Ballard, 1999). What this means is that the neuron's activity can be mapped to the other neurons' input, and this mapping is called

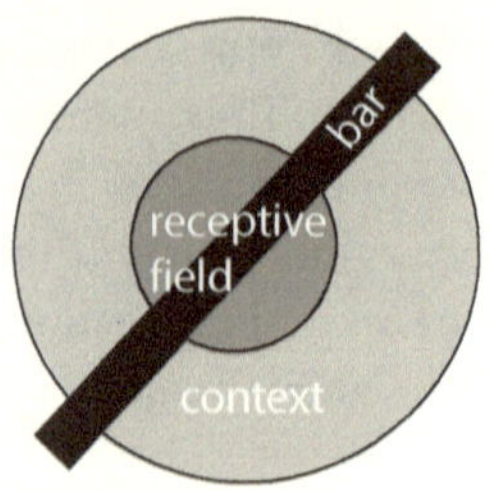

FIGURE 7.5. The pattern in the visual receptive field of a neuron (central disk) can be guessed from the context (outer annulus).

a "generative model." A hierarchical predictive coding model, such as in the seminal visual cortical model of Rao and Ballard (1999), is a (relatively abstract) connectionist model with feedforward and top-down connections. The ambition of that model was to explain the dependence of visual cortical responses on spatial context. Visual cortical neurons fire strongly to bars oriented in a certain way, but when the bar is lengthened, the response is inhibited. The idea of the model is that when the visual pattern within the neuron's receptive field can be predicted from the surrounding context, the neuron's response is suppressed (figure 7.5). This is a variation of efficient coding theory: the neuron should not waste energy encoding a stimulus that is already implied by the context.

In this model, the top-down connections are tuned so as to cancel as much as possible the feedforward input to the target neurons (figure 7.6). It follows that the transmitted feedforward signal becomes the mismatch between input and prediction. Here, the higher-order neurons encode a prediction of the feedforward input, that is, one can map their activity to the expected input. But this prediction is also realized in the synaptic machinery, namely as the total inhibitory current due to the top-down connections. Furthermore, it is then postulated that mismatches between feedforward and top-down signals drive synaptic modifications, to the effect that the mismatch is reduced. The presence of this mechanism justifies the use of the term "error"—something that the system tends to reduce.

On the physiological level, there are arguments in favor of such synaptic cancellation. Indeed, in vivo, cortical pyramidal cells maintain their membrane potential several mV below spiking threshold, so that action potentials are triggered when the potential transiently fluctuates above its mean (Rossant et al., 2011). This can be explained if excitation is canceled on average by inhibition, so that only fluctuations remain (Shadlen and Newsome, 1994). This had led to the consensual view that, in physiological conditions, neurons operate in a so-called balanced, or fluctuation-driven, regime. But this particular physiological regime is a feature of various unrelated theories: predictive coding theory, but also electrical homeostasis (avoiding epileptic seizures), binding by synchrony (Singer, 1999; von der Malsburg, 1981), synchrony receptive field theory (Brette, 2012), synfire chains (Abeles, 1991).

For the matter of this chapter, we must first realize that, despite its name, predictive coding theory is in fact not initially a theory of anticipation.

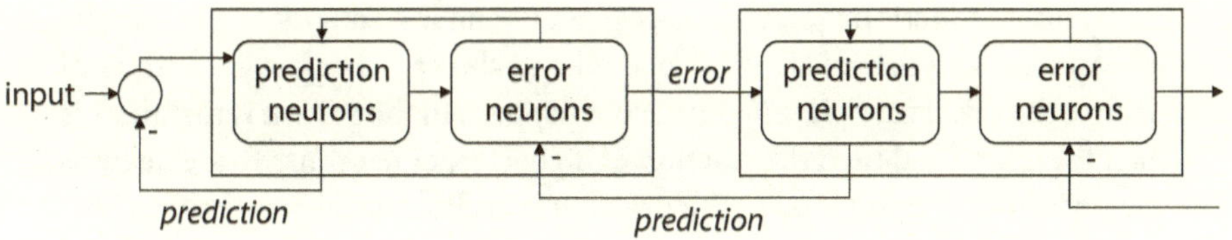

FIGURE 7.6. Hierarchical predictive coding model of the visual system (after Rao and Ballard, 1999).

Predictive Coding Is a Theory of Inference, Not Anticipation

There appears to be a confusion in much of the literature about predictive coding—namely, that it is depicted as being about predicting what will happen next (Clark, 2013). Of course, this is what the term "prediction" suggests, but as with many other words of scientific jargon (such as "information"), the confusion comes from the ambiguous use of that term.

To see this, we need to look at some details of the visual cortical model of Rao and Ballard (1999). In that model, the activity of neurons in a higher layer encodes a "prediction" of the activity in the lower layer, which could be the image. What this means technically is that the activity of those prediction neurons is passed through a certain function (which depends on synaptic weights), and the result is the predicted activity of the lower layer. In the model, the activity of the prediction neurons is *defined* implicitly as a function of the activity of the lower layer, as the activity that minimizes the prediction error. In practice, the image is presented, and then the activity of the prediction neurons is calculated through a gradient descent procedure based on the prediction error. So, to repeat: first the image is presented, then the activity of the prediction neurons converges to some value, which is then interpreted as a prediction of the image.

Thus, clearly, the "prediction" neurons do not anticipate the image, they react to it. What comes first is the image, and then the activity of the neurons forms a sort of compressed representation of the image. Predictive coding is usually not anticipatory. But then why is this called "prediction"? This unfortunate terminology comes from a confusion of perspectives with different temporalities. When we say for example that the central part of the bar (see figure 7.5) can be predicted from context, we mean that if we mask the center and just look at the surrounding annulus, then we can guess what is in the center. But in reality, the entire visual stimulus is shown at once.

Thus, when the observer describes the image and the activity of the higher-order neurons, she can point out that if she is just given the higher-order neural activity, she can guess what the image is, by passing it through the appropriate function. So, first she is given the neural activity, she makes the guess, and then the image is uncovered and she can check it. But that is of course the observer's temporality. What happens in the system is that there is first the image and then the reaction of higher-order neurons. Those neurons do not anticipate the image. The prediction is about what was there before, not about what comes next—which is obvious in retrospect, since the model was about representing static images.

Many modern predictive coding models share this feature. For example, in Denève's theory of spike-based predictive coding (Boerlin et al., 2013; Deneve, 2008), neurons spike whenever a prediction error (implemented as the membrane potential) becomes too large. Again, despite the term "predictive," the model is not anticipatory but reactive: technically, the model implements a greedy algorithm, meaning that it minimizes only the error on the current estimation, rather than the future error.

Thus, as initially construed, predictive coding theory is about prediction in the sense of the magician's predictions: you pick a card, and the magician predicts what card you *currently* have in hand. The temporality (pre-diction) is in the fact that you are going to show your card after the prediction is made, but it remains that you picked your card before. This is not a theory of anticipation, but rather a theory of inference. Indeed, predictive processing is sometimes described as finding the "causes" of the image, another confusion stemming from the use of statistical jargon designating the variables of a statistical model. The top-down connections are said to implement a "generative model," which means a parametric model of the image, and the model implements a procedure to determine the parameters, not to predict what's next.

Of course, predictive coding theory could be amended to be a theory of anticipation, and indeed there have been efforts in that direction (Rao et al., 2023). This requires the model to minimize the cancellation error between current top-down input and future feedforward input, and this is technically a rather different kind of model.

However, even if properly amended, predictive coding theory faces a more fundamental limitation as a theory of anticipation. All the intuitive appeal of predictive coding theory and predictive models in general stems from the idea that anticipation is often associated with an implicit prediction. But predictive models are about very special kinds of predictions, namely predicting a numerical value at a given time, and these are not typically the useful kinds of predictions.

Implicit Predictions in Anticipation

As we have seen in the preceding section and in chapter 6, perception and anticipation are tightly linked. This is also an intuition of predictive coding theory—for example, Clark (2013) writes that "to successfully represent the world in perception, if these models are correct, depends crucially upon cancelling out sensory prediction error." But unfortunately, the latter claim is not correct. For example, the fact that a sound source is in front is indicated by the fact that the two acoustic signals are identical. It is not necessary to predict the numerical values of the two signals to account for this fact. The same holds in general for invariant structure, sensorimotor relations, affordances or perceptual categories in general. When we see a hand protruding from a tree, we might predict that there is someone behind the tree. The anticipatory behavior consists in acting based on the prediction, for example by going around the tree. Here the prediction is not a particular visual pattern, a particular face, but a visual category, not something that can be "canceled out." It is also not about the value of a signal at given time in the future, but at an unspecified later time, conditioned on (hypothetical) action.

We have described a number of anticipatory phenomena at the start of this chapter, and most of them do not involve numerical predictions. For example, I take an umbrella because I predict that it might rain. But "rain" is not a particular pattern of sensory stimulation. It might be a category of visual, auditory, and tactile stimuli. What is relevant to the behavior is that rain makes you wet, but if I use this prediction and take my umbrella, then the prediction is *not* realized.

If I see a chair, I predict that I *can* sit. Thus, the prediction would be realized if I decided to sit, but maybe I do not want to sit, and yet I still see the chair. In all these examples, the kind of prediction (explicit or implicit) that underlies anticipation is not generically the prediction of a particular pattern of sensory inputs at a particular instant in the future. What is predicted is most often relational and typically rather abstract, such as a possibility of interaction. When it is predicted to occur is not generically the "next instant" but often an undefined future instant. In fact, the kind of predictions useful for action selection is generically not about what *will* happen, but about what *would* happen, if I did this or that action.

It is important to realize that in most cases, it is the observer that makes predictions, rather than the organism. For example, we have seen that eukaryotic cells anticipate membrane rupture by triggering repair events when the membrane is mechanically stressed, thanks to mechanoreceptors that open and let calcium enter the cell (see figure 7.1). Here, we might want to say that the cell predicts membrane rupture, in that it behaves *as if it knew* that the membrane might break. But of course, *we* are making the prediction. We could say that calcium concentration encodes the probability of imminent membrane rupture. But membrane stress encodes that probability in exactly the same way, even if the cell did not anticipate anything—if it were an inert bilipidic membrane, for example. The reason why we identify some physiological signal with a prediction is always in relation with particular actions taken by the cell (e.g., repair mechanisms).

But as we have seen, the encoding framework does not allow framing anticipation as an action that depends on a prediction. To circumvent this problem, a popular proposition called *active inference* is that anticipation consists instead in acting *so as* to meet a prediction.

Active Inference

Active inference extends predictive coding theory by stating that action is chosen so as to minimize "surprise," that is, to meet the sensory prediction. It is part of an ambitious unifying framework, the free energy principle, according to which the brain manipulates a probabilistic generative model of its sensory inputs, which it tries to optimize by either adjusting the model (learning) or by changing the inputs (action) (Friston, 2009, 2010).

At first sight, this proposition is appealing since it provides a principled way to account for agency and goals. It also has a certain psychological appeal. We tend to think that we act rationally as a function of objective information, but in truth, we know that much of our actions are chosen so as to reinforce our beliefs. For example, one would read newspapers that tend to have congruent political opinions. This is also a known phenomenon in philosophy of science (Feyerabend, 2010; Kuhn, 1962; Lakatos, 1976): experimental data are produced within a certain conceptual framework.

This important psychological insight is not new. It is the main focus of a famous psychology theory called *cognitive dissonance theory* (Festinger, 1957). In a field study called "When Prophecy Fails," Festinger et al. (1956) describe a small sect that predicts the end of the world for December 21st. When the day arrives and nothing happens, many believers do not stop believing. Instead, they make a new prediction, reach out to newspapers, and try to recruit new believers.

But while cognitive dissonance theory is a rich and complex theory that accounts for many aspects of human behavior, active inference fails to account for any significant fraction of that behavior. The problem is that, as we have seen, the predictions of predictive coding theory are numerical predictions of sensory signals, unlike what we usually call "prediction." The newspaper is a case in point: by picking the right newspaper, you expect to read something that comforts your system of beliefs, but you obviously do not know what exactly is going to be written there, and this is the whole point of reading. Thus, cognitive dissonance theory does not, in fact, stipulate that people try to match detailed expectations, but only that they try to avoid their abstract system of beliefs to be challenged.

Moreover, cognitive dissonance theory is not a universal psychological theory. For example, it does not account for how people build their systems of beliefs in the first place. Many psychological phenomena are obviously not covered by cognitive dissonance theory, and therefore even less by active inference. Why do people surf, for example? It turns out that this question is the basis of another major psychological theory, flow theory (Csikszentmihalyi, 2008), which addresses autotelic activities—things that we do for their own sake. Those are activities that achieve a balance between mastery and challenge, so that one is fully absorbed in the activity. Activities that are fully predictable are not enjoyable: they are boring. Thus, the idea that all behavior can be framed as meeting one's predictions trivializes psychological phenomena. It is not plausible.

This objection to active inference is not entirely new. It has been pointed out that if one wants to minimize surprise, then the simplest way would be to shut off stimulation altogether by going to the closest dark room, or simply by closing your eyes, and we obviously do not do that—this is called the "dark room problem" (Friston et al., 2012). The proposed solution is *priors*: you come out of the womb with a prior expectation of light, so that you seek light and you open your eyes. In this way, all behavior can be accounted for by picking the appropriate priors.

Of course, this move simply buries everything that is not accounted for by the theory into unexplained "hard-wired" behavior, which is why Bickhard calls it "epicyclic elaboration of hyperpriors" (Bickhard, 2016a). But even then, it does not work (see in particular Nave [2025]). If there is any possibility of learning at all, then when you blink, you should learn that closing your eyes makes the visual field dark, whatever your initial prior. After learning, surprise is minimized when you close your eyes, and therefore this is what you should do, according to active inference. Either there is no learning, which means that all behavior is hard-wired in priors, so the theory is empty, or learning is possible, and the theory is prone to the dark room problem.

The initial appeal of active inference is that it seems to solve the problem of action selection: just choose the action that meets your prediction. There is no need to account for goals, which is the major issue in philosophy of life. But this is an illusion, because then the problem becomes to choose the relevant prediction, and relevance is determined by goals.

Goals in Anticipation

Goals and Predictions

Consider the anticipatory posture of the runner in the starting blocks (see figure 4.9). The prediction is that the race will start soon, the goal is to arrive as quickly as possible at the finish line, and so the selected action is a change in posture. In this case, the action has no impact whatsoever on the prediction. Of course, the rhetorical trick of active inference is precisely to call the goal a prediction. So, the runner predicts that she will arrive quickly at the finish line. In this way, the person acts so as to meet her "prediction." But this is a rhetorical move with no explanatory power. It can work only as long as one decides to name as "prediction" exactly the goals of the person and forgets about all the other predictions that she might make. For example, we might grant that the runner predicts that she will win the race, although that is obviously false for many self-aware mediocre runners. But she would *also* predict that the race will start soon, that she might fall while running, that she might lose the race. The runner might also see the clouds and predict that it will rain. In any situation, you can predict many things: some events that you want to happen, others that you want to avoid, and yet others that you might not care about. Therefore, stipulating that you act so as to meet your predictions cannot work, since there are several predictions at any given time.

Anticipation appears to be action that fulfills a goal, and which depends on a prediction *relevant to that goal*. The relevant prediction is specified by the goal, and so it cannot precede goals, as in active inference. In fact, in many cases, there is no prediction to be found in the system. Rather, we describe the phenomenon of anticipation relative to a prediction that *we* make. For example, eukaryotic cells anticipate membrane damage by triggering repair processes when the membrane is mechanically stressed. There is no prediction in the cell, unless we want to call the repair processes themselves a "prediction." Therefore, I suggest that anticipation is not action based on prediction, but the *exploitation of regularities*. The term "regularity" points to an epistemic component of anticipation, but one that does not necessarily rely on a prediction—even though the observer can describe this regularity by making predictions. The term "exploitation" means that anticipation is always associated with a goal. But what are goals?

Goals as an Organization Principle

Goals are a tricky concept. When you take an umbrella, your goal is to not get wet. It is intuitively clear what this means: to not get wet is what you want, and you act accordingly. But when stress on the membrane triggers repair processes, does the cell *want* to avoid membrane rupture? This seems like a metaphorical use. Yet, we feel that the processes triggered by membrane stress are goal-driven, in the sense that they are organized in such a way as to achieve something useful for the organism, namely preparing for membrane repair. This is admittedly a weaker notion of goal, but it retains many of the features of the mental concept of goal, which are relevant to explain the broad class of anticipatory phenomena that we have discussed. It is not my ambition here to explain what an intention is, but to provide a useful concept to designate the fact that a system is directed toward some end. Thus, we will keep in the back of our head that there is a distinction to be made, about what it means to consciously intend to do something, but I will not address it.

When we attribute a goal to someone, we do so because it *explains* its behavior, but we are not in her head. In this sense, goals are an organizing principle. A goal cannot just be the effect of actions, otherwise the outcome of a failed action would qualify as a goal, and every observable behavior as well as involuntary twitches could qualify as anticipation. Thus, there must be a distinction between the effect of an action and what it is directed toward. This is tricky since we only see the outcome of actions, not what it *should* have been.

Consider the umbrella example. I notice clouds and I take my umbrella so as not to get wet. I anticipated in that I exploited the regular association of clouds and future rain, in order to achieve my goal of keeping dry. Suppose now that instead, when I notice clouds, I stop and stare at the clouds, and I always end up soaked. It seems that I do not anticipate the rain. But perhaps what actually happens is that I love standing in the rain, and so I actually do anticipate: I know that clouds are often followed by future rain, so I stop so as to get wet. This is troubling, since the same observed behavior might be anticipatory or not: it all depends on what my goal was (to get wet or to keep dry).

We can make progress by taking a more systemic view of behavior. How can we know if a person loves rain or not? If we cannot speak with her, we might guess by observing what she does when there is rain. If she just stays there, she might like it (or not care about it). If she immediately runs to a shelter, she probably does not like it. Thus, if we consider the entire behavioral repertoire associated with rain, it makes sense to say that the person who stays put when there are clouds but runs to a shelter when it starts raining acts so as to keep dry, but fails to anticipate the rain. To label this organizing principle as "goal" (to keep dry) is a bit more than a way to put some order in a set of observations. In this case, if we are right, it means that if we were to tell the person that it will rain, then she would run to a shelter or take an umbrella. In other words, a goal is an element of a theory of an individual's behavior.

Anticipation as Preparation

Here, we have characterized anticipation in relation with a reaction: some event A (rain) triggers an action (sheltering); a prior event B (clouds) correlates with the future occurrence of A, and triggers the same action (sheltering). Anticipation as prior reaction fits a number of anticipatory phenomena. For example, membrane stress triggers the same processes as would be triggered by membrane rupture, an event that typically follows membrane stress. In the same way, when your digestive system gets active around noon, it does the same thing as when it ingests food at other times. This happens when the circadian clock is in a state that correlates with the ingestion of food in the near future.

We can see that this notion of anticipation connects with the classical theory of conditioning. This is an interesting connection, but it does not address all cases of anticipation, such as the posture of the runner in the starting blocks, or chromosome decompaction that facilitates gene expression in daylight. To include these examples, we need to broaden our definition to understand anticipation not as prior reaction but as *preparation*.

The runner takes her position in the starting blocks, which will have the effect of increasing her chances to win the race, but the race has not even started. This postural change tends to improve her future performance, but the gunshot that starts the race does not trigger that posture: it triggers running. Thus, this anticipatory behavior is not a prior reaction as in conditioning, but a preparation in view of a goal. Here, the goal can still be viewed as an organizing principle: running the race explains why the person runs as fast as possible, and why she takes a particular posture before the race starts.

Consider the notion of anticipation in perception. When I hear a sound on the right, I might turn to the right to face the source of the sound. But this action occurs after the sound is perceived. In the same way, when I see a chair, I anticipate that I can sit on it, and this happens before I actually sit on it, and I might not decide to sit at all. To perceive the affordances of a chair is not to act accordingly, but rather to prepare oneself so as to act accordingly. This is a case of anticipation as preparation, in the same way as the runner takes a particular posture before the race starts. When I hear a sound on the right, the nervous system places itself in a configuration (change in structure or mode of activity) that will influence future behavior in an appropriate way.

To frame anticipation as preparation allows us to see the living organization as fundamentally anticipatory, and this allows us to ground goals and anticipation in the logic of the living.

The Material Grounding of Anticipation

So far, we have identified the goal of a system with a state it tends to in various situations. This is essentially a dynamical characterization: an attractor; or conversely, if we take the example of avoiding the rain, an unstable manifold. But there is nothing specific of living beings here. Consider for example a cupboard, standing vertically on the ground (see figure 4.8). If it is pushed a little, it will come back to its initial position. Thus, the system behaves in a way that is directed toward a goal, which is its upright position. This does not appear to be very different in nature from saying that a person's goal is to avoid the rain because she acts so as to avoid the rain in various situations.

However, there is an important distinction in living beings. The theoretical biologist Robert Rosen (1985) argued that anticipation is a constitutive fact of life, and this is also a core tenet of the enactivist tradition, which also tends to assign goals and intentions to all living beings (Thompson, 2010). The reason is that, as we have seen in chapter 2, living beings are organizations of processes that actively produce the conditions of existence of those very processes. This means that the output of a process is a condition for the future existence of that same process. In this sense, the process anticipates its future needs. Rosen expressed this idea in terms of Aristotle's causal analysis, by claiming that in a living organization, final causes are also efficient causes. A similar idea was prefigured in Kant's concept of organism as "natural purpose" (Thompson, 2010): something that is both cause and effect of itself.

An underappreciated fact in Rosen's theory as well as in enactivism is that the continued operation of the autopoietic organization does not derive simply from its closure, the fact that the processes loop back on themselves. It is obviously a necessary condition, but in no way a sufficient condition. In order for the processes to keep going, the specific material realization matters very much. (The failure to appreciate this point is what Zahnoun [2023] calls the "functionalist fallacy.") Each process must output the right amount of product, at the right time. This point is well known in dynamical systems theory, more precisely bifurcation theory, which identifies parameter regions associated with certain qualitative behaviors. It appears to be a blind spot in Rosen's theory because his approach is relational, that is, only concerned with closed diagrams of arrows connecting nodes, and not with the specific properties of what the arrows and nodes stand for. A simple example is temperature: a materially realized autopoietic organization, such as a living cell, maintains itself only within a narrow range of temperature, even though temperature essentially affects kinetic aspects of chemical interactions, and not the graph of interaction.

The corollary of this remark is that, in order for the autopoietic organization to exist (i.e., for the processes to run and continuously dissipate energy), each process must be adjusted in a particular way and have certain nontrivial properties, such that its output, when further processed by the organization, ensures the future conditions for the continuation of that process, in the right amount and at the right time. These nontrivial properties are related to the context formed by the entire organization, which specifies the regular relationship between the output of a process and its future input. In other words, the properties of processes embody knowledge about the system formed by the living organization and its environment. It is in this sense that the processes of a living system are anticipatory, which justifies searching for cognitive properties across all life, not only in nervous systems (Baluška and Levin, 2016; Lyon et al., 2021).

This is why we might talk meaningfully about the *function* of the digestive system: to extract nutrients and energy from food so that cells can keep on working, ensuring in particular the continued working of the digestive system itself. In the context of the organism, digestion is an anticipatory process. Note that the digestive system, or any set of processes in a living system, does not *have* a function, in the sense of a preexisting job description that evolution would have filled. Rather, as Bickhard (2009) points out, it *serves* a function within the organism: its operation is such that, in the context of the organism, it ensures the continued autopoietic operation of the system, in a particular way (by providing nutrients). It is mistaken to conceptualize function as deriving from evolution—that is, as what the processes have been "selected for"—because in order for some process to be selected by evolution, because of its usefulness, it must already serve a function within the organism. Clearly, parts of ourselves serve functions that have not been "selected for" by evolution: for example, the hand serves the function of typing on a keyboard, but it would be mistaken to say that the hand "has" the function of typing on a keyboard, as in something intrinsic that would preexist to the interaction with a keyboard. Thus, we can attribute functions to components of a living system, functions that they *serve* within the organism, without appealing to an external agency (as in machines) or to evolution.

These remarks allow us to characterize living processes as anticipatory, in a way that the cupboard is not. Indeed, the stability of the cupboard is not a condition for the cupboard's continued existence. We have seen in chapter 2 that a living system produces its own constraints (such as the vascular system), which allow its processes to continue running (Montévil and Mossio, 2015). This fits the view of anticipation as a form of preparation: the living system builds and maintains the vascular system, which then acts as a constraint to direct nutrients and oxygen to keep the living system running.

Thus, goals of living systems appear to be grounded in their own material condition of existence. Note that this is a useful but still incomplete characterization as it does not account, for example, for sexual behavior and art. In fact, it is a characterization that is not entirely specific of living systems. As pointed out by Villalobos and Ward (2015), it also applies to a candle flame: the vaporization and burning of wax in the flame produces heat, which liquefies the wax at the bottom, allowing it to then ascend by capillarity, making it available for burning. Thus, the burning process is anticipatory in the sense that we have discussed. This does not imply that the flame has intentions, and neither should it necessarily imply intentions on behalf of living systems.

Thus, we have characterized anticipation as a nontrivial property of dissipative systems, a class of systems to which living systems belong. Such systems are anticipatory in the sense that they have the nontrivial properties of producing the conditions of their continued operation, properties that depend on the organism and the ecosystem. This characterization of anticipation and goals does not confer intentions in a mental sense, but it has a number of desirable features: the "goal" is what the organization tends to (to maintain itself despite perturbations), and it does not derive from external agency; the "anticipation" may succeed or not, depending on some future outcome.

While this does not provide a clear distinction between nonliving dissipative systems and living systems, we can certainly sense that living systems achieve higher degrees of anticipation.

The Ladder of Anticipation

Reaction and Anticipation

Anticipation is often contrasted with reaction. For example, consider the membrane repair mechanisms (see figure 7.1). When the cell membrane is ruptured, various mechanisms are triggered so as to repair the membrane. This can be seen as a reaction to membrane rupture. The goal is to maintain membrane integrity, and actions are taken when that goal is not fulfilled anymore. In contrast, the repair mechanisms that are triggered when the membrane is merely stressed is an anticipation of membrane rupture, because actions are taken now, when there is no membrane rupture, which lead to membrane repair in the future.

As we have seen, the reaction to membrane rupture is also anticipatory, in that the repair processes have the highly nontrivial properties of ensuring the future existence of the living system. Thus, repair processes triggered by membrane stress appear to be a secondary kind of anticipation. The primary goal is to avoid death. The reaction to membrane rupture realizes the anticipation that preserving membrane integrity might avoid death. This anticipation creates a secondary goal, to preserve membrane integrity. The reaction to membrane stress realizes the anticipation that membrane stress might lead to membrane rupture, which is to be avoided. Thus, the repair processes triggered by membrane stress show a higher degree of anticipation, in that they anticipate an event that is the basis of another anticipation.

Thus, the distinction that we typically make between reaction and anticipation is relative to a secondary goal—for example, to preserve membrane integrity. In the reactive phenomenon, actions are triggered by a mismatch between the current state and the desired state (membrane integrity), that tend to bring the state back to the desired state. In the secondary anticipatory phenomenon, actions are taken that tend to avoid or reduce the mismatch between the future state and the desired state, and those actions are taken before there is a mismatch: they are triggered by other conditions. This appears more anticipatory in that it relies on an additional regularity between membrane stress and future membrane rupture.

This distinction is well-known in control theory, which distinguishes between two classes of control systems: feedback control and feedforward control, also called "model-based" control or planning. The feedback system measures the mismatch between the desired state and the actual state, then acts in the direction that reduces that mismatch. The feedforward system acts as a function of various measurements in such a way as to reach the desired state. As a result, in a well-adjusted feedforward system, there may be no mismatch at all.

A paradigmatic example of feedback control is the thermostat. When temperature is lower than the desired temperature, the heater is switched on. When temperature is adequate (or slightly higher), it is switched off. In this way, temperature is controlled. Now you might want to maintain the temperature only when you are home. You would then turn on the heater when you come back home, and it would take some time before the desired temperature is reached. A more sophisticated heater would be programmed with your usual schedule, so as to switch on the thermostat before you come home. A smart heater could then calculate when to start heating, based on the current temperature and the heating rate (in °C/h), so that the desired temperature is reached right when you come home. Here, the basic thermostat only reacts when there is a mismatch between actual and desired temperature, so as to reduce this mismatch, where the smart thermostat acts as a function of various cues, before there is any mismatch with the desired state, in such a way as to reduce future mismatch, or even avoid it entirely.

Degrees of Anticipation

We have contrasted reaction and anticipation, but as we have pointed out, this is a distinction relative to a secondary goal. Relative to the primary goal of self-maintenance, "reaction" is also anticipatory. Indeed, a reaction is also oriented toward a goal, and for a reaction to succeed, it needs to have certain properties that depend on regularities. The thermostat should start the heater when the temperature is lower than the target, rather than when it is higher. More generally, in feedback control, the feedback is set up in such a way that actions are expected to reduce the error. In contrast, feedforward control uses a model to output just the right action that cancels the error, possibly before there is an error.

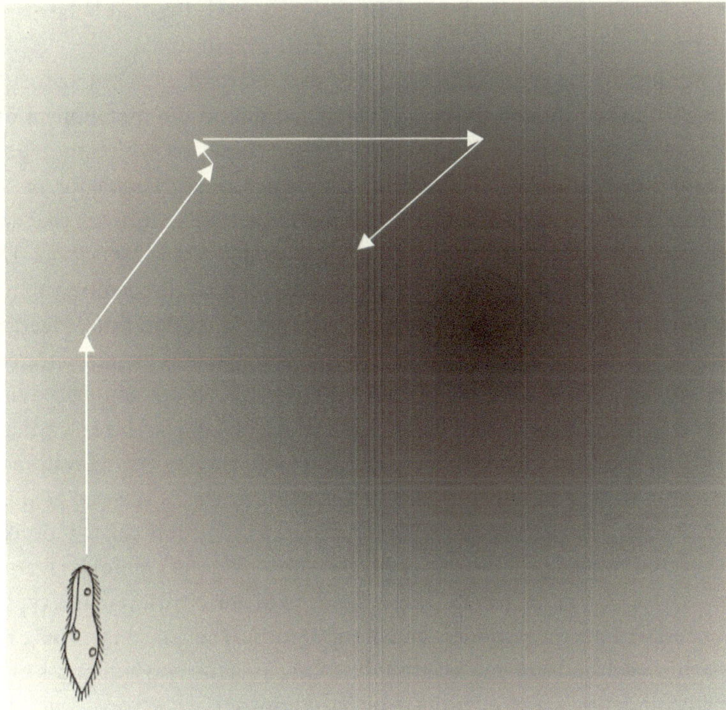

FIGURE 7.7. Trial-and-error chemotaxis in *Paramecium*.

There are also processes that are less anticipatory than feedback. Consider chemotaxis in *E. coli* or *Paramecium* (figure 7.7). When the concentration of the attractant increases, the organism keeps swimming in the same direction. When it decreases, it changes direction. In this way, it spends more time swimming in the right direction, and ultimately finds the source of attractant. The process is thus somehow goal-directed, but individual actions are not directed, they are random: it is a trial-and-error process.

There is still some form of anticipation at play here, since it is expected that the error will tend to decrease rather than increase by this process. There are various physiological signals in the cell, and the system is set up such that the right physiological signal triggers the directional change.

This also presupposes that the organism implicitly knows that a chemical signal is associated with something good like food, which is also anticipatory. A less anticipatory behavior would consist in only reacting to food. This is demonstrated in a model of bacterial chemokinesis (chemical modulation of speed) controlled by metabolism (Egbert et al., 2010), where a product of metabolism inhibits swimming. In this way, the organism stops swimming when it enters a region with food. In this way, goal-directed behavior gets rooted in material autopoietic processes of the organism.

Summing up, we can identify a ladder with increasingly anticipatory processes. At the bottom are the processes that maintain the metabolism of the autopoietic organization. Right above are processes by which metabolism regulates the organism's motility. Then comes exploratory behavior, or "trial-and-error." A mismatch with a desired state triggers an action, but that action is not directed in any particular way; it is essentially a random action. This is close to Darwinian ideas, where an appearance of goal-directedness emerges from the biased selection of random events. Above exploration is feedback: actions triggered by mismatch are directed in a way that is expected to reduce mismatch. This is the key concept of cybernetics, which also supports dynamicist views of cognition (van Gelder, 1998; Powers, 1973). Above feedback is anticipated reaction as in conditioning: the system takes the relevant action in response to a prior cue, before any mismatch with the desired goal is detected. A sophisticated version is planning: here, the system takes actions that are expected to lead to a desired state, but those actions are not triggered by mismatch, but by other events. This is the kind of anticipation that is typically associated with internal models and computation. There is in fact some room above this level, such as imagination: one imagines the consequences of various possible actions, and possibly selects the appropriate action—but it is not obvious what this might mean in physiological terms.

What changes along the ladder of anticipation, and in what sense is there an increase in anticipation?

Knowledge and Robustness

Intuitively, it seems that what changes when the degree of anticipation increases is that implicit knowledge gets incorporated. For example, the thermostat's design incorporates the knowledge that activating the switch when the error signal is on should make that signal disappear. The cat that turns its gaze toward a source of sound "knows" that contracting its muscles in a certain way should make its eyes turn to a certain direction.

Of course, the cat does not literally know that, and it is the thermostat's designer who knows about its causal organization. But we can rephrase this idea by pointing out that, when a system incorporates knowledge, its success becomes conditional on that knowledge being true. For example, consider a phonotactic device that turns toward a sound source. To do this, it subtracts the sound intensity from two microphones and feeds the result into a motor, so that the device turns toward the louder ear until it faces the sound source. For the device to work, there are at least two conditions. First, the microphones must be arranged in such a way that the motor will run in the right direction, not the opposite direction. If the microphones are swapped, then the device turns away from the sound source. Second, the two microphones must transduce sound in exactly the same way. If one microphone has a higher gain, then the input to the motor will be biased and the device will not face the sound source. However, the device would still work correctly if the microphones were nonlinear (e.g., saturating), but identical.

Now consider a more sophisticated device that first subtracts the two sound intensities (in decibels), then maps the result to the sound's direction using a table, and then calculates the number of motor steps to make. This is more anticipatory than the previous device, since the behavior does not rely on continued interaction with the world to succeed. But for it to work, more stringent conditions must be satisfied—namely, the two microphones must be identical linear transducers with fixed specifications matching those of the table, and the motor step must have a certain size.

Thus, a system anticipates when it works conditionally with certain regularities. A more anticipatory system has more regularity conditions, which means more possible cases of failure. Thus, anticipation goes hand in hand with fragility. For example, the ant, with its sophisticated anticipatory navigation system, fails to return to its nest when it is displaced by the scientist: this is precisely how we know that this behavior is anticipatory. Anticipation exploits regularities, but also depends on them.

From this remark, we can draw two important conclusions. The first conclusion is that the ladder of anticipation is not a ladder of progress. Even though we tend to find higher degrees of anticipation when we go from single cells to multicellular organisms, in particular primates, more anticipation is not unconditionally better. An anticipatory system is both powerful and fragile, as it relies on regularities that might break up. Therefore, anticipation must be conceived not as progress, but as specialization: the system gets specialized to a particular context that exhibit certain regularities. It follows that there is no reason why higher steps of the ladder should entirely replace lower ones. Indeed, we find trial-and-error as well as feedforward and feedback control in the behavior of the same species. The less anticipatory aspects of behavior are not evolutionary relics: they are functional, because they apply to broader conditions. As we have seen, we also find anticipatory processes of different kinds in single cells. For multicellular organisms, we have seen in chapter 2 that development, which at first sight appears to be genetically planned, actually includes trial-and-error processes, with cells that divide, migrate, and then get pruned. This holds for the embryo, for the motor system, for the brain.

The second lesson is that since anticipation is contingent on some context, an anticipatory system is necessarily a flexible system, one that is not rigidly specified. How then can anticipation be framed as computation, which is a complete formal specification of steps? This paradox comes from the dualistic nature of computation, where the program, a rigid formal specification, is written by a programmer, who is in charge of making the specification, by observing and exploiting regularities. When there is no external specification, anticipation must emerge, which means that the system must change. This is what is usually called "learning."

Learning

Evolutionary Emergence of Anticipation

Consider again the membrane repair processes. These are presented within an evolutionary framework (Brunet and Arendt, 2016): first there were repair processes triggered by the entry of calcium in the cell; then mechanoreceptors appeared by mutation and natural selection, which let calcium enter the cell when the membrane is stressed, triggering the repair processes in advance. There is a regularity between membrane stress and future membrane rupture, and processes get selected because of this regularity. In this sense, the regularity causes the phenomenon, which is why we call it anticipatory.

However, selection does not act on individuals. The regularity is not observed by the individual, which would then anticipate future events: rather, the regularity acts as a constraint on the kinds of individuals that we can observe in the long term. This leads to an important subtlety about anticipatory phenomena, which I will explain with a thought experiment. Consider a diverse population of rabbits that react to a certain musky odor in various ways: some hide, others freeze, others seek the source of the odor, others do nothing. Now a wolf appears, who turns out to have that odor. Only the rabbits who hide survive. As a result of selection, all surviving rabbits now hide whenever they smell the wolf. They show an anticipatory behavior, in the same sense as cells anticipate membrane rupture when their membrane is stressed. But selection has not changed the individuals, and yet, we could not have possibly said that those individuals anticipated the wolf before the predator existed. Therefore, in this case, anticipation is not an intrinsic property of the organism. Put in a different context, the same rabbit is not anticipatory anymore.

We previously characterized anticipation as a phenomenon in which the organism exploits a regularity. However, in our evolutionary example, individual behavior does not depend on the regularity. Individuals do not adapt in this example (unlike in Lamarckism): simply, those that are not adapted to the new regularity disappear. Although selected individuals can be said to exploit the regularity in the sense that their structure is adapted to it, they do not act *as a function of* the regularity. The unit that exploits regularities in the stronger sense of acting appropriately as a function of regularities is the population undergoing natural selection, not the individual. As we saw in the previous chapter, evolution does not create knowledge, and genomes do not store information, for the simple reason that selection does not change individuals: the rabbit who survives the wolf is the same rabbit who initially hid before wolves appeared, and therefore it does not know that there are wolves in the world.

Thus, by considering the emergence of anticipation, we have uncovered an important distinction. Some processes can be called anticipatory in that they exploit regularities, but this is relative to a particular context. Even an autopoietic organization with an unlimited input of free energy works only within a certain range of temperature (usually quite narrow). But for a system to be considered intrinsically anticipatory—that is, it acts appropriately as a function of regularities—there must be both anticipatory processes, which exploit regularities, as well as other processes that adjust those anticipatory processes. The latter might be called learning processes. Therefore, we may submit that an anticipatory *system* is a set of anticipatory processes together with learning processes.

In the Darwinian example earlier, the learning process occurs at the population level, and therefore the anticipatory system would be the population, not the individual.

We noted earlier that a candle flame is anticipatory in the sense that its processes ensure their own future existence, in the same way as the autopoietic organization of living systems. But flames do not learn, and therefore the candle flame does not constitute an anticipatory system, nor does it belong to a wider anticipatory system. In contrast, living cells belong to an intergenerational anticipatory system. In fact, modern living cells also adapt to the conditions of their continued existence within their lifetime, including bacteria (Lyon, 2015), and therefore, are anticipatory systems in their own right.

The Scale of Learning

We have previously described a scale of sophistication in anticipation, whereby candle flames exhibit a primary form of anticipation, whereas the ant's navigation abilities display more sophisticated forms of anticipation. We can now distinguish another scale, which is a scale of emergence.

An anticipatory system does not know the future. Rather, it acts as a function of regularities it has been exposed to, assuming that these regularities will continue to hold. Therefore, anticipation necessarily relies on previous exposure to the regularity, whether across generations (through evolution) or within the individual's lifetime (learning and adaptation). Thus, to identify a system as anticipatory is to point at "learning" processes, which somehow incorporate regularities in the organism's structure. To be more accurate, a learning process is a process that changes the structure in such a way that the organism's actions *would have led* to more desirable outcomes (or more likely so). Under the assumption that the same causes produce the same effects (which is contextual), this leads to anticipatory behavior.

Thus, it seems that each anticipatory process must be tied to a learning process, which gives rise to it. But what gives rise to the learning process? Previously, we have noted that classical conditioning is a case of secondary anticipation: the dog learns to salivate when it hears the assistant's footsteps, by associating the footsteps with a preexisting anticipatory behavior, salivation when food is presented. Thus, learning may consist in associating new conditions with prior anticipatory processes. But this seems to lead to an infinite regress.

To address this issue, we can take some inspiration in theories of the origin of life. Kauffman considers a random set of polypeptides (Kauffman, 1986), where each of them catalyzes certain reactions. When the set becomes large enough, an autocatalytic network forms among a subset of the molecules (this is called a *phase transition* in physics): each molecule is produced by a catalysis allowed by another molecule. Thus, the catalytic processes produce the substances necessary for their own future continuation. We then have an example of anticipatory processes without an associated learning process. This works because the autocatalytic network incorporates regularities that it also creates (namely, the dynamics of the chemical processes). No learning process is necessary because the network only anticipates internal regularities, not external regularities that could change independently of the system. In this way, we avoid an infinite regress.

If we then consider various autocatalytic networks, those may or may not persist depending on environmental conditions. We then obtain a process of a natural selection, which shapes anticipatory processes depending on external regularities. Of course, an important ingredient in the natural selection of living beings is reproduction, but the origin of reproducing cells is an open question that I will not discuss here.

The crucial point is that natural selection, as a learning process, is not itself anticipatory. The key elements of Darwinism are blind variation and selection. The effect of a particular variation is not "known" in advance—the variation process is independent of the regularities. If it were, we would have to explain where that knowledge comes from, and this would lead to an infinite regress. Thus, primary learning processes must ultimately be a kind of trial-and-error process. In Darwinian evolution, the notion of error is not externally defined. Rather, it is an existential concept: "good" is whatever continues to exist; "bad" is what ceases to exist. This existential concept then conditions derivative errors implicit in anticipatory behavior.

One simple implication of Darwinian theory is that anticipatory processes must co-exist with nonanticipatory processes. The biological organism is never fully adapted, since otherwise there would be no space for further adaptation. Unlike machines, which are conceived according to a design, the possibility of emergence of anticipation implies that, in biological organisms, many things are not part of the design. This is also a lesson we can draw from monsters (Blumberg, 2010): life is exploratory.

Within the individual (rather than across generations), we can distinguish two kinds of learning processes, acting on different timescales. The most well-known kind is associative learning: a certain action leads to a desired effect in a certain identified situation, and a new causal link is produced, from that situation to that action. This is the case of conditioning, for example, or of learning to ride a bicycle.

There is another kind of learning process that is important in perception and sensorimotor interaction. It is related to a subtle concept that ecological psychologists call the "current future" (Turvey, 2018), which is essentially the continuation of the present.

The Current Future

An example is the motion of a ball that has been thrown: the immediate past trajectory of the ball somehow specifies where the ball will be in the near future. Another example is in hearing. Vowel and musical notes have a periodic acoustic wave, and the period is associated with the pitch of the sound (how

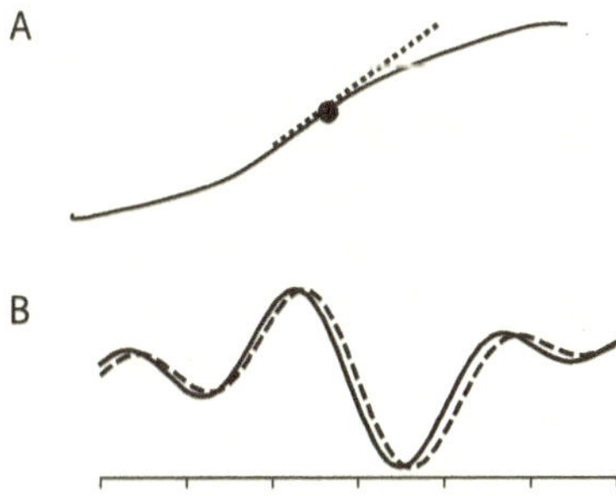

FIGURE 7.8. Anticipation by continuation. A: Anticipating a time-varying quantity using the tangent. B: Anticipating with a delayed filter (dashed curve: input signal $x(t) = \sin(6.2t) + \sin(4.4t) + \sin(2.8t)$; solid curve: output prediction $y(t)$ with $a = 5$, $K = 3$, $\tau = 2$).

high or low its sounds). Because it is periodic, the past of the waveform specifies its future.

This concept is related to Gibson's invariant structure, which is a law that governs sensory inputs (or sensorimotor relations). If this invariant structure extends temporally, then it describes the past as well as tentatively specifies the future. For example, if we identify a signal as periodic based on a finite observation, then we can also anticipate its future—and of course, we might be wrong. In the same way, if we identify a trajectory as parabolic, then its immediate past specifies its future—assuming, again, that the trajectory remains parabolic. The process by which the structure is identified can be seen as a learning process.

A simple example is when one is asked to tap along to a regular rhythm. People tend to do this by tapping slightly before the next beat, and they may also continue to tap when the rhythm stops (Aschersleben, 2002). This requires a process by which the immediate past shapes action so that the action's effect matches some criterion. What kind of process is that? One of the perhaps counterintuitive insights of ecological psychology is that this kind of anticipation (which Turvey and colleagues call "strong anticipation"; Stepp and Turvey [2010]) does not necessarily require structural changes, like a change in synaptic weights in connectionist theory. For example, I can anticipate the future value of a signal $x(t)$ by using the formula $x(t+\tau) \approx x(t)+(x(t)-x(t-\tau))$ (graphically, by continuing the curve of $x(t)$ with the tangent; figure 7.8A). To make this prediction, I am informed by the past but I do not need to tune any parameter. What is less well known is that certain kinds of simple dynamical systems can perform much more sophisticated continuations, as shown by Voss (Voss, 2000, 2016a,b). For example, suppose the signal $x(t)$ is fed into the following causal linear filter:

$$\frac{dy}{dt}(t) = -\alpha y(t) + K(x(t) - y(t-\tau))$$

It turns out that, under broad conditions, the filtered signal $y(t)$ is approximately proportional to the value of x at a *later* time, $x(t+\tau')$ (with $\tau' < \tau$) (figure 7.8B). This may seem surprising since y is a causal filter, but there is no magic. Just like the continuation of a curve using the tangent works approximately as long as the curve is sufficiently smooth, the continuation of a signal with this anticipatory filter works on a limited region of the spectrum. In practice, this means that the output of the filter converges to the anticipated signal $x(t+\tau')$ after a certain time, which can be seen as the learning phase. But this learning occurs without structural changes, and its effect fades away as soon as the signal changes.

Thus, we have distinguished three different timescales for learning processes involved in anticipation. One is the "current future": anticipation as continuation of the current dynamics, a property displayed by certain dynamical systems. At the timescale of the individual life, we find learning processes that rely on different kinds of association, such as conditioning or affordance learning. Ultimately, these processes are rooted in the trial-and-error logic of natural selection, which operates on an intergenerational timescale.

With these remarks in mind, we can now examine the relation between computation and anticipation.

Computation and Anticipation

Consider a washing machine. It is designed in such a way that its processes lead to the output of clean clothes: it is goal-directed. Arguably, its processes are anticipatory, but does the machine anticipate? It is the designer who is aware of the physical regularities, and who incorporated this knowledge in the machine. Thus, it is actually the designer who anticipates. If the machine goes wrong, the machine does not fix its own processes; a person does. With the distinctions we have made, the anticipatory system, which acts appropriately *as a function* of regularities, is not the machine, but the machine plus designer. The machine itself is blind.

In the same way, consider the view that brains are computers optimized by evolution. We might want to see a program as an anticipatory process, but who anticipates? With computers, it is the programmer, the person who writes the program, plans the sequence of computational steps, and checks whether the program is right. If we substitute evolution for the programmer, then the learning process is outside the individual, and so anticipation lies at the population level, not the individual. A program itself does not anticipate. A program could be a random sequence of instructions, and there would be no anticipation there. Again, this is a consequence of the dualistic nature of computation, with an external agency giving meaning to its inputs and outputs and deciding on what the computation is supposed to do. In brief, a computation is not intrinsically anticipatory, at least not generically.

Remember that the representational–computational model of the mind was motivated by the anticipatory behavior of animals, such as the ant's navigation abilities. But unfortunately, computation is not a good model of anticipation. Computation, which is the application of formal rules expected to give a correct result, is logically secondary to the establishment of those rules, by processes that necessarily involve variation and interaction with the world. Therefore, to the extent that cognition is anticipatory, it is necessarily embodied. But computation is by definition disembodied, since it is formal. Therefore, cognition cannot be just computation.

Connectionism allows for self-modification of the system, by processes usually termed "learning rules." However, it is often underappreciated that those learning rules are themselves anticipatory, and that anticipation is not grounded within the system. The most successful learning rule in modern connectionist networks is gradient descent. Gradient descent consists in changing the parameters (synaptic weights) in a direction where error decreases. How does the system know which direction to choose? It is given by the programmer (nowadays, by automatic program differentiation). In addition, the effect of a synaptic change on error depends on the detailed properties of the entire neural circuit downstream of the synapse. Thus, the error is backpropagated from the output layer to the inner layers of the network, and each step requires anticipating the effect of a synaptic weight change on the activation of the postsynaptic neuron. It also requires communicating that effect to other neurons, which implicitly presupposes a rigid relationship between error and whatever signal carries it. Thus, error backpropagation relies on seeing neurons as rigid prespecified structures, except for their synaptic weights. As we have seen, neurons (cells in general) are not like that.

Thus, what makes backpropagation biologically implausible is not the lack of dendrites and other structural details, but the lack of grounding of these anticipatory properties, which require learning processes. There are a few propositions in the theoretical neuroscience literature for how cells might learn the right structural modifications (e.g., Harris, 2008; Seung, 2003), and those follow a completely different logic than function differentiation and communication, namely an exploratory logic.

In this chapter, I have tried to characterize the phenomenon of anticipation, as it appears in its diversity in biological systems. But the question of *how* the organism anticipates remains largely open. Neurocomputationalism answers this question by describing algorithms, such as learning algorithms, and then asks how the organism *implements* the algorithms. This leads us to the concept of *implementation*. As we will now see, this terminology introduces a biased view of the organization of brain processes.

8

The Organization of Brain Processes

IN THE MAINSTREAM view of brain and mind, biology is "implementation." Implementation is an engineering concept. To build a machine, one starts by asking what it should achieve, then describes how it works at an abstract level, and finally wonders about the parts and materials that meet the functional specifications. In computational neuroscience, this corresponds to Marr's three-level analysis (Marr, 1982). One starts by describing the abstract problem to be solved (the computational level), and then the important abstract variables and the algorithm that the brain uses to solve the problem (the algorithmic and representational level), and finally, the implementation level describes how brain processes might realize the algorithm. In cognitive science, this is the mainstream doctrine known as *functionalism*: what matters for mental phenomena is the causal organization of functional states, and this organization can be realized in many ways—this is the "multiple realizability" of mental states. (See Zahnoun [2023] for a critique of functionalism.) As Putnam put it: "we could be made of Swiss cheese and it wouldn't matter" (Putnam, 1975).

One of the difficulties with Marr's three-level approach—that is, the engineering view of brains—is that only the highest and lowest levels have a direct empirical grip. The "computational" level is constrained by behavior or psychological phenomena, while the "implementation" level is constrained by physiological or anatomical observations. But the "algorithmic" or "representational" level is a theoretical construct with no independent access to reality. This level does not follow logically from the computational level (there are many ways to solve problems), and depends on "implementation" for its empirical assessment. Yet, it is that intermediate level that specifies the functional organization of the brain, considered as an "information-processing device." Therefore, applying the engineering mindset to the study of the living runs a high risk of inserting prejudices about how brain processes are organized. As we will see, this risk is indeed realized in typical modeling approaches.

We will start with one of the prejudices explicitly endorsed by Marr and his followers, that the brain is an "information-processing device" (*Information*

Reductionism). As we have seen in chapter 6, that terminology is committed to the substance view of information, the epistemic phlogiston that gets processed through the successive stages of a factory chain. This is not how the brain is organized. There is no information *in* the brain for someone to read; there are brain processes by which the organism gets informed. This leads us to a fundamental misconception of the algorithmic view of the brain, namely that the activity of neurons can be identified to a computational state (*Neural Activity*). But an activity is not a state, let alone a computational state: this is a basic confusion between substance and process that lies at the heart of most neurocomputational approaches. Finally, the engineering mindset leads to a flawed view of multicellularity (*The Nature of Multicellularity*). The brain is made of neurons, but no one actually made a brain from neurons: unlike machines, a multicellular organism is not an assembly of components with the right specifications, organized into neatly separated modules, but a cellular colony that develops. In other words, brains are not "implemented."

Information Reductionism

Perception as an Input–Output Process

A computation has an input and an output. Accordingly, in neurocomputational models of perception, the problem of perception is typically framed as one of mapping input data, such as an image or sound waves, to categories (face label) or properties (sound direction). Indeed, we see an image and then we might say who we saw. With an engineering mindset, we might imagine a machine that captures a photograph, then transforms it into a symbol that stands for the identity of a face. But the recognition of a person is a mental phenomenon, not an output symbol. In fact, an image is also a mental phenomenon, and not an input.

Indeed, as we have seen in chapters 3 and 4, there are no images as the input to the visual system, at least not in the sense of the images that we see. Rather, the organism is coupled to an optical scene through mobile organs, the eyes, which actively explore it in a feedback loop. The fastest eye movements occur unconsciously on a 10 ms timescale over very small angles, when we fixate an object (Martinez-Conde et al., 2004; Rucci et al., 2007). When these movements are prevented, we cannot see. Therefore, movement is constitutive of perception, it does not occur only after perception has taken place. Thus, perception is not a projection of an image onto a central screen, but an exploratory activity that involves a dynamical coupling between environment and organism (Ahissar and Assa, 2016). This circular relation between the environment and the organism was pointed out more than a century ago by Dewey (1896), in a criticism of the concept of "stimulus": "the motor response determines the stimulus, just as truly as sensory stimulus determines the movement."

Thus, when we see a still image, the actual sensory input is not still at all. It is the still image that is a mental construction, and not the input, which is dynamic. In the same way, one of the properties of images such as photographs is translation invariance: translating a photograph does not change the perceived identity of the person in the photograph. This has motivated a powerful technical trick in artificial neural networks known as convolutional networks (LeCun and Bengio, 1998): connections between layers are enforced to be translation invariant (via "shared weights"), which drastically reduces the number of adjustable parameters. But translation invariance does not exist at the level of the retina, for the retina is not homogeneous. The eye has inhomogeneous optical aberrations, the fovea has a much finer spatial resolution than the periphery and different color sensitivity, and the visual field is sampled by a mosaic of cells of different types, meaning that two neighboring cells typically have very different properties (see O'Regan and Noë [2001]). There is no translation invariance at the interface between the nervous system and the optical environment. Again, this invariance appears at the psychological level: if the photograph is moved, one sees the same person. But this does not occur physiologically by having shared synaptic weights, or translation-invariant weights.

Thus, what is typically taken as input is in fact the result of a process. We realized this only because we took interest in the biological level, the "lowest" level, rather than treated it as implementation detail. Thus, biology constrains not just how "brain algorithms" are implemented, but the very formulation of the problem.

The second error with the "information processing" view is to take the psychological phenomenon of perception as a sort of machine output: there is a component of the machine whose state stands for face identity, or sound direction. Implicit in this perspective are one unsupported assumption and one fallacy. The fallacy is the well-known homuncular fallacy: an output symbol needs to be interpreted, and therefore presupposes a cognitive agent. We will examine this fallacy shortly, in the context of modeling. The unsupported assumption is that the brain is strictly modular, with the activity of some neurons acting as a computational variable standing for some feature (face identity, sound direction).

Computational Modularity

When we say that the brain computes sound direction, it is implied that there is some place in the brain where the state of a neuron or group of neurons corresponds to sound direction, like a computational variable. In the case of sound localization, these neurons are usually assigned to the neural structures in the auditory brainstem, such as the inferior colliculus and the olivary complex, with an anatomical circuit going from the cochlea to the brainstem through the auditory nerve. Computing something involves transforming the input into an output, and that output is in the brain. The output of this circuit can then be used by other circuits further down the road from perception to action, for various tasks: turning to the source of the sound, fleeing from that source, paying attention to that sound, talking about it ("it is on the left"), and so on. Thus, there is a functional justification to that particular arrangement: the computational work is factored into estimating sound direction and doing various things that depend on sound direction.

In terms of the organization of brain processes, this means that the process that outputs sound direction does not depend on what is done with that output. This is what an engineer would typically do: to build a robot, one would design perception modules, typically one per sensory modality, planning modules, and motor control modules (but see Brooks [1986] for an alternative way of building robots). The reasons for this modularity are cognitive and social. Cognitively, this allows one to focus on one simple problem at a time, following Descartes's rule of analysis (Descartes, 1637): to divide the problem into smaller, more intelligible, pieces ("to divide each of the difficulties I would examine, in as many pieces possible that would be required to better solve them" [translation mine]). Socially, it also allows work to be divided between different people, or even different companies: one in charge of each module. These modules can then be reused in different contexts, different machines that combine modules in new ways. For this to work, the input–output properties of modules are well-specified—for example, by stating that a module's output represents sound direction. What the output represents should not depend on the context in which the module is embedded.

Thus, there are excellent reasons for engineers to use modular design. But brains are not made by people, let alone by human societies. However appealing the functional argument for computational modularity might be, it is hard to reconcile with the known physiology and anatomy of nervous systems. We have seen in chapter 5 that there is no neuron or group of neurons whose activity depends on one particular feature or a few features. Indeed, careful analysis shows that this is not even possible in principle, since features are contextual: they depend on both the scene, whose structure constrains possible features, and on the task, which defines what features are relevant. Rather, any given task enrolls a wide set of brain regions that interact, some of which are specialized to certain sensory modalities, with no fixed semantics associated to any neuron's activity. The same brain region can be recruited in a variety of tasks (or "reused"; see Anderson [2010]). Arguably, this could be seen as a form of modularity, but in the sense that the same person can do different things depending on its skills, not in the computational sense of the composition of input–output mappings.

Anatomically, the topology of brain connectivity is clearly not linear but entangled (Pessoa, 2022), with a few exceptions. It is true that the retina sends many axons to the rest of the nervous system, while the converse is rare. But in central regions of the brain, bidirectional connections are common. The connectivity of brains is not feedforward, from sensory neurons to motoneurons. Surely, there is a "feedforward" path from retinal ganglion cells to the thalamus, then to the occipital cortex, which then diverges between the dorsal pathway (parietal cortex) and ventral pathway (inferotemporal cortex), then to other structures such as the prefrontal cortex, involved in executive control. But with the exception of retinothalamic projections, all these connections are mirrored by abundant "feedback" connections, in the opposite direction (Gilbert and Li, 2013; figure 8.1).

Given these anatomical facts, it is not reasonable to believe that we can pull out a circuit and expect that it will work essentially as it does when it is embedded in the brain. Concretely, this implies that the causal organization of the brain cannot be neatly split into distinct processing modules, as the input–output processing view would suggest. Despite the common Cartesian rhetoric, which sees those feedback pathways as "top-down" instructions, mechanistically we have a system made of coupled components. Locally, the timescale of this coupling can be as short as a millisecond. For example, many neurons are coupled by electrical synapses, also called gap junctions. Unlike chemical synapses, which are directional and involve a transmission delay as neurotransmitters move across the synaptic cleft, electrical synapses are bidirectional and essentially instantaneous: the two connected neurons are electrically coupled. In this case, it is not possible to model the activity of neurons as a sequence of updates, as in an algorithm. Rather, neural activities are co-determined.

In brief, neurons are coupled together and with the body and environment. They are organized in such a way that the nervous system cannot be neatly split into independent modules. To be sure, there is some amount of specialization in the brain: some brain areas are involved in certain broad categories of tasks rather than others. But these areas cannot be seen as components with fixed input–output specifications.

We now turn to the nature of models of perception, when perception is conceived as the output of symbols.

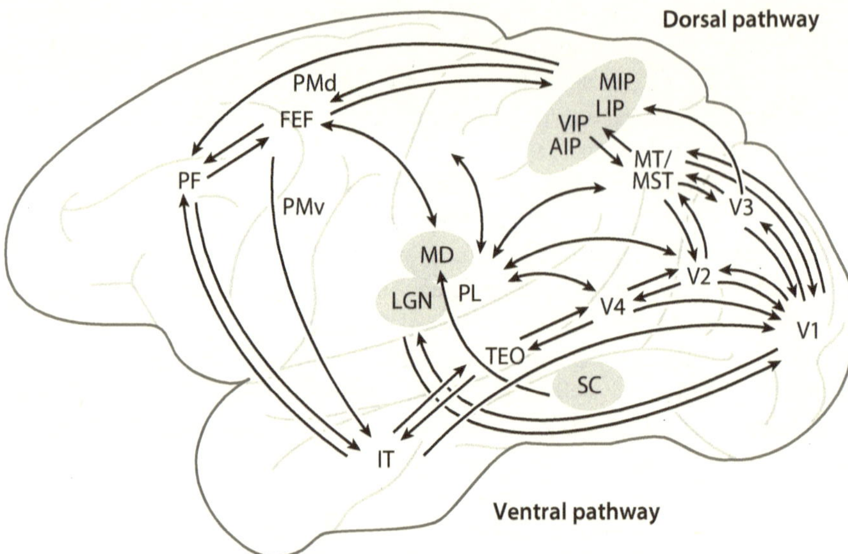

FIGURE 8.1. Simplified connectivity of the visual system. Adapted from Gilbert CD, Das A. 2021. In Kandel et al., editors. Principles of Neural Science. McGraw Hill.

Homuncular Modeling

Suppose I am sitting next to a blind friend. I am looking at a collection of photographs. Whenever I see David Marr, I tap on my friend's shoulder. Does my friend see David Marr? Obviously not: she feels a tap on her shoulder. In the same way, the fact that a neuron spikes when David Marr is in sight does not explain why I see David Marr. Thus, visual perception cannot be identified with the production of a symbol that stands for a meaningful category of images. This is what artificial neural networks do, but this is not perception. Perception is a mental process, not a symbol.

Yet, the conventional approach to neural modeling of perception is to frame it as an input–output process, where the output is a code for the perceptual symbol (face identity, sound direction). Implicitly, it is assumed that this output is connected to some other neural circuits, such that it *then* triggers perception, or some relevant behavior, but those circuits are not explicitly modeled.

Concretely, this means that neural models of perception are typically hybrid models: one part may be a biological model, whose components are supposed to correspond to components of the brains, such as neurons; another part is a formal mapping that is used to bridge the domain of neural activity to the domain of objects and features. This formal mapping is, by construction, realized by the observer, not the brain, and as we have seen both for computation (chapter 4) and anticipation (chapter 7), this means that part of the work is actually done by the observer. Thus, this hybrid model is homuncular: part of the model is realized not by the brain but by a person—in fact, not a little person but a well-identified grown-up, generally a human with higher education in science.

Teller called the decoding mapping a "linking proposition" (Teller, 1984), a statement that relates perceptual states to physiological states. As she pointed out, linking propositions are often not explicit in the scientific literature. But sometimes they are, and one can then see their problematic nature. One example is the concept of the "ideal observer," which figures in many neuroscience studies (Macmillan and Creelman, 2005). In the context of an experiment where the firing rate of a neuron (or group of neurons) is measured, the ideal observer is a theoretical construct that performs the task of interest with optimal accuracy based on the observable. Thus, it is an explicitly homuncular concept. For example, as we have seen in chapter 6, one can measure the response of a binaural neuron to spatialized sounds and ask whether an ideal observer of the neuron's firing rate could tell apart sounds coming from different directions. On this basis, it has been concluded that the firing rate is so sensitive to sound direction that "it might not be necessary to pool the outputs from many neurons to account for the high accuracy with which human observers can localize sounds" (Shackleton et al., 2003; Skottun, 1998). What is extraordinary about this statement is that something about perception can apparently be deduced from the physiological measurement of a single neuron, without considering at any point the system in which this neuron is embedded.

The implicit assumption is that the human brain implements something like an ideal observer of the measured neuron. To examine this assumption, it is useful to spell out what the ideal observer is supposed to do exactly. Suppose the subject is asked which of two sounds came from the left. The ideal observer must answer based on the firing rate of the measured neuron. What does it do in practice? When the first sound is presented, the ideal observer stores the number of spikes produced by the neuron in a window of a fixed duration after the sound, a duration chosen by the experimenter (figure 8.2). It ignores all spikes produced before and after that window. Then the second sound is presented, upon which it stores again the number of spikes produced by the neuron. Then it retrieves the two stored numbers, compares them while ignoring everything else (e.g., the activity of all other neurons), and decides to push one of two buttons.

When it is spelled out in this way, it is not at all clear how the brain might realize the ideal observer. The tacit assumption is that the brain has been "optimized by evolution," and so it should do the optimal thing. But when we

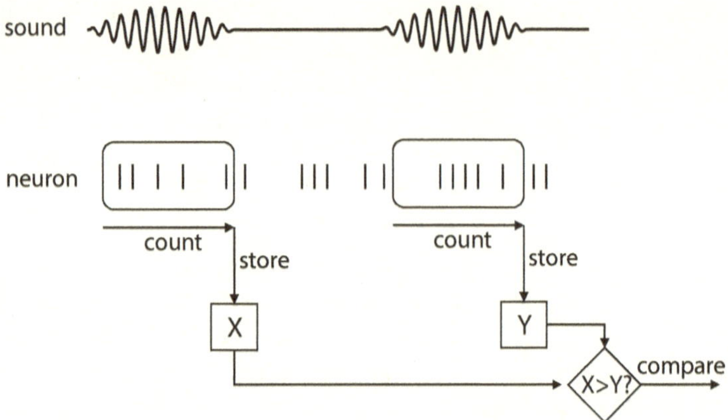

FIGURE 8.2. An ideal observer of a neuron reacting to two sounds. The observer counts spikes in experimenter-specified windows, stores them, and ultimately compares the stored numbers.

examine what this means, we realize that the ideal observer uses information that is available only to the experimenter: what part of the auditory stream is considered as "the stimulus" and which parts of neural activity are to be stored and compared, the knowledge that the same sound is played at different positions, the neuron's tuning curve *for that sound*, and so on. In addition, the ideal observer uses nothing *more* than the information available to the experimenter. In the preceding quote, it is acknowledged that pooling the response of other neurons would make the discrimination task easier, by reducing noise. But the ideal observer does not do that, as it only considers that specific neuron being measured by the experimenter. In other words, the "ideal observer" is not the best thing that the brain can do, but the best thing that the experimenter can do. Therefore, not only is the ideal observer not explicitly modeled, but it cannot possibly be a biological model—it cannot be "implemented."

The problem comes from considering the activity of some neurons as a kind of symbolic output, which is homuncular—someone has to interpret the symbols, but this critical part is not part of the biological model. This state of affairs causes major epistemological problems, since it makes it virtually impossible to empirically assess models. For example, some claim that sound direction is "represented" by the average activity of neural populations (the "hemispheric model" [McAlpine et al., 2001]), others by the detailed pattern of activity. But the difference between those models is only the speculative interpretation different scientists make of the *same* data: the observer can infer sound direction from average activity, or from detailed patterns of activity (although in this case, there is a *theoretical* argument against the hemispheric model [Goodman et al., 2013]).

Thus, typical models in computational neuroscience combine a biological part and a homuncular part. The biological part has components corresponding to neurons or brain regions, and an input and an output in the neural domain (e.g., firing rates). The homuncular part consists of formal mappings between things in the brain and things in the world or in the mind. Only the biological part is the focus of the model and of the argumentation, while the homuncular part is left unexplained, and in practice is performed by the scientist, with the information available to her.

In terms of modeling, we have already pointed out the solution to this problem: embodiment. For example, a model of sound localization should be a model of an orientational response (e.g., orientation reflexes), including the motor part. The model would then explain not how the scientist can infer sound direction from the activity of some neurons, but how an animal orients its head to a sound source. In other words, the appropriate modeling approach is systemic rather than reductionist—"models that behave" (Gomez-Marin, 2017).

What Kind of System?

When we try to explicitly model the physiological system responsible for behavior, another issue with the information processing view appears: information is atemporal. The same information is produced whether the process is fast or the output has a latency of one second. In essence, the information processing view is a bureaucratic model of the brain (Brette, 2019b): neurons are seen as filling and exchanging some kinds of forms that get processed in due time. A bureaucratic system is a system where information can be decoupled from action. There is no such decoupling in neural activity: an action potential cannot be suspended for future use; it is a transient electrical event that impacts connected neurons and vanishes.

The brain is a dynamical system, not a bureaucratic system, and time matters very much. For example, in sound localization models, behavior is based on sound direction codes. Concretely, this means that the activity of auditory neurons is observed for some time, and then the summed activity is mapped to a sound direction estimate. That is what the experimentalist does, but neurons do not register spikes over time intervals delimited by a clock until they decide to do something with them. Each neuron's spike has an essentially immediate effect on the neuron's targets—hence the terminology "action potential." For example, if spikes of auditory neurons directly control the contraction of neck muscles, then the head turns as soon as a sound is presented, which in turn changes the auditory input. The observer can still measure the summed neural activity and find that it correlates with sound direction, but the behavior is actually based on feedback, not on reading the activity over the entire period. The information processing view assumes that the brain is decoupled from the environment, but this is not generally the case.

If instead we want to account for ballistic movements triggered by sounds, then we need to pay attention to many aspects beyond the mere production of "information." For example, what triggers the initiation of movement? How are the spikes of auditory neurons temporally integrated, and how does this temporally extended activity result in appropriate movement? When an auditory neuron responds to a sound, do the first spikes influence the later spikes through the feedback connections that are known to exist? None of these crucial questions can be answered by considering measured spikes as "information"; we need to pay attention to the dynamics and organization of the system.

In this respect, conventional terminology in neuroscience is rather misleading. The visual "system," for example, is a set of anatomical structures from the eye to the occipital cortex that are important for seeing. This might be a useful anatomical grouping, but not a functional system. To explain any visually guided behavior, one needs to consider a system much wider than the visual "system." In the same way, one can understand how blood circulates throughout the body by considering the cardiovascular system, while the arteries and veins of the leg do not constitute a "leg circulatory system": we cannot account for circulation if arteries and veins are not connected.

Another major prejudice of the information processing view is the primacy of the stimulus: if what the brain does is to transform an input into an output, then neural activity is just responses to stimuli. This view denies a major aspect of life: the autonomy of the organism.

Autonomous Activity

Sensory neuroscience is focused on the concept of stimulus. A stimulus is whatever makes the organism react. Thus, it is a behaviorist notion, where behavior is conceptualized as responses to stimuli. In the context of an experiment, the stimulus is typically a change in sensory signals that is of interest to the experimenter. For example, an oriented bar is flashed in front of the eyes, or a sound is produced. The experimenter then measures behavior or neural activity, and these observations are called "responses" to stimuli. This terminology is used not just for sensory neurons such as photoreceptors, but also deep inside the brain for cortical neurons. Even in consciousness studies, cognitive neuroscientists propose that the electrophysiological signal measured after the presentation of a noticeable stimulus (as opposed to a subliminal stimulus) is a signature of consciousness (Mashour et al., 2020), as if consciousness was also a response to a stimulus—as if the person were not conscious of something before the stimulus, or when the stimulus is not noticed. Just as behaviorism sees behavior as responses to stimuli, cognitivism sees mental states, including conscious ones, as responses to stimuli (specifically, outputs of computations).

Thus, neural activity is typically conceptualized as responses to stimuli, as the output of some processing of sensory inputs. It follows that when the neural activity that follows the presentation of a stimulus varies over trials, this variability is framed as "noise." Indeed, it is often said that neural responses are typically noisy, because neural activity is not always the same when the animal is in the same experimental configuration.

But this comes from framing neural activity as responses. If we presented various pictures to a person and measured her heartbeats, we might find that the heart beats slightly faster when certain pictures are presented, but the beats would occur at random times relative to the stimulus. Yet, heart beating is not "noisy": rather, the heart has an intrinsic ongoing activity, generally regular, which is perturbed by the presentation of the stimulus. Indeed, heartbeats are not responses to the picture presentation; the heart beats also when the experimenter is away.

The same is true for neurons in the brain, including sensory neurons. Neurons spike whether there is a stimulus or not. In the neuroscience literature, this phenomenon is called "spontaneous activity" or "resting state" activity (Deco et al., 2011). Studies of this activity show organized patterns, with some similarity to the patterns observed in stimulus-induced activity. For example, place cells of the hippocampus are active during rest and sleep, and show the same patterns of activity as observed during awake behavior (spatial navigation), although at a different pace (Ólafsdóttir et al., 2018). It has been known for a long time that the brain has a highly organized spontaneous electrical activity, as manifest in the oscillations measured by electroencephalograms at various frequencies (Buzsáki, 2011). Oscillations are also measured when various stimuli are presented, but they are usually not tightly time-locked to the stimulus. Thus, neural activity is not simply responses to stimuli, and it follows that variability in neural activity for the same experimental configuration is not necessarily noise. Neurons are active, this activity is self-organized, and it changes when the environment changes.

How could it be otherwise? The mind does not turn off when the experimenter goes away, or while the subject is waiting for the stimulus—waiting is a mental activity too. Unless we believe that the mind is entirely unrelated to the brain, the brain should not turn off either. Why then is the activity of neurons framed as responses to stimuli? To make this consistent with agency and the autonomy of the living, we must postulate that the autonomous organism is somewhere beyond the reach of experimental measurements. In other words, the input–output processing view of cognition is deeply committed to Cartesian dualism: what we measure are automatic responses, and the organism is somewhere else, making sense of neural responses and commanding neurons.

If instead we want to develop a naturalistic view of brain function, then we must reject the idea that neurons transform input stimuli into output representations. The brain has an organized endogenous activity, and it is coupled to the environment. This implies that a "stimulus" does not produce a "response": rather, it perturbs the ongoing dynamics of an otherwise autonomous system. Tuning curves are a partial and contextual measurement of this perturbation, not a measurement of an input–output mapping. This motivates what Buzsáki calls an "inside-out" approach to studying the brain (Buzsáki, 2019).

This leads us to a fundamental misconception at the heart of neurocomputational modeling, namely the idea that neural activity is a sort of state—namely, a computational state.

Neural Activity

Processes and States

Computationalism describes cognition as the manipulation of mental states, identified to computational states, in explicit analogy with the formal states of an automaton. The brain is said to implement these computations in the sense that physical states of the brain can be mapped to computational states in the formal model of computation, such that brain processes can be interpreted as computations, as we have seen in chapter 4. Typically, these physical states are the activity of some neurons. For example, in artificial neural networks, the computational variables are called "neural activations," and correspond essentially to firing rates.

This conceptual framework is usually unquestioned, because it is often thought that "we can view every physical system as a computer" (Shagrir, 2006): we just call firing rates "computational states" and brain processes "computations." However, the fact that it was not so trivial to build actual computers should give us pause. There are indeed mistaken assumptions in the neurocomputational framework, which conflates activity with state.

Remember that classical neurocomputational models of visual perception transform an input image into a mental symbol, such as a face label. This is because psychologically, we see a stable image and we are led to believe that there must be a stable physical entity in the brain that stands for that image, a "neural representation," which is then further processed (Ramsey, 2007). But when we looked at the biology of seeing, we realized that there is nothing like an image in neurons of the retina: the visual field is dynamic and the activity of neurons is transient, made of electrical spikes. The stability of the visual world is a property of perceptual activity, not an intrinsic property of raw perceptual material.

Consider the following thought experiment inspired from the TV series *Bewitched* (Brette, 2019b). In that series, Samantha, the housewife, is actually a witch. She can twitch her nose and everyone freezes except her. Then she twitches her nose and everyone unfreezes, without noticing that anything happened. For them, time has effectively stopped. What happens during that time is not a continuous experience of the same mental state. Rather, there is no experience, no mental image at all, and it is as if time had jumped from one instant to another. This is because an experience is not a thing but an activity (Erdin and Bickhard, 2018; Zahnoun, 2020). Despite the common terminology, a "mental state" is nothing like a physical state.

As the concept of information illustrates (chapter 6), the neurocomputational framework sees brain processes as operating on some raw cognitive substance, which is stable by default and has informational content—that is, which has the properties of mental experience. This perspective on phenomena where the properties of the phenomenon are traced to the properties of an underlying substance is called *substance metaphysics*. It is to be contrasted with *process metaphysics*, where the properties emerge from the processes. This distinction goes back to Aristotle and beyond, and while *substance metaphysics* remains common in biology and psychology, it has been progressively replaced by *process metaphysics* in most other sciences, as explained by Bickhard (2009). As we have seen, in the seventeenth century, burning used to be explained by the release of phlogiston in the air. To complete the cycle, phlogiston is later captured by growing trees. Thus, the theory postulates a stable fire-like element and explains the phenomenon as a displacement of that element. It was later superseded by a theory that described combustion as an out-of-equilibrium process where molecules change through chemical reactions that release heat, sustaining the continuation of the process (a positive feedback loop). Heat used to be described as a self-repellent fluid called caloric, which would then tend to flow from hotter to colder bodies. In modern physics, heat is not a thing but a property of thermodynamic processes. Heat does not only move with the fluid (the process of convection), it is also energy transferred by the process of conduction (through the collision of molecules), and by the process of electromagnetic radiation—two phenomena that are not properly captured by the concept of a moving thing. Since the twentieth century, we know that atoms are also not "things," but systems of moving particles, and in quantum field theory, particles are also not things but particular phenomena in unlocalized dynamic quantum fields.

In biology, vitalism (in its simplest form) used to explain life as the result of a particular substance infusing organisms. As we have seen in chapter 2, life is now theorized in terms of self-sustaining organizations of out-of-equilibrium processes, and these organizations change through developmental and evolutionary processes. In the eighteenth century, preformationism held that development is the unfolding of a preformed miniature version of the animal contained in the embryo. It has been replaced by the dynamical processes of development, where the organism is shaped by the interaction of genetic processes with the growing body and environment. Yet, naïve substance views remain surprisingly common in biology. The popular view of the genome as a map of the organism is indeed not very different from preformationism. In cognitive science in particular, cognitive functions are sometimes attributed to genes, such as the "gene of language," as if a protein could contain knowledge of grammar (or any kind of knowledge). In evolutionary psychology, one speaks of the innate behaviors of adult humans, such as the language faculty. This paradoxical terminology ("innate" should at least apply to newborns, but babies do not speak) means that psychological faculties are "already there" from the beginning, ready to be uncovered, again a preformationist notion that is contradicted by modern developmental biology, as is explained by Blumberg (2006). Here, substance metaphysics fuels a false dichotomy between genes and environment ("nature versus nurture"), as if there were no other option for adult behavior to already be *somewhere*: either it is already there in the genome, or it must result from "acquisition" from some other source. A snowflake has a very characteristic shape, yet that shape is neither inscribed in the water nor acquired from the environment: it results from a growth process, which in certain conditions, lead to the formation of snowflake, with a shape that is both characteristic of snowflakes and at the same time unique to any individual snowflake, because of varying initial and environmental conditions (figure 8.3).

In these examples, the major conceptual weakness is that the essential properties of the phenomena are already postulated properties of the substance itself, and therefore they cannot be explained—an organism is alive because it has a vital substance; the grown adult is already in the embryo from the beginning. Unfortunately, neuroscience and cognitive neuroscience remain largely committed to substance metaphysics, despite being blatantly misfit to biological phenomena, which are out-of-equilibrium processes. Thus, people can perceive because they have representations, which are things with perceptual content, which the brain can then "interpret." Organisms can anticipate because they have programs (literally: something written in advance), either genetic or cognitive.

FIGURE 8.3. Snowflakes photographed by Wilson Bentley (1902).

The implementation framework of neurocomputationalism makes two invalid assumptions based on substance metaphysics. The first assumption is that firing rates are states of the brain. But firing rates are activities, not states. The second assumption is that physical states of the brain are computational states. But as we will see, computational states are states with special properties, which is why building a computer is a technical challenge, and not just a rhetorical stance. Let us first clarify the concept of state in modeling.

What Are States?

States are key elements in the scientific description of systems, in particular in physics. The state of a system is its configuration at a given time. For example, consider an articulated puppet, which is animated by the puppeteer using threads (figure 8.4). Its state is the configuration of its different elements. It

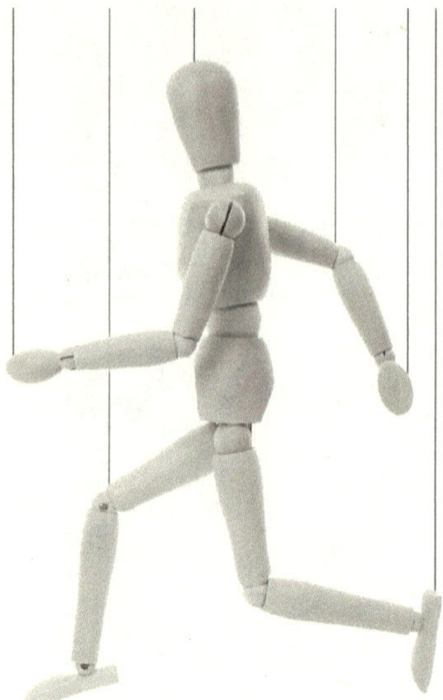

FIGURE 8.4. State of a wooden mannequin.

can be characterized by the angles of all articulations, for example, or by the position of the contact points of all threads. Once we have described this state, we have specified everything there is to know about the current shape of the puppet, but also about possible interactions with the puppet. For example, we know what will happen if we pull on a given thread.

In a dynamical system, the state is what gets updated by the dynamics. For example, the state of a pendulum is its angle and angular speed. Once these two values are known, the laws of mechanics can predict their values at any future time point (given the structure of the system). The system might very well be noisy, in which case the temporal evolution of the state might only be known in probability, but the state is still all there is to know about the system at this time.

Clearly, the concept of state is a modeling concept. For example, the pendulum might be noisy because of turbulences in the air. We could decide to model those turbulences as noise with certain characteristics, in which case the pendulum would be described by a stochastic dynamical system, and its state would be entirely characterized by its angle and angular speed. Or we could be much more ambitious and describe turbulences with Navier-Stokes equations. In this case, the state of the system would include fluid properties over the entire volume of the system, and the angle and angular speed would be state variables within a broader set of variables characterizing the state of the system.

In the same way, the state of a gas would be its pressure, volume, and temperature, if we decided to model it macroscopically with the ideal gas law ($PV = nRT$). But we could also decide to model it microscopically, with the position and momentum of each molecule and Newton's laws. Thus, the state of a system implicitly refers to a model. The laws may not be fully specified, but a set of quantities gets called a "state" to mean that the laws describe how these quantities change through time or interaction, or how they are related to each other at any given time. Within a model, the state of the system characterizes the system *at a particular instant.*

Thus, not all measurements are states. For example, we could measure the time it takes for an athlete to run 100 meters. This is not a state. Modeling the brain comes with identifying states of the brain. In general, the basic variables of neurocomputational models correspond to firing rates of neurons. Unfortunately, these are not states.

Firing Fates Are Measurements, Not States

Typically, the state of neurons is characterized in terms of their activity. For example, in formal neurons, a graded number represents the "activation" of a neuron. In layered formal networks, the activation of a neuron is a function of the activations of neurons in the previous layer. Experimentally, the activity of neurons is reported as *firing rates,* which is the average number of spikes produced by a neuron per unit of time (Brette, 2015). For example, in the 1920s, Edgar Adrian examined the electrical responses of nerves to stimuli, such as touching the skin. He observed that the nerve would fire electrical impulses ("action potentials" or "spikes") at a frequency that increased with the strength of the stimulus. In practice, he would count the number of spikes produced over some duration after stimulus onset and divide by the duration (figure 8.5).

But just like the state of a runner cannot be the time it takes her to run a hundred meters, the state of a neuron cannot be the number of spikes that it produces during some time interval. A firing rate is a statistic of electrical events over some time window chosen by the experimenter. In contrast, the membrane potential is a state variable: it is something that can be measured with an intracellular electrode at any instant, such that its temporal evolution can be described by the laws of electrophysiology.

It might be objected that, in practice, physical states are always measured over extended periods. There is no such thing as a truly instantaneous measurement. But in order to measure a firing rate, the observation period must be fairly long. Since the firing frequency of neurons rarely exceeds

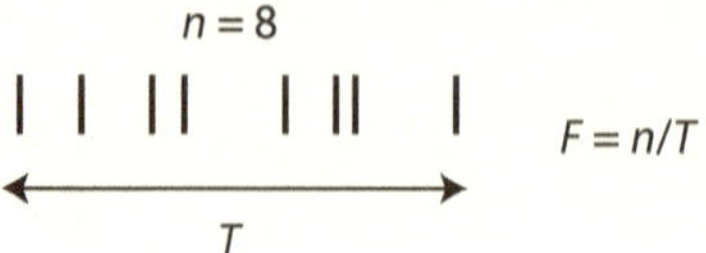

FIGURE 8.5. The firing rate of a neuron: average number of spikes per unit time.

100 Hz, and is more often in the 1 Hz range, it takes at least several tens or hundreds of milliseconds to be able to define a firing rate. But neurons interact with each other on a timescale of milliseconds—that is, within the observation period. In addition, the impact of spikes on a neuron also depends on the temporal arrangement of spikes on a timescale of milliseconds. That is, spikes received by a neuron have much more impact if they arrive synchronously than if they are scattered over tens of milliseconds (Brette, 2015; Rossant et al., 2011). Therefore, one cannot define a state variable from a measurement that extends over more than tens of milliseconds.

In the modeling literature, there are two common strategies to address this issue and use firing rates as state variables. One is to count spikes produced by many neurons, so that a shorter time interval can be used. The other one is to consider firing rates as probabilities of firing.

Macroscopic Neural States

As we have seen in the previous chapter, one strategy is to look not at individual neurons, but at populations of neurons. In this way, it is possible to measure average firing rates over short periods of time (figure 8.6). There are essentially two ways to do this. One is to group nearby neurons with similar properties into a single unit. This approach, often called "neural mass" models (Freeman, 1975), has a long tradition in theoretical neuroscience (Wilson and Cowan, 1972). It has some appeal because the description unit corresponds to the spatial unit of measurement in brain imaging (functional magnetic resonance imaging, intrinsic imaging, calcium imaging, electroencephalography, etc.). However, the choice of grouping is somewhat arbitrary, as there is no guarantee that neural properties are somewhat spatially continuous. In the hippocampus of rats, for example, place cells of nearby neurons can vary widely. Grouping them into a single unit would impede any functional description. Development does not generally lead to locally homogeneous structures—for example, in the retina, cells of different types are interlaced in a mosaic pattern.

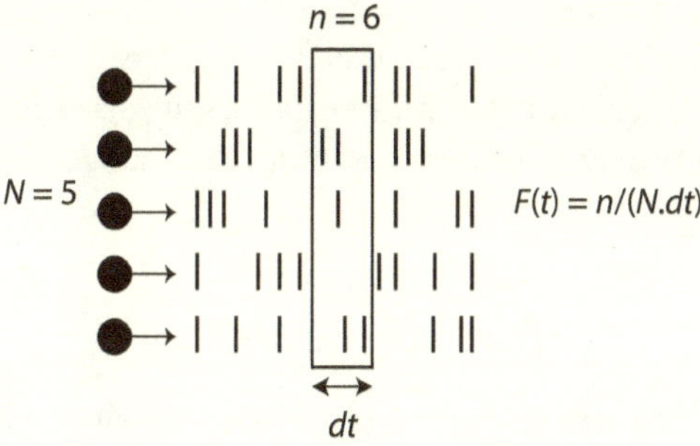

FIGURE 8.6. The firing rate of a neuron population: number of spikes averaged over neurons and a short temporal window. Original figure by Brette R. 2015. https://doi.org/10.3389/fnsys.2015.00151 / CC BY https://creativecommons.org/licenses/by/4.0/.

A more empirical approach that we have seen in chapter 5 is based on neural manifold analysis. It consists in reducing the activity of all measured neurons to a few main dimensions. In practice, each manifold coordinate is a function of the firing rates of all neurons calculated over (typically) 50 ms bins. Thus, it is similar to population rates described earlier, except the averaging is more sophisticated. Manifold analysis is more empirically grounded than neural mass models, but as we have seen in chapter 5, manifold dimensions change depending on the task. Therefore, manifold dimensions may be useful data analysis tools, but they are not states. In addition, in practice those variables are still defined by grouping spikes spread out over several tens of milliseconds, and we have seen that this is a problem because neurons are highly sensitive to temporal relations between spikes within such a timescale.

There are additional reasons why grouping neurons is not an adequate strategy to define neural states. Remember that states do not simply summarize observations; they are supposed to support inferences about the system, according to some model. In the context of dynamics, the future state is supposed to be determined (to some extent) by the present state. If we want to define states by grouping the activity of different neurons, then the interaction between two neurons should depend essentially on the group they belong to. In particular, synaptic weights should be defined at the group level, not at the level of individual pairs of neurons. In other words, we must dismiss the individuality of neurons and treat it as irrelevant noise. But this is a problem because neurons *are* individuals: cells with their own processes and individual history. Synaptic plasticity, for example, is primarily defined at the cellular level, not at the level of a population, let alone of a virtual population. In particular, the cellular individuality of plasticity processes implies that the particular groupings obtained from manifold analysis have no reason to be stable. Therefore, manifold dimensions are not states, only descriptive variables that summarize activity over some period of time. The same remark applies to neural masses.

Firing Rates as Probabilities

A different approach is to see firing rates not as counts but as probabilities. For example, we might say that the firing rate of a neuron is 10 Hz, at a precise instant t, and this would mean that over a small interval $[t, t+dt]$, the neuron might fire with probability 10 Hz times dt. This is similar to the population rate illustrated in figure 8.6, except we average over multiple realizations (trials) rather than over different neurons. Mathematically, this means that spike trains are modeled as stochastic point processes with a time-varying rate.

Empirically, experimenters often report time-varying firing rates obtained by averaging neural activity over many repeated trials with the same experimental condition (stimulus). Such measurements are called post-stimulus time histograms, or PSTHs, and they define a time-varying quantity. However, again, this is a statistic and not a state. At a given time, the neuron spikes or does not spike, the value of the PSTH has no physical existence, nor does it have any causal influence. Again, the mean number of spikes observed over many trials on a short time window is not a state: it is a statistic.

Neural firing is not like radioactivity: spikes are not emitted randomly. Neurons spike when their membrane potential exceeds a precise threshold. The threshold might vary, but these are mostly history-dependent variations, not noise (Platkiewicz and Brette, 2011). A neuron that receives the same time-varying input repeatedly through a current electrode would fire at very similar times (Brette and Guigon, 2003; Mainen and Sejnowski, 1995). Of course, there is stochasticity in the cell. For example, an incoming spike triggers the release of a random quantity of neurotransmitter at the synapse. This contributes some variability in the spike train produced by the receiving cell, although this is reduced by synaptic integration, but it remains that the output is still a spike train, not a continuous variable. As long as the response of neurons depends on the actual timing of spikes, this output cannot be summarized by a firing probability.

Nevertheless, there is a strong tradition in theoretical neuroscience that attempts to justify the use of firing probabilities as states, because theoretical analysis is much simpler with continuous variables than with spikes. The most sophisticated approaches are inspired from statistical mechanics. In statistical mechanics, individual particles may follow deterministic laws (Newton laws),

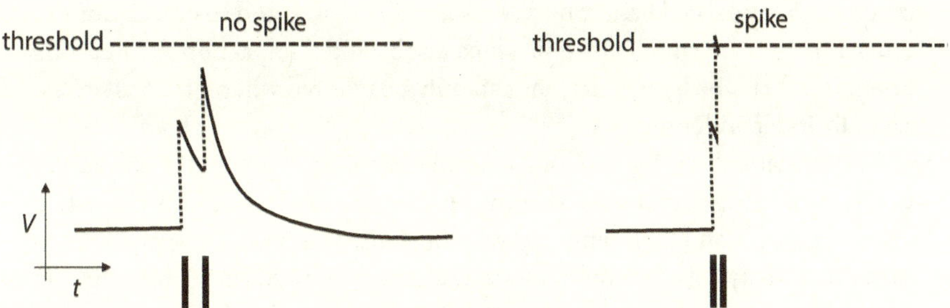

FIGURE 8.7. Coincidence detection: two spikes arriving simultaneously are more likely to trigger an output spike (right) than two spikes arriving at different times. (V is membrane potential.)

but if one assumes that their states are randomly distributed and independent, then it is possible to calculate equilibrium distributions and macroscopic laws. This works for gases, but neurons are not independent. (See Brette [2015] for a more detailed argument.)

Indeed, in theoretical work aiming at reducing spike-based models to rate-based models (Brunel, 2000), a key assumption is that neurons are randomly connected with low connection probability (sparse networks), which ensures that neurons are approximately independent. But empirically, the connectome does not have this property. For example, neurons are often bidirectionally connected or form clusters (Perin et al., 2011; Song et al., 2005). When neurons are not independent, the output firing rate is not a function of input firing rates anymore. A simple example is a neuron receiving spikes from two neurons within 1 second (figure 8.7). Each input neuron fires just one spike, thus, with a firing rate of 1 Hz. If the two spikes arrive at different times, the membrane potential does not reach threshold and no spike is triggered. If the two spikes arrive simultaneously, the threshold is exceeded is a spike is triggered. Thus, the output rate may be 0 or 1 Hz depending on the fine temporal relationships between the two input spike trains.

In more realistic cases, neurons receive spikes from thousands of neurons, but it remains that neurons are highly sensitive to fine correlations between their inputs (Rossant et al., 2011). It is possible to estimate this sensitivity to correlations quantitatively with a relatively simple technical argument (Brette, 2012). In the fluctuation-driven regime that characterizes neurons in vivo, the firing rate of a neuron depends on both the mean and variance of its inputs. The mean does not depend on correlations, but the variance scales with $N + cN^2$, where N is the number of inputs and c is their pairwise correlation (between 0 and 1). It follows that correlations tend to be the dominant term when they are of order $1/N$ or more. This is extremely small: for example, a cortical pyramidal cell has around $N = 10{,}000$ synapses, and so correlations of order 1/10,000 are sufficient to produce a substantial impact on the neuron's firing. In other words, correlations can only be ignored when neurons fire essentially independently.

In summary, the firing rate of a neuron cannot be expressed as a function of the rates of input neurons. The output of a neuron depends on individual incoming spikes and their temporal relationships. This makes it invalid to consider the "activity" of a neuron as a state: the activity of a neuron is what a neuron does (the spikes it produces), it is not a state. In particular, the "activation" of formal neurons (e.g., in deep learning models) does not correspond to any neural state.

At the biophysical level, neural networks are modeled as hybrid dynamical systems (Brette et al., 2007). These are particular kinds of dynamical systems that combine continuous dynamics and discrete changes. Individual neurons can be modeled with state variables such as membrane potential and ion channel opening, which are typically governed by differential equations. But interactions between neurons occur mainly through spikes—that is, when the membrane potential undergoes a strong fluctuation. The rate of those interaction events is what we call the activity of the neuron. It is not a state but a statistic of interaction events. It is an important statistic, since as a first approximation, it measures the energy spent by neurons (which is mainly related to synaptic transmission [Attwell and Laughlin, 2001]), but it is not the dynamical variable of the model.

However, it is possible to define state variables for neurons, such as their membrane potential. Membrane potentials are well defined and electrophysiologically measurable at any instant (with some caveats). But are those *computational* states?

States for Computation

One of the greatest philosophical challenges to explain what it means for a physical system to compute is to avoid triviality. For example, Putnam (1991) and Searle (1990) argued that virtually anything can be said to implement any computation, if we simply ask that there should be a mapping from physical states to states of a computational model. This difficulty has led Shagrir (2006) to propose that "to be a computer is not a matter of fact or discovery, but a matter of perspective." This is a rather paradoxical take, given how much effort and ingenuity it has taken to actually build computers. Alan Turing and John von Neumann are celebrated as geniuses for their insights that led to the development of computers, and yet, all they had to do was apparently just to look at a rock in the right way.

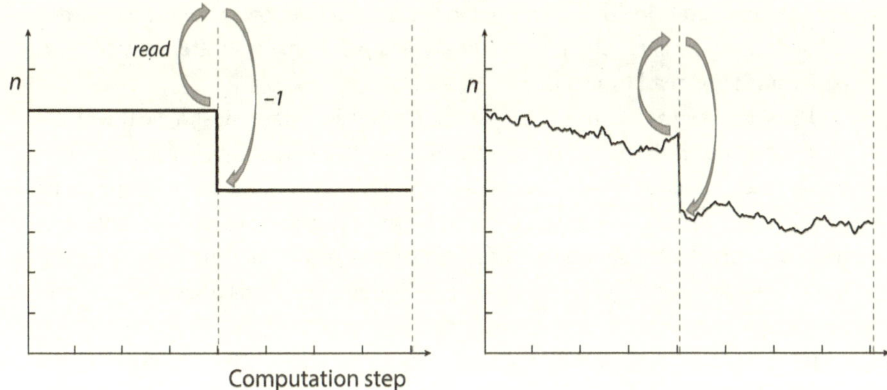

FIGURE 8.8. To serve as a computational state, a physical state must be stable between computations (left); otherwise the computation is not correct (right).

Let us take a step back and look at what problems people actually had to solve in order to build computing machines. Consider a simple algorithm: the calculation of the factorial of an integer *n* (the product of all numbers from 1 to *n*). It can be described as follows:

(1) m: = 1
(2) m: = m*n
(3) n: = n−1
(4) go to (2) unless n = 1
(5) return m

The instruction a: = b means that the value of b is assigned to the variable a. A system that implements the preceding computation would have a state defined by three values: m, n, and the line number. Once we know the state at a given time, we can predict how the computation will proceed—that is, the future state at any time. Thus, this is indeed a state of the computational model.

But these states have very particularly properties. Consider line 3, for example. In that line, the state gets updated in such a way that the new value equals the previous value minus one. For this to work, the physical state that stores the value of n must not change between explicit updates, which may happen at arbitrary time intervals. This means that the state must be stable by default, and only undergo discrete changes under specific conditions, at times of computations (figure 8.8, left).

This is not the case for all physical phenomena. In many cases, the physical state changes, often decays back to some thermodynamical equilibrium, and is subject to thermal noise (figure 8.8, right). This makes it impossible to carry out the computation. This is implicit in the notion that a computation is something that can be carried out *with a pen and paper*. What is written stays on the paper until it is read or modified.

Thus, to serve as a computational state, a physical state must have the properties of symbols on paper. It must be a stable state of the physical system, and updates are transitions between stable states. This is not a usual property of physical systems. The climate system or the solar system do not have this property. An electrical circuit made of resistors and capacitors does not have this property. Crucially, the membrane potential of neurons does not have this property.

It is because computation requires very particular kinds of physical states that it took much effort to actually build computers. It was not sufficient to simply look at any physical system and call it a computer. In early computing machines, such as Babbage's analytical engine (described in 1837), stable states were provided by the mechanical state of rigid objects, such as the discrete angle of a gear, with instructions encoded in the configuration of punched cards (Babbage, 2010). In the twentieth century, computers were built with electrical circuits, but unlike mechanical properties of rigid objects, electrical phenomena are typically transient. Thus, engineers designed electrical circuits with two stable states, the "flip-flop" circuit, by arranging transistors in pairs connected by a positive feedback loop. In this way, a binary computational state could be physically implemented—that is, a stable physical state that can be switched between two values. Before transistors, older computers used vacuum tubes, but the principle was the same: to make a bistable system so that the physical state of the circuit can act as a computational state.

The same holds for less conventional computers such as parallel computers and even quantum computers: computational states are encoded by stable states of the physical system, which is engineered in such a way as to exhibit such kinds of states. Quantum computers use quantum superpositions of binary states called qubits (quantum bits), and the technical challenge with quantum computers is precisely to make it so that qubits are stable. This is why it is actually not that easy to build a quantum computer.

As we have seen in chapter 2, living systems are out-of-equilibrium processes, they are generically unstable. The membrane potential of a neuron, for example, is highly fluctuating, and the state of neurons does not, in general, have the properties of flip-flop circuits. Of course, in a computer, the same could be said of the microscopic state of the transistors: atoms jitter, and their energy increases with temperature. The key to building a computer is to arrange those elements in a set of stable circuits, such that the macroscopic state of a circuit (on/off) depends only on the macroscopic state of other circuits, and not on their microscopic states. In this way, the computer can be consistently described in terms of stable macroscopic states—that is, computational states. But are neurons arranged in such a way?

Computational States in the Brain

When a neuron receives an action potential from another neuron, the membrane potential increases or decreases. This can be seen as a discrete transition in the state of the neuron. But the effect of this action is not stable: it typically decays within milliseconds. Therefore, membrane potentials are not computational states.

As we have seen, computational states can be built on top of dynamics. In the flip-flop circuit, the binary state is made of two attractors of a dynamical system. In the navigation system, one influential theory of head direction cells is that they form a "ring attractor" (Hulse and Jayaraman, 2020): the network is arranged with a topology of a ring, such that at any given time, cells placed about a particular position in the ring are continuously active in the absence of movement, and this stable position gets shifted by sensory input (figure 8.9). Much effort has indeed been devoted to demonstrating that the activity of head direction cells is best understood with dynamical concepts, as an attractor of the dynamical system constituted by the interacting neurons (Chaudhuri et al., 2019).

The analogy then is that neurons are like single transistors, and they are arranged in such a way as to form a system with attractors, so that they display stable states. In this way, computational states are built out of the dynamics of neurons.

However, although we can find stable circuits in the brain, the whole brain is not a collection of stable circuits, unlike a computer. In the computer, interactions occur only between equilibrium states of dynamical systems (flip-flop circuits), which is why the computer can be said to implement computations, while safely ignoring the microscopic dynamics of individual electronic components. This is not the case in the brain. First, neural phenomena can often be described in dynamical terms, but not always in terms of attractors (Vyas et al., 2020). More critically, neural interactions do not generally occur at equilibrium. Unlike in the computer, where the state of transistor is essentially piecewise constant at the timescale of interactions between circuits, in the brain there is no separation between the timescale of dynamics and the timescale of interactions. In fact, in many cases, the dynamics of neurons exhibits power-law dependence on time, meaning that there is no privileged timescale (Drew and Abbott, 2006).

Furthermore, neural attractors are usually described at a population level (such as the ring of neurons), but as we have seen in chapter 5, these are often

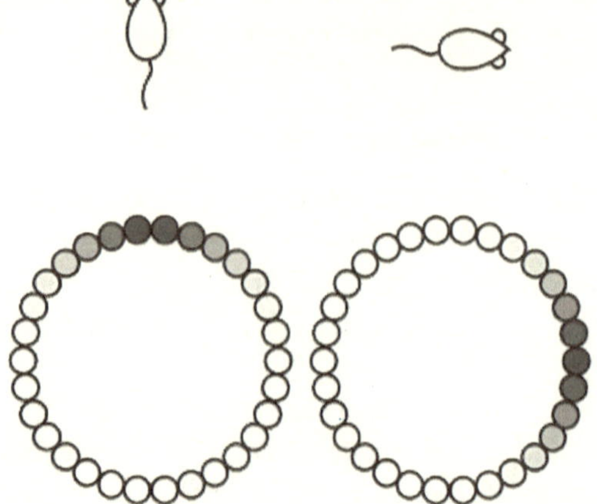

FIGURE 8.9. Ring attractor theory of head cells. The activity of head direction cells forms a stable bump, which shifts with the mouse's head position. Original figure by Wang C, Zhang K. 2019. https://doi.org/10.3389/fncom.2019.00096 / CC BY https://creativecommons.org/licenses/by/4.0/.

transient coalitions of neurons that depend on context (what the organism is doing). This implies that the state of individual neurons cannot be abstracted away, unlike the state of electronic components composing a flip-flop circuit. This is why neural populations do not form computational units, or units at all.

The impermeable boundary between dynamical and computational levels that characterizes computers corresponds to the hardware/software distinction in the computer metaphor. As we have seen in chapter 2, there is no such distinction in living systems. This is why theoretical neuroscientist Tony Bell (1999) argued that the brain is not a computer: "a computer is an intrinsically dualistic entity, with its physical set-up designed not to interfere with its logical set-up, which executes the computation. In empirical investigation, we find that the brain is not a dualistic entity."

Thus, states of the brain are not computational states. As we have seen, an alternative theoretical framework is the idea of the brain as a control system. However, brain states are not control states either.

States for Control

As we have seen in chapter 4, an alternative to the computational view of brain is the cybernetic concept of feedback control, which acknowledges the coupling relation between organism and environment (see *Information*

Reductionism, earlier in this chapter). Control systems can be computational systems, but they can also be dynamical systems with no computational states (van Gelder, 1995). This perspective has gained some momentum in neuroscience (Cisek, 2019). In psychology, Perceptual Control Theory (Powers, 1973) claims that behavior is the "control of perception": actions are chosen so as to control the perceptual state. However, if perception is an activity and not a thing or a number, then it is not so clear what "control of perception" might mean.

It is true that much behavior can be described in terms of control theory. For example, many animals actively stabilize their gaze during locomotion, by moving their eyes or their head. The characteristic head-bobbing movement of pigeons is such that the head remains stable while the pigeon walks, except at discrete instants, where the head abruptly moves forward (Friedman, 1975). Another example is the standing posture of humans (or in fact, pigeons), which is mechanically unstable, but actively controlled so that the person does not fall.

For an external observer, these behaviors are clear cases of feedback control, as described by control theory. There is a controlled variable, which is for example the angle of the body with the vertical, and a target value, which would be zero degree. When the body angle deviates from the vertical, a force is applied so as to reduce the difference between desired and actual angle. The same might be said of visual stabilization, based for example on the position of some fixated object.

However, these measurements are made by the observer: when the measured body position is shifted from the vertical, it comes back to that initial position. The observed phenomenon is then called "control of the body position." The problem is that the exact same observation could be done on a cupboard (chapter 4). It is also the case that the cupboard's motion tends to reduce the error between target angle and actual angle, by applying an adequate torque. Yet, we would not claim that the cupboard controls its position. In both humans and cupboards, what we observe is a stable fixed point, or an attractor, of a dynamical system. But we call the phenomenon "control" only in the case of humans, because the stable point results from the action of the nervous system, in other words because there is an agent responsible for making that point stable.

What does this agent do? In control theory, as classically construed, the system essentially mirrors what the observer does: it makes measurements. Then on the basis on those measurements, the system takes actions that tend to reduce the difference between measurement and target value. What this means technically is that the system has a state variable in a fixed relation with the measured quantity: an encoding. The state variable can then be compared to a target value, which is in the same relation with the physical quantity. But what evidence do we have that the brain controls its body in this way?

From a subjective perspective, it does seem that the perceived visual scene is stable, or that the perceived body position is stable. This is why Powers describes behavior as the control of the perceptual state. But this terminology is rather misleading. As we have seen, a perceptual "state" is not a state at all in the sense of modeling, in particular in the sense of control theory. We might feel that the visual scene is stable, but the state of neurons is not stable at all. In fact, as we have just discussed, even the visual input to the retina is not stable, because of miniature eye movements. But these are not perceived, which fosters the illusion that the state of neurons is stable. Therefore, it is not clear at all what it might mean to control a "perceptual state." At least, this phenomenon does not have a simple relation with control theory.

At the root of this misunderstanding is the idea that sensory neurons might be engaged in measuring the environment—temperature, light or sound intensity, stretch, and so on. This idea is implicit both in the mainstream information processing view and in the control theory view. But neither neural states (e.g., membrane potentials) nor neural activity are measurements, that is, in a fixed relation with the measured quantity. Sensory neurons do not respond to constant signals by a constant state. For example, when we touch an object, mechanoreceptors in the skin produce spike trains, that is, highly fluctuating signals, while the pressure is constant. There is no constant state, and therefore no state that could act as a measurement of pressure. Similarly, when we hear a tone (a beep), we perceive a stimulus of constant pitch and loudness. Yet, neither the acoustical wave nor the state of auditory neurons is constant: neurons fire spikes.

One might object that at a larger temporal scale, the firing rate of sensory neurons might reflect the loudness of the sound. But even that is not true. Many sensory neurons adapt, which means that their firing rate decreases when a constant stimulus is applied, and more generally depends on the history of previous stimuli. For example, the firing rate of auditory nerve fibers responding to a sound of given intensity depends on the statistics of sounds heard before, in such a way that the firing rate varies broadly over that set of sounds (Wen et al., 2009). Clearly, neither the state of auditory nerve fibers nor their average activity can act as a measurement of sound intensity.

This fact raises considerable issues for both computational and control views of perception and behavior. Consider for example the problem of localizing sounds based on sound intensity differences between the two ears. Both computational and control views start from the premises that sound intensity is measured at both ears. These two measurements are subtracted, and then either sound direction is inferred from the result (computational model), or the head turns in the direction indicated by the sign of the result (feedback control model). In both cases, there must be a fixed, static relation between sound intensity and the measurement. If the auditory neurons adapt on each side (which is the case), neither of these models will work.

Furthermore, measurements have an additional crucial property. A thermometer gives a measurement of temperature not just because the reading varies systematically with temperature, but also because we expect to read 100°C when we immerse the thermometer in boiling water, at any time and with any thermometer. That is, we have agreed on a particular relation between temperature and the reading on the thermometer. Without this, we cannot compare the readings of two thermometers. In the same way, in order to compare sound intensity at both ears, there must be two states acting as measurements of sound intensity, that is, in the *same* relation with sound intensity. But it is implausible that auditory neurons on opposite sides of the brain respond in exactly the same way to the same sound presented at their corresponding ear, given that their properties result from development, not from an engineering plan.

Thus, the organism does not measure the physical dimension and compare it to a target value, as in control theory. Rather, the nervous system is coupled to the body and environment in such a way that the physical dimension is stable, without having a correspondingly stable state in the biological system. The same remark holds for actuators: a stable muscular contraction is maintained by spike trains produced by motoneurons; it does not correspond to a stable state of a motoneuron. Each spike triggers a particular temporal pattern of contraction, and these patterns interact nonlinearly.

We now turn to the relation between individual neurons and the brain.

The Nature of Multicellularity

Levels of Organization

In the computer metaphor, the brain is seen as a *parallel* computer, with its neurons acting as tiny individual processors. More generally, a machine is made of components that are assembled. There are usually hierarchical levels of organization: the machine is made of parts, and each part is made of components. In the modern electronic computer, these levels might be the whole computer, the processors, the flip-flop circuits, and the electronic components (e.g., transistors). In the same way, a common trope of modern neuroscience is that the brain can be modeled at different scales, and then the challenge is to "bridge scales." These scales are the whole brain, neural populations (as measured, e.g., in functional imaging), individual neurons, and molecules.

An implicit assumption is that each level can be modeled independently, and the upper level can be described as an assemblage of elements of the lower level. This is again the engineering bias: a machine is assembled from parts, each part has a specification in terms of how it may interact with other parts, and the logic of the machine may be functionally described without going beneath the level of parts—this is the concept of "implementation." As we have seen, this is what we have in an electronic computer: flip-flop circuits are built out of dynamical electrical components, and interact only at times when circuits have reached a stable state. As a result, microscopic states of distinct flip-flop circuits do not interfere: they are only coupled by their macroscopic binary states. It follows that a set of flip-flop circuits can be modeled as coupled binary elements. The microscopic detail of the flip-flop circuit—that is, the lower level—matters only if one wants to understand how the specifications of the circuit as a binary device are achieved.

However, the scales that brain modeling is supposed to bridge are scales of measurements, not levels of organization. What levels of organization actually exist in the brain?

An obvious answer is that neurons might be considered as the components of the brain. Indeed, histological work in the nineteenth century, in particular by Ramón y Cajal (Ramón y Cajal, 1899), has shown that the nervous system is made of individual cells. This is known as the *neuron doctrine*. There are exceptions, such as the squid giant axon, which is a syncytium resulting from the fusion of hundreds of axons (Young, 1936), but this is the general rule for the brains of most animals. Because cells are bounded by a membrane, coupling between cells is essentially mediated by interactions at the membrane. Thus, whatever is going on in the neuron could in principle be treated as a black box, and the interaction of neurons could still be modeled. In other words, in multicellular organisms, there are two clear levels of organization, the cell and the organism.

How are neurons coupled? To a large extent, this coupling is mediated by electrical events, action potentials or "spikes," and this coupling is directional. That is, a spike goes from a presynaptic neuron to a postsynaptic neuron (with a chemical intermediary), and never in the opposite direction. These electrical events are essentially stereotypical ("all or none"), although there are subtleties (Zbili et al., 2020). Finally, at the timescale of behavior—that is, of seconds—only electrical interactions between neurons are fast enough to make a difference at the organism scale, at least for animals. These remarks justify the view that a neuron can be modeled as mapping input events to output events, with no direct interaction between neurons beneath the level of events (Brette et al., 2007). There are important qualifications to this general picture, which we will discuss shortly. But as a first approximation, the electrical level of the cellular membrane seems to qualify as a genuine organization level. Are there levels below or above that level?

First, can there be higher levels of modeling, between the level of electrical events and the behavioral level? As we have already discussed, the answer is generally negative, because groups of neurons do not form units with well-defined states. A brain area does not communicate with another brain area through a fixed set of input and output slots, at least not at a coarser grain than the individual neurons. Rather, neurons from a brain area are flexibly coupled with some particular neurons in another brain area. Brain areas may be specialized to some extent, but in terms of coupling, the brain is fundamentally entangled (Pessoa, 2022). Brain areas are anatomical levels of description, not strictly hierarchical levels of organization like in machines.

What about levels below the cellular membrane? Are there intermediate levels between the lower level of molecular interactions and the higher level of the electrical membrane? In electrophysiological modeling, this is what the Hodgkin-Huxley formalism suggests. It is known that individual ionic channels switch on and off stochastically as a function of membrane potential, but as they are independent, the law of large numbers applies and all microscopic ionic currents of the same type can be lumped into a single macroscopic current (say, the sodium current), which varies smoothly with the membrane potential. Therefore, it is possible to describe the electrical phenomenon in terms of macroscopic currents while ignoring microscopic currents, which are independent stochastic realizations of the macroscopic quantity. This establishes the level of macroscopic currents as an intermediate level between single molecules and the cellular membrane—although note that this is not an intermediate scale.

However, this is a modeling convenience that works only for some particular cellular phenomena and timescales, such as the propagation of an action potential across a homogeneous axon. Indeed, this description ignores chemical phenomena, which are local, unlike electrical phenomena. For example, in *Paramecium*, a mechanical stress on the membrane opens local mechanoreceptors, which produces a local ion flux through the membrane (Eckert, 1972). Since *Paramecium* is isopotential, this ion flux leads to a change in potential on the entire membrane, which may trigger an action potential and a reversal of ciliary beating (the "avoiding reaction"), as we have seen in chapter 2. However, the effect of the mechanical stress can depend on where it is applied. For example, a strong mechanical stress triggers the ejection of little needles called trichocysts in the direction of the stimulus, which is thought to be a defense mechanism (Knoll et al., 1991). Physiologically, this is mediated by a local calcium influx, which triggers a local molecular pathway leading to exocytosis (Plattner, 2020). Thus, this influx has simultaneously local and global effects.

Furthermore, the global effect (action potential) works by feedback from the global scale of the electrical membrane to the local scale of calcium channels in the cilia (a case of "downward causation," as discussed in chapter 2). In fact, we know that genetic phenomena involve single molecules (DNA strands), which implies that they are stochastic, while affecting the entire cell (see chapter 2). Molecular and membrane levels are entangled, not hierarchical, and there is no intermediate level (e.g., lumped currents).

Thus, there are no universal levels beneath the cell in the same sense as the computer has the level of connected components (processor, memory, hard drive, etc.) and then the level of logical gates and flip-flop circuits. Microscopic states of flip-flop circuits do not interact, except through the macroscopic binary states, but single molecules of the cell do. In summary, there are two distinct organization levels in the brain: the cell and the organism. This simply reflects the fact that multicellular organisms are organisms composed of autonomous living units.

Even then, these levels are not impermeable, unlike the organization levels of a machine. I have mentioned earlier that there are qualifications to the claim that neurons can be modeled as mapping input spikes to output spikes. For example, the circulation of current in a neuron generates an electrical field outside the membrane, and this can in turn affect neighboring neurons—these are called ephaptic interactions (Anastassiou et al., 2011). It is in fact this field that is measured with an electro-encephalogram (EEG). Neurons can also interact by direct coupling of their membrane through "gap junctions," essentially holes in the membrane. This coupling mediates ohmic currents, but also allows the transport of various molecules. Neurons also produce nitric oxide, a gas that then diffuses across membranes and binds to receptors inside other neurons, thus, not at the membrane (Prast and Philippu, 2001). This gas modulates neurotransmission, and therefore electrical function.

In contrast, levels of a machine are supposed to be independent. The engineer implements a logical scheme (the higher level) by using components (the lower level) with particular specifications. As far as the higher level is concerned, the components are entirely specified by their description. This strict separation between levels is a desirable engineering feature for cognitive and social reasons, as I discussed in the context of computational modularity. For cognitive reasons, because in order to build things from components, one must think of components abstractly, in terms of its interactions with other components. The specifications of one component should preexist to the machine being built, and the abstraction we make when we design should exhaust all the possible modes of interaction with other components. There is also a social aspect to the strict separation of levels, namely that if one person builds a machine from components made by other persons, for example electronic components produced by a factory, she needs to know how those components will interact, and generally she needs to know it in advance. This is made possible by building components in such a way that its properties can be specified and communicated, so that components cannot interact except through the specified variables. The specification forms an impermeable level that wraps the component. These cognitive and social aspects also subtend the hardware/software distinction in computers: this strict separation of levels allows different people to build computers and to write programs, and it allows running different programs without making a new machine.

This sociocognitive motivation does not exist within a living organism, since there is no designer, let alone a group of designers. There is no strict separation between levels, as Anthony Bell (1999) has pointed out, as he criticized

> the prevalent tendency to view biological organisms as machines in the exact technical sense in which computers are machines, i.e. in the sense that they are physical instantiations of finite models which do not permit physical interactions beneath the level of their machine parts (e.g. the logic gate) to influence their functionality.

Cells are living entities, not components. Behavior is not "implemented" with cells in the sense of engineering, any more than human history is implemented with people. A multicellular organism is made of cells, but not in the same sense that a machine is made of parts: cells are individuals that develop, they are not literally "made."

The Problem of Individuality

A specific epistemological issue in biology is the problem of individuality. When we study the movement of rigid objects in the air, the physical model has variables such as the position and speed of the center of mass, which can be mapped to observables of any such object. But in the case of brains, this is rarely the case. The neurons of a neural network model do not correspond to specific neurons of a person or animal—different people do not even have the same number of neurons. There are exceptions, such as models of the microscopic worm *C. elegans,* with its 302 identified neurons (within a given genetic line) (Izquierdo and Beer, 2016), or models of unicellular organisms such as *Paramecium* (Elices et al., 2023). But even then, the matching is only anatomical, for each individual has its own life history, which makes individuals different.

What then is the epistemological nature of a model's "neuron," if it does not correspond to any actual neuron?

A possible answer was provided by Chomsky in the context of linguistics: "A theory is not a taxonomy, an organization of lots of data. It is a rigorous description of a possible human language" (Chomsky, 2002). Thus, paraphrasing Chomsky, a brain model is a *possible brain*. This is what the strange terminology "biologically plausible model" means (why not just good model?): the model *could be* biologically realized.

But what is a "possible brain" exactly? The engineering mindset gives it a particular meaning: a possible brain is a brain that *you* can construct from a set of biological components, namely neurons. This is indeed the concept of implementation in computer science: an algorithm can be implemented with a given language by combining instructions from that language. It is in this sense that an algorithm can be "implemented with" neurons, or that the brain "computes with" neurons. In other words, "biological implementation" is the art of finding biological processes that can be assembled so as to execute the algorithm. This notion of implementation raises some obvious epistemological issues that I will address with a few examples.

Computing with Neurons

As we have seen in chapter 1, the binary neuron model of McCulloch and Pitts (McCulloch and Pitts, 1943) triggered considerable interest because they demonstrated that any logical function could be obtained by combining those formal neurons in the right way. In the same vein, modern formal neural networks, which replace binary activity by a graded number and the threshold by a nonlinear transformation, are universal approximators (Hornik et al., 1989), which means that it is possible to build a neural network that approximates any nonlinear function from n to m variables with an arbitrary precision.

Thus, one can "implement" any function with formal neurons. This makes it possible to compute with formal neurons. However, in this notion of implementation, neurons are understood as building blocks, elements that are assembled together according to an arbitrarily plan. This is the notion of program in the computer metaphor; the program is the set of parameters of the network (synaptic weights, thresholds). But there is no one to set the program, which is why the brain is not a computer, just like a pile of transistors is not a computer (Brette, 2022a), as we have seen in chapter 4. Thus, synaptic weights and other properties of neurons are not parameters but variables, subject to physiological and developmental processes. The status of "parameter" we give to various biological properties is only a heuristic distinction of timescales, not a distinction between what is constrained by the model and what is arbitrary.

A multicellular organism is not an assemblage of cells according to a plan: it is a result of development by division from a single cell. Thus, a possible brain is not an arbitrary assembly of neuron-like elements, but a possible result of biological development. Not all assemblages of neurons are in fact "biologically plausible" in this sense.

A common proposition to bypass this biological fact is that parameters are tuned by evolution.

Evolution Is Not Parameter Optimization

Ironically, evolution is often summoned to save the engineering perspective on life. That is, it is argued that evolution tunes model parameters, since evolution optimizes biological structures. With this perspective, the range of possible brains is exactly the range offered by universality theorems, namely: any computable mapping. Therefore, any assembly of neurons that achieves something useful is biologically plausible. That is the idea of the brain as a sort of computer tuned by evolution. This is a rather convenient take, since it saves much tedious empirical work.

But this is not true. Being a multicellular organism means that the organism develops by division from a single cell. All cells have the same genome. What exactly can evolution tune? If a mutation occurs in a gene, then the corresponding protein might change. This might change its properties, such as the electrical properties or the permeability of voltage-gated ionic channels, or it might change its localization, or possibly its expression. But this change will occur in all cells that express the protein. Evolution does not tune the properties of single cells individually. It could not tune synaptic weights, for example. This should be obvious since in general, even the number of cells and synapses differs between genetically identical individuals.

Mutations can also alter the expression of genes. For example, the spiking threshold of a neuron depends on the number of sodium channels in the axonal initial segment (Goethals and Brette, 2020). This number is not encoded in any specific region of the genome—it differs between cells anyway. The expression level of a protein is a dynamical systems property that depends on *transcription networks* (Alon, 2019). The expression of a gene can be modulated by a protein called *transcription factor* (as well as by various noncoding DNA regions), which itself is expressed from a gene that can be modulated, and so on. Transcription factors can also be regulated by various cellular events (mediated by second messengers such as calcium or cyclic nucleotides). The number of channels of a given type in a cell, when it is stable, is then a dynamical equilibrium of the transcription network, coupled to the cellular environment. This equilibrium depends on both the genome and the particular history of the cell—including a history of cellular differentiation; it is not encoded in any particular gene.

Cells of an animal are individuals with the same genome, and therefore mutations do not alter cells independently like components of a machine. Mutations also do not alter elements of the cells independently. For example, synapses of the same type are made of essentially the same proteins. Therefore, their expression cannot be independently altered by mutations. It follows that the strength of an individual synapse cannot be tuned by evolution. Synapses differ within a cell because of a different local environment and history (in particular, a history of electrical activity), not because of a different genetic code. Again, evolution does not tune the individual components of an organism. Rather, it alters developmental and physiological processes, resulting in changes in dynamic equilibria. These equilibria also depend on the environment. In the same way, the differences between individual cells of an animal are not due to differences between corresponding parts of the genome, but to constraints exerted by the genome on developmental processes, leading to heterogeneous development of the organism (see chapter 2).

With the engineering perspective, one sees component properties (e.g., cellular properties) as parameters, and the organism as an assembly of components with the right properties. But those parameters are not the knobs that evolution turns. They are dynamical variables that change on a timescale that might be slower than behavior, but much faster than evolution. In modeling terminology, these are slow variables, not parameters.

On the Concept of Parameter

Thus, a possible brain is not an arbitrary assembly of neural material. A possible brain is one that can plausibly develop and maintain itself. This changes the perspective on the concept of parameter. The elements that are fixed within the lifetime of an organism or of a cell are, to a first approximation, the coding sequences of proteins and their regulatory sequences, and those are shared between cells—even this view of fixed coding genes is being seriously challenged (Shapiro, 2022). The rest is not a "parameter" of the organism.

Let us examine the implications in the seminal Hodgkin-Huxley model of action potentials in the squid axon. This model has many parameters. It includes a sodium current and a potassium current, flowing through two types of ionic channels. To a first (rough) approximation, the parameters of the channels are linked to their structure, encoded in the corresponding genes. But the number of channels is not encoded anywhere. The axon morphology, which determines various electrical parameters, is not encoded anywhere either. Rather, morphology and channel numbers are dynamical variables resulting from a system of various kinds of interactions (chemical, electrical, and mechanical). Yet, successful action potential initiation and propagation relies on those variables having the right magnitude. The model does not explain how this comes to be the case. The remaining challenge, then, is to explain what kind of system allows regulating the number of channels in such a way that the axon remains functional in various circumstances (such as different cells, individuals, or species).

More generally, if the proper functionality of a multicellular model (such as a neural network) relies on the specific individual values of cellular parameters, then "biologically plausibility" requires explaining how those parameters may plausibly acquire the right values, given that they cannot be individually tuned through evolution. In connectionism, this is the focus of learning algorithms. Indeed, synaptic weights cannot be independently tuned by evolution, and therefore, in modern connectionism, they are tuned by learning processes, also called *synaptic plasticity* in neurophysiology. However, synapses are not the only plastic elements in a cell.

Everything Is Soft

Formal neural networks raised considerable interest when Rosenblatt introduced an algorithm to train a single-layer neural network, called a "perceptron," based on examples of correct input–output associations (Rosenblatt, 1962), and this interest was revived when techniques adapted to multilayer perceptrons were introduced (Rumelhart et al., 1986). Thus, learning processes are, reasonably, a major focus of neural network theory. However, in connectionism, only synaptic weights are considered plastic (and the activation threshold, which is formally treated as an extra weight). Connectionism is then a variation on the computational distinction between software and hardware: software is the set of synaptic weights, executing on a fixed hardware. The activation functions are fixed; sensory transduction ("encoding") is fixed; learning processes are fixed. It is indeed the core assumption of connectionism, the one that underlies the claim that "you are your connectome" (Seung, 2012), that what matters functionally is the set of connections between otherwise identical elements.

However, in a cell, everything is soft, not just synaptic strength. The expression of all proteins is regulated by transcription networks (Alon, 2019). Proteins such as ionic channels can be quickly altered as a function of activity by phosphorylation and other molecular processes (Levitan, 1994). The axonal initiation site, where action potentials initiate, can move, extend or shrink depending on activity (Grubb et al., 2011). Axonal delays can also change with activity, for example through changes in myelination (Fields, 2015). All of these changes have major impact on electrical function.

These are well-known neurobiological facts. Why then are they essentially ignored in connectionism? One reason can be traced back to the introduction of the binary neuron model of McCulloch and Pitts (1943). As we have seen in chapter 4, there were prior neuron models in the 1930s introduced by Rashevsky, and the main innovation of McCulloch and Pitts was to discard time. In this way, the state of a neuron becomes a function of the states of other neurons at the previous iteration. As there is no time in the model, the effect of a neuron's activity on a target neuron can be parameterized by a single parameter, the synaptic weight. And since neuron states are binary, excitability can be parameterized by a single parameter, the threshold. Thus, all structural changes, such as protein number and phosphorylation state at various locations in the cell, can be lumped into modifications of synaptic weights and threshold.

This is a result of the formalism, not an empirical finding. In reality, time exists and synaptic responses have kinetics. Structural changes at the synapse may result in changes in release probability, unitary current, short-term dynamics (whether the synapse facilitates or depresses). Structural changes on the dendrite where the synapse sits can result in changes in the timescale of the synaptic response. Structural changes in the axon may result in changes in propagation delays.

Modern artificial neurons have graded rather than binary states, but this then comes with an additional assumption, that all neurons have the same activation function—the function that maps the summed activity of input neurons to the neuron's activity. As we have seen in the preceding section, there is no such thing as an activation function, since there is no such thing as a graded state of activity: activity comes in spikes, and there are graded measurements of those spikes over some temporal window, not graded states. But even broadly construed, the mapping from input spikes to output spikes differs between neurons. Thus, either the network's functionality should be robust to such differences, or those differences should result from learning processes.

These facts impose strong constraints on the kind of learning processes that can be deemed biologically plausible. For example, most connectionist algorithms rely on the calculation of the gradient of the activation function relative to every synaptic weight. When we look in detail, we find that the learning process consists in a modification that is local, at a particular synapse, but it depends on the cell's "activation function," which depends in a precise way on structural properties distributed across the cell—the axonal initial segment, the cell body, all the dendritic pathway from cell body to the synapse, which are both synapse- and cell-specific. In other words, not only each cell but also each synapse must have its own learning process. Since all cells have the same genome, this means that the learning process at each synapse must itself be learned specifically. In contrast, the learning processes of deep neural networks are based on formal differentiation, that is, on prespecification based on the known structure and properties of the network. This is not biologically plausible. This remark comes in addition to the neglect of dynamical aspects in connectionism.

Thus, unlike machines, functional properties of an organism cannot rely on rigid specifications of its components. Beyond "synaptic strength," virtually everything in the cell is subject to regulation processes. The engineering perspective sees evolution as an optimization process that tunes the components of the living machine, but evolution cannot and does not tune cellular properties individually. Evolution is not an optimization perspective—optimization is on the contrary a finalist perspective. Evolution is a historical perspective: to each current organism corresponds a path of functional organisms that lead to it.

The Evolutionary Perspective on Multicellular Organisms

Darwin was inspired by the artificial selection of domesticated animals and plants. Thus, he described how animals evolve into other animals (or plants into plants). But a very important implication of evolutionary theory is that, since there were only unicellular organisms 2 billion years ago, every single cell of an animal must descend from a unicellular organism. As discussed in chapter 2, it is thought that all animals descend from a marine protist living in colonies, which evolved from optional multicellularity to systematic multicellularity. Thus, the evolutionary story of animals is not that of a tiny protist that has gradually morphed into a large organism, but rather of a colony of protists that has bound and then further differentiated.

What this means is that, contrary to the preconceptions of connectionism, individual cells such as neurons should not be conceived as elementary components with simple stereotypical input–output properties, but as organisms that live in community. These organisms are siblings, individuals with the same genome but their own histories, which interact with and adapt to their local environments. Action potentials are not symbols but activities of those living forms.

9

Living Brains

Brains Beyond Machines

Cognition as a Biological Phenomenon

Cognition is a biological phenomenon. Yet, most theoretical frameworks in neuroscience are not rooted in biological concepts but in engineering concepts, in particular computer science. Every aspect of behavior is considered as the result of computation. The brain itself, if not a computer, is at least an information-processing device, a machine that transforms data into actionable information. Functionalism, the mainstream doctrine of cognitive science, asserts that cognition is just computation and what matters for computation is just those aspects of biology that are machine-like. This is a stunning assertion, since the very concept of computation is a series of operations with a goal, according to which the computation may succeed or fail, and machines have no goal of their own. Cognition is about anticipation, and for this reason, the roots of cognition are to be found in the specificities of life, not the specificities of machines.

The machine view of life is so ingrained in scientific culture that it seems very difficult for many scientists to imagine that living organisms could be something else than machines, or that brains could be something else than computers—"what else could it be?" seems to be a valid scientific objection (van Gelder, 1995). As I pointed out in chapter 1, molecular biologists and cognitive scientists appear to have replaced God with the engineer. The Church has disappeared, but otherwise there is a stunning intellectual continuity: God was a watchmaker, and now life is clockwork. Is there really nothing in the world apart from human artifacts and magic? Couldn't life be its own thing, rather than a kind of clock?

There is no doubt that engineering concepts and analogies can sometimes be useful (Sterling and Laughlin, 2017). But those have limitations, since living organisms are not engineered. In the same way, it can be useful to liken an atom to a solar system, with electrons orbiting around the nucleus. But particles are not actually solid spheres in motion, and eventually the mechanistic view of physics of the seventeenth century had to be abandoned. In 1690, Huygens insisted that light, just like all natural phenomena, must be explained mechanically, because otherwise we should give up any hope of ever understanding physics (Huygens, 1920). He was wrong. Yes, quantum physics is hard and counterintuitive, because microscopic objects do not behave like the macroscopic objects of our daily experience. But it explains the microscopic world, while classical physics does not.

It appears that molecular biology and cognitive science have entirely skipped the intellectual revolutions of electromagnetism in the nineteenth century and of quantum physics in the early twentieth century and are still entrenched in Huygens's seventeenth-century mechanistic view of the world. In order to understand life and brains as such, rather than as some kinds of engineered devices, we must first reject the injunction of computational orthodoxy, which declares a priori that biology is essentially irrelevant to understand cognition. Let us first study biology and only then declare what features are relevant.

What Is Life?

What do we know today about living organisms, in particular animals? First, a living organism is not a *thing*. It has no material persistence. It is an organization of out-of-equilibrium processes that keeps itself running by exchanging energy and matter with the environment. Thus, like machines, it is an organized system—that is, an arrangement of parts that achieves something. But unlike machines, the parts are changing, have no fixed specifications, and have no existence prior to the organism—they develop. This implies that cataloguing the parts and their properties is not a valid modeling approach, since the properties of parts depend on the integrated level (the "whole"). Such catalogues may be useful resources, but they cannot be used by themselves to model the brain, let alone understand it. In other words, "bottom-up" approaches cannot work, precisely because living organisms are not really machines. The structure of living organisms, and brains in particular, must be understood as a result of development, not assembly.

Furthermore, living organisms are fundamentally coupled with the environment, as a condition for their existence. This means that the right level of systems modeling is the brain, body and environment, which are coupled. The brain is organized with respect to the body and environment; without body and environment, the brain is only structured, not organized. The brain cannot be understood as an isolated physical object, independently of what it contributes to the life of the organism (that is, the "top" of top-down approaches).

The life of the organism is an empirical constraint that is just as much valid as the empirical constraints of structural measurements.

Living organisms are constitutively precarious, they must harvest energy in their environment in order to continue existing as out-of-equilibrium processes. This is why living organizations are fundamentally goal-driven and anticipatory. More precisely, a living organization is anticipatory because the correct output of the processes is a condition for the future continuation of those same processes. This is the primary form of anticipation upon which more sophisticated anticipatory phenomena are based. Machines, in contrast, have no goals: their engineers do—whether a calculator outputs the expected result or not makes no difference to the machine itself.

A living organism is autonomous. This means two things. First, each living organism is an individual with its own organization, its own way of harnessing energy in the environment. This organization, this individual logic, changes through life and evolution. It is indeed a condition of evolution that a structural change can be something else than an error, in contrast with machines specified by plans. This implies in particular that brain models cannot be based on statistical measurements of different individuals, because this destroys organization—indeed, an average brain cannot have memories, skills or personality.

Second, the relation between environment and living organism is not one of command. Behavior is made of actions, not reactions: actions are taken as a function of their expected consequences, in relation with the individual logic of the organism—for example, whether the action is expected to provide food to the organism. Therefore, it is a mistake to interpret all neural activity unrelated to the experimenter's stimulus as "noise." Brains and organisms are active by default, and they are coupled to the environment. Sensory coupling can change that activity, but it does not drive it. The very idea of a "receptive field" of neurons is misleading, or the idea that neurons are "information-processing devices" mapping an input to an output. Indeed, to make this terminology consistent with empirical facts, neuroscientists had to invent "extra-classical effects" (outside the receptive field), "top-down modulation" (changed by context, including the organism's goal), and "representational drift" (changing over time). Once all these qualifications are taken into account, we find that neurons are coupled to the environment, they do not respond to it. The primacy of the stimulus is inherited from behaviorism, which sees the brain as a tunable input–output machine, but despite the "cognitive revolution" that was meant to overthrow behaviorism, it is fully (and mistakenly) preserved in neurocomputationalism.

For the same reason, perception is not "information processing," a mapping from input data to symbols. For example, there are no images prior to their processing, like in artificial neural networks that detect objects. Visual perception results from active vibratory scanning of the optical environment, that is, it is an activity of the coupled brain–environment system.

Biological autonomy is another reason why the computer is a poor analogy of the brain. A computer is a programmable machine. Not only is the brain not a machine, but neither animals nor humans are programmable in any sense. Animals cannot follow arbitrary instructions; they can be trained or guided. Humans of a certain age and education can follow arbitrary instructions, but will they? Humans can follow instructions, machines will. This is indeed what the phrase "I am not a machine" means in ordinary language. Genes are not instructions either, but more like boundary conditions: they strongly influence or constrain the process of development, but as do other factors. The body's shape cannot be formally derived from the sequence of nucleotides, unlike the output of an algorithm.

What Is an Animal?

An animal is made of cells, not in the sense that a machine is made of parts, but rather in the sense that a society is made of people. Cells are living entities, and the animal develops by successive divisions and differentiation from a single cell. In the course of development, this colony of cells builds a local environment that we call the body of the animal, which also typically hosts other symbiotic life-forms, such as gut microbiota.

Animals have not evolved by progressive complexification of unicellular organisms. Rather, colonial unicellular organisms (protists) have progressively evolved transient multicellular stages that then became permanent. In other words, the animal is a colony of protists that has settled. This means that cells are genetically identical individuals living within a collective organization. The relations between these cellular individuals are not encoded in the genome, and indeed if some external condition changes, or because of the stochasticity of normal development, cells can organize differently in a way that often remains functional. Since cells are individual biological entities, they are also autonomous, they have an individual history, and their structural properties depend on the environment they grow in, which is the organism. Thus, cells are very unlike the parts of a machine, or the components of a computer.

A crucial consequence is that the properties of neurons are not formally specified. This applies not just to "synaptic weights" (an abstraction with only vague biophysical counterpart), but to the entire structure of the neuron. The distinction between hardware and software is a crucial feature of computers, since it is what gives meaning to the very notion of a program, but this distinction does not exist in biological organisms. Without formal specification, there cannot be computation. Indeed, the output of a computation is formally defined, by (a small set of) preexisting rules defined on symbols. Without formal rules, instructions do not correspond to a computation but to a recipe, whose output is empirically defined, not formally defined. Thus, neurons do not compute because their actions do not occur within a formal space. It follows that the activity of the brain cannot reasonably be framed as distributed computation.

This is a critical feature that lies at the core of modern artificial neural networks. Indeed, their learning algorithms (backpropagation and variations) are based on the fact that the models are differentiable computations, so that the effect of a potential change in synaptic weight can be precisely computed, rather than just observed. This cannot work with biological neurons. Computations can in some cases be differentiated, because they are formally defined; recipes cannot be differentiated.

Neurons are not formally specified because the brain is not engineered. It is this formal aspect of engineering that is mistakenly taken on board by neurocomputationalism, and lies at the core of many of its conceptual issues.

Against Reverse Engineering

When neuroscientists try to interpret the brain's activity as a form of computation, or as a form of prediction, or as information, the methodological approach is to "reverse engineer" the brain—that is, to find the formal specification that the machine realizes. This is the standard concept of "implementation": the activity of neurons is mapped to symbols (the neural code), such that brain processes map to symbolic rules. These rules are the formal specification of the machine.

As we have seen in detail in chapter 5, there are many issues with neural codes, both empirical and conceptual. In particular, many empirical findings concur to demonstrate that the activity of neurons is not formally specified at all. The concept of a mapping from neural activity to a formal symbol can be retained only if it depends on both context ("top-down modulation") and time ("adaptation," "representational drift"), and allows for some variability ("neural noise"); but then that is not a mapping at all.

More deeply, the core issue with the concept of implementation is that it is an interpretative framework. It is the observer who comes up with an interpretation of what the brain must be doing, of what the brain's activity must *mean*. But this is a reverse-engineering methodology that can only work if the brain is indeed engineered, if each component is there for a *reason*. As we have seen repeatedly, it does not work with brains. In chapter 3, we have seen that virtually any connectome can be interpreted as implementing any information-processing task, by simply choosing an interpretation for its inputs and outputs. In chapter 4, we have seen that a pile of paper with a switch can be interpreted as calculating noncomputable mappings. In chapter 5, we have seen that the activity of the retina can be interpreted as representing the identity of faces—since there is a mapping from the retinal image to face identity. In chapter 6, we have seen that this leads to an engineering view of information as what can be *potentially* decoded by an external observer, with the paradoxical property that cognitive processes can only degrade information (this is known as the data processing inequality). In chapter 7, we have seen that the input of any calculation can be interpreted as a prediction of its output—the formal mapping being precisely that input–output mapping. All of this is possible under the concept of implementation because the formal mapping from neural activity to symbols is not realized in the observed brain, but in the observer's mind.

Thus, the implementation model, based on encodings, cannot coherently explain what it means for the brain to compute; it cannot give a coherent account of mental representation; it cannot explain how perception can provide information about the world; it cannot explain how organisms anticipate.

This is not so much of an issue for reverse engineering actual machines, because the engineer would not allow arbitrary mappings from the machine to its formal specification, but *reasonable* mappings. For example, she would choose the same symbols for inputs and outputs; she would give a single function and fixed properties to any single spatially delimited component; she would choose simple mappings that she can easily describe with words, and such that the user can input symbols with simple actions and easily read output symbols. All of this is implicit when we try to understand how a machine works. But none of this applies to brains, which are not designed for someone else's use, which are not assembled from a plan, and whose main structural logic is "whatever works."

The implementation model of cognition does not work because there is no formal specification of the brain, and no inner interpreter of neural activity as symbols. How then are we to make sense of the brain's activity?

Embodiment

Again, we must remind ourselves that cognition is a biological phenomenon, not a formal procedure. To be sure, humans have designed formal procedures that attempt to mimic cognitive processes, but in any case, the primary phenomenon is biological. Many authors have argued that cognition cannot be understood independently of the body, or even that the body contributes to cognitive phenomena: cognition is embodied (Chemero, 2011; Gallagher,

2006; Pfeifer and Bongard, 2006; Zahnoun, 2023). But although this is now a familiar theme in cognitive science and philosophy of mind, it has had very little influence on neuroscience, in particular modeling.

What remains underappreciated is that considering cognition as embodied solves the problem with the implementation model of the brain. If cognition is conceived not as a formal procedure occurring in an imaginary symbolic space but as a phenomenon of the brain–body–environment system, then there is no need for interpretation anymore. For example, instead of looking for neurons whose activity can be interpreted as a computational state storing the direction of a sound source—which does not exist, we ask how the nervous system contributes to orienting the organism toward a sound source.

Of course, this is a difficult scientific task, because the object of the model is now extended from the sensory neurons that fire when a stimulus is presented to the subset of the nervous system involved in the behavior, together with relevant parts of the body (such as muscles). But the reason for this difficulty is the same reason why trying to make sense of neural activity by some kind of symbolic interpretation is invalid: what a part of the brain does depends on the broader system in which it is embedded, and indeed that system might be complex, and it might not conform with engineering prejudices. If the coupling between brain, body, and environment is indeed complex, if the organization of the nervous system is indeed entangled rather than linear, then these are precisely reasons to dismiss the interpretative methodology and adopt a systems view of brain and cognition.

Clearly, engineering concepts should be used with care when applied to biology, since living organisms are not engineered. The theme of embodied cognition points toward a systems view of brain, cognition, and behavior, in contrast with the cognitive reductionism of neurocomputationalism (all cognition can be reduced to elementary computations). But what kind of systems are cognitive biological systems?

Cognitive Biological Systems

Dynamical Systems

A defining feature of biological systems is that they are organizations of out-of-equilibrium processes, which must actively maintain themselves. A biological system is not stable like a rock is stable. Its material content changes, but some features of its organization may remain. Many aspects of biological organisms are not stable but cyclic, such as the cell cycle. But there is a form of stability, in that the organism often returns to its usual cycle after a perturbation. These are concepts of dynamical systems theory: stable cycles, attractors, and so on.

Thinking about life requires a change in perspective from substance metaphysics to process metaphysics (Bickhard, 2009, 2024; Jaeger and Monk, 2015; Nicholson and Dupre, 2018). Computation is based on substance metaphysics. A computational state is a property that is fixed by default, until it is explicitly modified through the algorithm. This is a key property that allows the algorithm to store and retrieve values, a prerequisite for calculations. Living organisms, and brains in particular, follow the exact opposite logic. As we have seen in chapter 8, the activity of neurons is not a computational state; it is indeed an activity: a storm of electrical events triggering various chemical changes. In some cases, attractors or cycles can form in the cooperative activity of neurons. Thus, a form of stability can arise from the interaction of processes, but this is not the default.

To build an electronic computer, stable states are constructed out of dynamical systems (flip-flop circuits), such that the dynamical nature of the underlying hardware can be entirely ignored: computation occurs as transitions between computational states, entirely shielded from the dynamics of electrons. Such shielding does not exist in brains. Thus, biological cognition cannot be reduced to elementary computations, supposedly implemented by neurons. Rather, computation is an elaborate form of cognition.

Of course, in mathematical terminology, computation is a (certain kind of) discrete dynamical system. But what this means is precisely that the mathematical "time" has little to do with time, but with the number of iterations of a formal operation. The same computation occurs whether it runs fast or slow, but you cannot interact with another dynamical system, such as the body and environment, unless the time of the cognitive system is the actual physical time.

This difference between computation and dynamics has been appreciated by a number of authors in cognitive science (Beer and Gallagher, 1992; van Gelder, 1995, 1998; Juarrero, 2002). In fact, in theoretical neuroscience, dynamical concepts are familiar conceptual tools, as we have seen in chapter 5. Unfortunately, the tension between dynamical and computational concepts often goes unrecognized. This is due to a loose use of computational terminology, where a phenomenon is deemed "computational" if it appears to be involved in solving a problem. But in that broad sense, "computation" does not imply computational representations or states, and certainly not algorithms. On the contrary, what empirical research has shown is that neurons can form transient context-dependent coalitions, not computational units shielding the presumed algorithmic logic from the underlying activity of neurons.

One symptom of this confusion is that in many dynamical models of the brain, the descriptive variable that undergoes dynamical changes is itself the activity of neurons (the firing rate), which is already a measurement of dynamical events. To a first approximation, nervous systems are special kinds of dynamical systems, namely continuous dynamical systems that interact through timed events (spikes). What we call "neural activity" is a description of these events, which mediate coupling between cellular dynamical systems. Neural activity is not the object of dynamics but a statistical description of coupling events.

But cognition is not just dynamics. The climate system, for example, is not cognitive. Biological dynamical systems are special in another crucial way: they are anticipatory.

Anticipatory Processes

In noncomputational accounts of cognition, the classical paradigm is feedback control, as we have seen in chapter 4 (Pezzulo and Cisek, 2016; Powers, 1973; van Gelder, 1995). Control is indeed an appealing conceptual framework, since it is about goal-driven dynamics and benefits from a large theoretical literature. There is no doubt that control theory can be useful to analyze some behavior, but it also suffers from important limitations.

First, feedback control is just one case of anticipatory phenomena, as we have seen in chapters 4 and 7. Some interactions with the environment are not based on error correction, such as Ariadne's thread; many are exploratory, based on trial and error. Much behavior is not based on interaction but on anticipatory association, such as when taking an umbrella on a cloudy day. Anticipation is rich and diverse. Broadly, anticipatory systems are systems that exploit regularities.

The control paradigm is also restrictive in a more subtle way. Just like computation, control is an engineering concept, and this comes with a conceptual burden. Remember the cupboard: when it is perturbed within a certain range, it comes back to its original position. Mathematically, it could be formalized in terms of control theory, with a target position and an error that gets reduced by the cupboard's dynamics. But this is not a classic case of control because no measurement is involved. Nothing within the cupboard measures its own angle, and the discrepancy with the target angle; rather, it is the coupling between cupboard and environment (in particular, gravity) that depends on the angle, a relational property of the cupboard-environment system.

When one tries to keep a horizontal arm, for example, both motoneurons and proprioceptive neurons are continuously active. Thus, while the body is in a stable configuration, the membrane potential of neurons fluctuates all the time. The neurons are not in a stable state that would correspond to a measurement of position or error. It is the coupled system consisting of the nervous system and body (and gravity) that is such that the body is stable, and it does not necessarily involve minimizing a physiological quantity. It is convenient for the engineer to separate the measurement system, which explicitly represents the distance of the system to the target, from the actuation system, which acts so as to reduce that distance. But there is no reason why this convenience should apply to a biological system, as a long as the coupled system has the desired property of being stable in the relevant dimension. Thus, Powers's claim that cognition is the "control of perception" (Powers, 1973) is very misleading. Perception is not a state in the sense of a measurement, but an activity, and therefore it cannot be controlled in the sense of control theory.

We meet again the concept of embodiment. Anticipation is not a property of an isolated system. It is realized in the coupling of the system with its environment. This does not require targets and errors to be encoded in the biological system. In particular, it does not require predictions, a highly problematic concept when applied to entities that do not speak, like neurons. As we have seen in chapter 7, anticipatory phenomena are very diverse. Some involve feedback, others involve exploration, association, continuation, or planning. Ultimately, anticipation is grounded in the self-sustaining nature of the living organization—processes whose proper outputs condition the future existence of those same processes. Very rarely, anticipation relies on prediction.

Thus, biological systems are dynamical systems that display embodied anticipation, realized in their interaction with the environment. Cognition appears in the degrees of sophistication and specialization of anticipatory phenomena, which must be associated with learning processes.

This characterization applies to all living organisms, but cognition is traditionally assigned to animals. What is special about animals?

Multicellularity: Cognition as Collective Behavior

A blind spot of both classical computationalism and dynamicism is the multicellular nature of animal cognition. Connectionism acknowledges that brain function is due to the interaction of many neurons, but it sees neurons as simple input–output devices rather than living entities. As we have seen, neurons are not stereotypical devices but living individuals, even though they have the same genome. In particular, they are not formally specified, they are not assembled according to a plan, and they are themselves anticipatory systems. How can we then understand the relation between neurons and the brain, if not as components of a machine?

Paul Cisek has defended the usefulness of taking an evolutionary perspective on animal cognition (Cisek, 2019). Here I will take it a step further and consider the transition from unicellularity to multicellularity. As we have seen in chapter 2, the last eukaryotic common ancestor (LECA) was probably a marine protist with a flagellum, which displayed many of the cellular structures and metabolic pathways found in today's eukaryotes—both protists and animals (Koumandou et al., 2013). That protist must have been able to survive in the fluctuating environment of the sea, to feed, avoid dangers and so on. A likely scenario for the transition to animal multicellularity is that colonies of protists transitioned from optional multicellularity to permanent multicellularity. Thus, animals are not sophisticated protists. Rather, they are somehow *made of* protists. Cells in general and neurons in particular are not at all like simple input–output devices: they are autonomous life-forms that have adopted a colonial sedentary lifestyle within an environment that they have built—namely, the body.

This gives us a different perspective on animal behavior: it is a case of collective behavior. This perspective differs in crucial ways from the idea of the brain as a parallel computer with neurons implementing distributed computations.

First, neurons are not elementary input–output devices but living individuals. What they do is not formally specified and it changes over time. Brain function cannot rely on a formal specification of its components, as in distributed computation. The electrical activity of neurons, that is, spikes, is not a computational variable. It cannot be, first because spikes are events, not states, and second because there is no formally specified "format" for such variables. Rather, spikes are actions on other neurons. This makes brain function not a formal procedure but an interactive activity.

These individuals are themselves biological systems, which are anticipatory, adaptive, and autonomous. This is known from the literature on protist behavior. For example, *Paramecium*, a well-known ciliate, has a rich behavior (Brette, 2021). It is an excitable cell that produces action potentials under various circumstances. The action potential then triggers an *avoiding reaction*, where the ciliate swims backward and then changes direction and swims forward again. Using this avoiding reaction, it can avoid obstacles, locate sources of food as well as mates by chemotaxis, regulate its temperature by thermotaxis, attach to surfaces. This behavior also relies on spontaneous activity: in the absence of stimulus, *Paramecium* also occasionally displays avoiding reactions due to membrane potential fluctuations (Moolenaar et al., 1976), and this property determines its exploratory behavior. Both its physiology and behavior change with circadian (~24 h) (Hasegawa et al., 1997) and ultradian (~1 h) (Kippert, 1996) rhythms. Thus, it has an autonomous activity with an internal organization, as well as goal-directed interactions with the environment. Remarkably, these abilities adapt to environmental changes. For example, a change in the ionic composition of the swimming medium shifts the membrane potential, which can strongly perturb the swimming behavior, but after a couple of hours, *Paramecium* swims normally in the new medium, thanks to an appropriate regulation of membrane permeabilities that restores the resting potential (Oka et al., 1986). When left for a couple of hours in a heated medium, its physiology shifts so as to recover normal behavior and most remarkably, the new temperature becomes the preferred temperature (Nakaoka et al., 1982). This behavioral adaptation is accompanied by electrophysiological changes (Martinac and Machemer, 1984). A paramecium trapped in the dead-end of a narrow capillary can escape by altering its normal avoiding reaction: after a minute, the ciliate swims backward for an unusually long duration (Kunita et al., 2014). There is even a reported case of classical conditioning, where *Paramecium* is trained to react to sound by an avoiding reaction, by pairing sound with an electrical shock (Hennessey et al., 1979). Another well-known ciliate, *Stentor*, shows habituation to mechanical stimulation with many of its classically known features (Wood, 1973). All these capacities are robust since they continue working as the cell grows and divides. Such adaptive abilities are also found in other species including bacteria (Lyon, 2015).

Of course, this behavior is much less sophisticated than animal behavior. In particular, sensorimotor behavior remains rudimentary. Many protists are photosensitive, but none can visually recognize individuals. No protist can perform dead reckoning. Yet, it is behavior: autonomous, goal-directed, adaptive, individual, interactive. A striking observation is that the single cell is not an elementary processor, performing simple formal operations like additions: it is an elementary adaptive system.

This distinguishes the collective behavior perspective from both classical computationalism and connectionism. In contrast with computationalism, elementary cognition is not elementary computation. The cellular basis of cognition, which I shall call *proto-cognition*, is not formal but dynamic and exploratory. The anticipatory and adaptive properties appear to precede the formal properties of computation, which might emerge from the collective activity of cells. Anticipation is not based on computation—it is the converse. In contrast with connectionism, the elementary components of the brain are not simple calculators, but (relatively) simple adaptive systems. This makes it impossible to see cognition as a distributed formal procedure, such as formal differentiation. In addition, the fact that there are many species of protists with diverse behavior shows that the radical emphasis on the connectome is exaggerated. Surely, the specific way individuals interact is very important in collective behavior, but what it is that interacts obviously also matters.

Second, this perspective also casts a new light on how cognitive work is distributed among neurons and brain areas. In collective behavior, individuals and local groups may specialize. They may be exposed to certain aspects of the world and interact preferentially with certain groups of individuals. But it does not mean that they implement a fixed operation, like a part of a machine in charge of a specific operation. Remember from chapter 5 that properties of neurons in the primary visual cortex depend on behavioral context (Gilbert and Li, 2013). Their activity appears to depend specifically on those object properties that are relevant to that task. This contradicts the idea of neural activity encoding some property of the world (the "tuning curve"), but it makes perfect sense if this activity is understood as the collective behavior of neurons engaged in solving visual problems.

Collective behavior allows some differentiation of individuals, but not necessarily a strict functional modularity. Thus, the same brain areas can be involved in various cognitive tasks, including evolutionarily new ones such as reading (a property called "neural reuse" [Anderson, 2010]), and generally any cognitive task engages a large web of brain areas (Pessoa, 2022).

Third, unlike behaviorism, cognition as collective behavior allows for internal states, or rather internal activity, to play a large role in cognition. In fact, in this respect, both classical computationalism and connectionism did not succeed in parting with behaviorism's primacy of the stimulus, despite the initial motivation of the cognitive revolution. Indeed, in both frameworks, perception remains a transformation of sense data to mental states, identified to machine states, and cognition is seen as "information processing," with a clear directionality from stimulus to mental states and commands (Hurley's "classical sandwich" [Hurley, 2001]). There is no spontaneous activity in either computationalism or connectionism that is not noise. Yet, the electrical activity of the brain is known to be organized by endogenous rhythms at various timescales (Buzsáki, 2011, 2019). Again, this does not fit well with computationalism, which interprets brain oscillations as epiphenomenal, but it makes sense as a sign of the coordination of cellular behavior.

Fourth, unlike the parts of a machine, the individuals that contribute to collective behavior develop and self-organize. This shifts the focus from the structure of the adult neural network (i.e., the connectome and other catalogues of parts) to the organizing processes that take place over the course of development and beyond. As a rule of thumb, since all neurons have the same genome, any structural parameter that differs between neurons (or perhaps, between neuron types), must result from developmental processes rather than genomic specification. Any stable cellular property corresponding to a structure subject to turn-over, that is, most cellular components apart from DNA, must be an attractor of an underlying process. It is the logic of these processes that underlies the collective organization.

This perspective leaves space for theoretical approaches that are deemphasized in the computational perspective. These are, for example, approaches inspired from statistical physics, where collective phenomena emerge through phase transitions—like when a solid turns into liquid, and suddenly follows new macroscopic laws. Other approaches include Darwinian or more generally exploratory processes, as we have mentioned in chapter 4, where a problem gets solved through an undetermined path rather than by a series of formal operations. Others might focus on the formation of transient coalitions of neurons or rhythmic activity, as we have mentioned in chapter 5. None of these approaches can exist if cognition is just the application of formal rules on symbols, or a series of matrix products.

Is there anything special about nervous systems, compared to other organs? After all, the liver is also a collection of interacting cells. Apart from its coupling with motor and sensory systems, the nervous system does obviously have something special: nerves. Axons and dendrites allow neurons to interact with specific cells at distant places, in a coordinated way—the same spike is propagated toward the axonal targets of a given neuron. In contrast, cells in other parts of the organism interact mostly locally and not very specifically. In the brain, it is thought that the neural network has a "small-world" structure (Liao et al., 2017)—that is, that one can go from one neuron to any other one by passing through a small number of synaptic connections. This would be impossible if the brain were a paving of spherical cells interacting with their neighbors. This network property allows brain-wide coordination—that is, it allows global brain properties to immediately feedback on individual neurons (in other words, downward causation, as we have seen in chapters 2 and 3).

Outlook

In neurocomputational theory, living organisms are machines and the brain is a computer. There is something to these engineering analogies, but they have been radicalized to an unreasonable extent. Cognitivism explicitly claims that biology is irrelevant to cognition, and neurocomputationalism sees biology as "implementation" of cognitive algorithms—that is, as an afterthought. Such dramatic claims are not the conclusion of a careful study of biology: they are theoretical speculations on the nature of cognition, a biological phenomenon. When we do pay attention to biology, we find tremendous differences with engineering concepts. Not just structural detail—the fact that there are dendrites and so on—but fundamental conceptual differences in the way neurons work, interact with each other, and are organized.

How much do we understand about brains? The truth is we know little about how to relate mental activity with the activity of neurons. The neurocomputationalist paradigm consists in mapping neural activity to a formal symbolic space (encodingism), but this approach is deeply flawed, because the exact same activity can be interpreted in an infinite number of ways. It is flawed to the extent that it cannot provide a coherent account of computation, mental representation, anticipation (incorrectly framed as prediction), or information (incorrectly defined as what can be potentially decoded by an observer). Embodiment sidesteps this issue by shifting the focus from cognition to cognitive behavior. That is, instead of asking how neurons implement cognition, we ask how neurons and the body meaningfully interact with the environment.

Once we realize that neurons are not like parts of machine but living entities, it becomes clear that much remains to be understood about neurons and brains. What are the anticipatory properties of neurons and how do they adapt to changing conditions, according to what logic? How can the interaction of neurons, which are not formally specified, lead to adapted behavior at the organismic scale? How do they self-organize in such a way that the organism is functional? How can such dynamic and adaptive entities retain stable collective functionality over long timescales? Cellular biology has changed considerably over the past decades (Ball, 2023), and much remains to be discovered about how cells work. This calls for humility and curiosity regarding biology, rather than disdain.

I proposed to see the brain's activity as the collective behavior of a colony of living entities, rather than a distributed computer. This is more a perspective than a theory—namely, it is a biological perspective. Nonetheless, it does suggest certain theoretical directions.

First, theoretical neuroscience has heavily borrowed from computer science, and relatively neglected the related theoretical subfields of biology, such as theoretical and systems biology. This includes many relevant topics such as the metabolic and genetic bases of cellular adaptation, the endogenous dynamics of cells, the laws of development, anticipation in cellular processes (such as circadian rhythms), as well as more fundamental topics, such as how a system of processes can be self-sustaining. All these topics are a source of inspiration for theoretical neuroscience.

Second, if we see the brain's activity as collective behavior, then a fundamental issue is to understand the nature of the interacting units. Cells, and neurons in particular, are anticipatory self-organizing systems. Because of connectionism, much theoretical effort has been devoted to understanding synaptic plasticity, but similar questions arise when we turn to cells: how does a neuron build and maintain an excitable system? What is the logic of intrinsic plasticity—that is, the plasticity of everything else than synapses? These are themes of theoretical cellular neurobiology, currently a small subfield of theoretical neuroscience.

These questions are particularly difficult to address in the cells of metazoans, because the organism is composed of many cells, which do not behave in the same way when isolated (or cannot be isolated). Thus, a possible source of inspiration is to look at the behavior and physiology of protists, or more generally of unicellular eukaryotes, since those are simultaneously cells and organisms, and have a visible behavior since they are motile. As I outlined for the ciliate *Paramecium* (Brette, 2021), protists can be studied with a neuroscience lens, to understand the biological basis of their proto-cognitive properties.

Third, to understand the collective behavior of neurons demands specific theoretical developments. These can build on the solid theoretical corpus of dynamical systems theory, as well as on the physics literature on collective properties (such as statistical physics). However, there is a peculiarity in biological systems, which is normativity: the system achieves something for the organism. We have seen that normativity can be addressed through a fundamental property of biological systems: anticipation. The theoretical challenge, then, is to relate the anticipatory properties of the system with the more rudimentary anticipatory properties of its constitutive units.

Given how open these research directions are, the spectacular claims of neurocomputationalism appear rather premature. Will we be able to upload people's minds to simulations? For sure, if that happens, it will not be a continuation of current research, and I gave reasons to doubt that it would be possible even in principle—let alone in practice. Are artificial neural networks on their way to quickly reach "artificial general intelligence"? None of the artifacts made by humans displays autonomy or intrinsic goals. No symbolic machine, as impressive as it may be, can understand what a symbol stands for without embodiment—a banana is indeed more than its syntactic and statistical relations to other words. If we take autonomy, goals, and adaptation seriously, then what we call "artificial intelligence" is not a first step toward human intelligence any more than a dishwasher is a first step toward general human-like behavior, even though dishwashers are already better than humans at washing the dishes. Probably, it is a first step toward better dishwashers.

When the brain was a computer, the understanding of brains, mind, and behavior was supposed to be a matter of decades, if not years. Just the time to decipher the brain's algorithms. This recurrent promise has been broken repeatedly (Dreyfus, 2012). But when we realize the obvious—that cognition is a biological phenomenon—suddenly the frontiers of knowledge appear to be wide open.

REFERENCES

Aabakken L. 2011. How to diagnose dyspepsia. In Duvnjak M, editor. Dyspepsia in Clinical Practice. New York: Springer, pp. 53–60. doi:10.1007/978-1-4419-1730-0_6.

Abeles M. 1982. Role of the cortical neuron: integrator or coincidence detector? *Isr J Med Sci* **18**:83–92.

Abeles M. 1982. Role of the cortical neuron: integrator or coincidence detector? *Isr J Med Sci* **18**:83–92.

Abeles M. 1991. Corticonics: Neural Circuits of the Cerebral Cortex. Cambridge: Cambridge University Press.

Abraham TH. 2002. (Physio)logical circuits: the intellectual origins of the McCulloch–Pitts neural networks. *J Hist Behav Sci* **38**:3–25. doi:10.1002/jhbs.1094.

Ackley DH, Hinton GE, Sejnowski TJ. 1985. A learning algorithm for Boltzmann machines. *Cogn Sci* **9**:147–169. doi:10.1016/S0364-0213(85)80012-4.

Adams RA, Shipp S, Friston KJ. 2013. Predictions not commands: active inference in the motor system. *Brain Struct Funct* **218**:611–643. doi:10.1007/s00429-012-0475-5.

Adrian ED. 1928. The basis of sensation. New York: W. W. Norton.

Adrian ED, Zotterman Y. 1926a. The impulses produced by sensory nerve endings: part II. The response of a single end-organ. *J Physiol* **61**:151–171.

Adrian ED, Zotterman Y. 1926b. The impulses produced by sensory nerve endings: part III. Impulses set up by touch and pressure. *J Physiol* **61**:465–483.

Ahissar E, Assa E. 2016. Perception as a closed-loop convergence process. *eLife* **5**. doi:10.7554/eLife.12830.

Alon U. 2019. An Introduction to Systems Biology, 2nd ed. Boca Raton, FL: Routledge.

Anastassiou CA, Perin R, Markram H, Koch C. 2011. Ephaptic coupling of cortical neurons. *Nat Neurosci* **14**:217–223. doi:10.1038/nn.2727.

Anderson ML. 2010. Neural reuse: a fundamental organizational principle of the brain. *Behav Brain Sci* **33**:245–266. doi:10.1017/S0140525X10000853.

Araque A, Navarrete M. 2010. Glial cells in neuronal network function. *Philos Trans R Soc B Biol Sci* **365**:2375–2381. doi:10.1098/rstb.2009.0313.

Aronov D, Nevers R, Tank DW. 2017. Mapping of a non-spatial dimension by the hippocampal/entorhinal circuit. *Nature* **543**:719–722. doi:10.1038/nature21692.

Arthur W. 2021. Understanding Evo-Devo, Understanding Life. Cambridge: Cambridge University Press. doi:10.1017/9781108873130.

Aschersleben G. 2002. Temporal control of movements in sensorimotor synchronization. *Brain Cogn* **48**:66–79. doi:10.1006/brcg.2001.1304.

Attwell D, Laughlin SB. 2001. An energy budget for signaling in the grey matter of the brain. *J Cereb Blood Flow Metab* **21**:1133–1145.

Aury J-M, Jaillon O, Duret L, et al. 2006. Global trends of whole-genome duplications revealed by the ciliate Paramecium tetraurelia. *Nature* **444**:171–178. doi:10.1038/nature05230.

Babbage C. 2010. Babbage's Calculating Engines: Being a Collection of Papers Relating to Them; Their History and Construction, 1st ed. Cambridge: Cambridge University Press.

Baldwin JM. 1896. A new factor in evolution. *Am Nat* **30**:441–451.

Ball P. 2023. How Life Works: A User's Guide to the New Biology, 1st ed. Chicago: University of Chicago Press.

Baluška F, Levin M. 2016. On having no head: cognition throughout biological systems. *Front Psychol* **7**. doi:10.3389/fpsyg.2016.00902.

Barlow H. 1961. Possible principles underlying the transformations of sensory messages. In Rosenblith W, editor. Sensory Communication. Cambridge, MA: MIT Press, pp. 217–234.

Barlow HB, Fitzhugh R, Kuffler SW. 1957. Change of organization in the receptive fields of the cat's retina during dark adaptation. *J Physiol* **137**:338–354. doi:10.1113/jphysiol.1957.sp005817.

Bar-On YM, Phillips R, Milo R. 2018. The biomass distribution on Earth. *Proc Natl Acad Sci* **115**:6506–6511. doi:10.1073/pnas.1711842115.

Barry C, Hayman R, Burgess N, Jeffery KJ. 2007. Experience-dependent rescaling of entorhinal grids. *Nat Neurosci* **10**:682–684. doi:10.1038/nn1905.

Beer RD, Gallagher JC. 1992. Evolving dynamical neural networks for adaptive behavior. *Adapt Behav* **1**:91–122. doi:10.1177/105971239200100105.

Bell AJ. 1999. Levels and loops: the future of artificial intelligence and neuroscience. *Philos Trans R Soc B Biol Sci* **354**:2013–2020.

Benichoux V, Fontaine B, Karino S, Joris PX, Brette R. 2015. Frequency-dependent time differences between the ears are matched in neural tuning. *eLife* 10.7554/eLife.06072. doi:http://dx.doi.org/10.7554/eLife.06072.

Benichoux V, Rébillat M, Brette R. 2016. On the variation of interaural time differences with frequency. *J Acoust Soc Am* **139**:1810–1821. doi:10.1121/1.4944638.

Benichoux V, Tollin DJ. 2018. These are not the neurons you are looking for. *eLife* **7**:e39244. doi:10.7554/eLife.39244.

Berg HC. 2008. *E. coli* in Motion. New York: Springer Science and Business Media.

Bernstein N. 1967. The Co-ordination and Regulation of Movements. New York: Pergamon Press.

Bialek W, Tishby N. 1999. Predictive information. *ArXiv E-Prints* cond-mat/9902341. doi:10.48550/arXiv.cond-mat/9902341.

Bickhard MH. 2009. The interactivist model. *Synthese* **166**:547–591. doi:10.1007/s11229-008-9375-x.

Bickhard MH. 2015. What could cognition be if not computation . . . or connectionism, or dynamic systems? *J Theor Philos Psychol* **35**:53–66. doi:10.1037/a0038059.

Bickhard MH. 2016a. The anticipatory brain: two approaches. In Müller VC, editor. Fundamental Issues of Artificial Intelligence, Synthese Library. Cham: Springer International Publishing, pp. 261–283. doi:10.1007/978-3-319-26485-1_16.

Bickhard MH. 2016b. Inter- and en- activism: some thoughts and comparisons. *New Ideas Psychol* **41**:23–32. doi:10.1016/j.newideapsych.2015.12.002.

Bickhard MH. 2024. The Whole Person. London: Academic Press.

Bizzi E, Cheung VC. 2013. The neural origin of muscle synergies. *Front Comput Neurosci* **7**:51.

Blumberg MS. 2006. Basic Instinct: The Genesis of Behavior. Basic Books.

Blumberg MS. 2010. Freaks of Nature: What Anomalies Tell Us about Development and Evolution, illustrated ed. Oxford: Oxford University Press.

Boerlin M, Machens CK, Denève S. 2013. Predictive coding of dynamical variables in balanced spiking networks. *PLoS Comput Biol* **9**:e1003258. doi:10.1371/journal.pcbi.1003258.

Bohr N. 1958. Atomic Physics and Human Knowledge. New York: Wiley.

Bolz J, Gilbert CD. 1986. Generation of end-inhibition in the visual cortex via interlaminar connections. *Nature* **320**:362–365. doi:10.1038/320362a0.

Braitenberg V. 1984. Vehicles: Experiments in Synthetic Psychology. Cambridge, MA: MIT Press.

Brenner S. 1973. The Genetics of Behaviour. *Br Med Bull* **29**:269–271. doi:10.1093/oxfordjournals.bmb.a071019.

Brette R. 2004. Dynamics of one-dimensional spiking neuron models. *J Math Biol* **48**:38–56. https://doi.org/10.1007/s00285-003-0223-9.

Brette R. 2012. Computing with neural synchrony. *PLoS Comp Biol* **8**:e1002561.

Brette R. 2015. Philosophy of the spike: rate-based vs. spike-based theories of the brain. *Front Syst Neurosci* 151. doi:10.3389/fnsys.2015.00151.

Brette R. 2016. Subjective physics. In El Hady A, editor. Closed Loop Neuroscience. London: Academic Press, pp. 146–170.

Brette R. 2018. The world is complex, not just noisy. *Behav Brain Sci* **41**:e227. doi:10.1017/S0140525X18001292.

Brette R. 2019a. Is coding a relevant metaphor for the brain? *Behav Brain Sci* 1–44. doi:10.1017/S0140525X19000049.

Brette R. 2019b. Neural coding: the bureaucratic model of the brain. *Behav Brain Sci* **42**:e243. doi:10.1017/S0140525X19001997.

Brette R. 2021. Integrative neuroscience of Paramecium, a "swimming neuron." *eNeuro* **8**: ENEURO.0018-21.2021. doi:10.1523/ENEURO.0018-21.2021.

Brette R. 2022a. Brains as computers: metaphor, analogy, theory or fact? *Front Ecol Evol* **10**:878729. https://doi.org/10.3389/fevo.2022.878729}

Brette R. 2022b. Does the present moment depend on the moments not lived? *Behav Brain Sci* **45**:e43. doi:10.1017/S0140525X21001904.

Brette R, Guigon E. 2003. Reliability of spike timing is a general property of spiking model neurons. *Neural Comput* **15**:279–308.

Brette R, Rudolph M, Carnevale T, et al. 2007. Simulation of networks of spiking neurons: a review of tools and strategies. *J Comput Neurosci* **23**:349–98. doi:10.1007/s10827-007-0038-6.

Brezina V. 2010. Beyond the wiring diagram: signalling through complex neuromodulator networks. *Philos Trans R Soc B Biol Sci* **365**:2363–2374. doi:10.1098/rstb.2010.0105.

Brooks RA. 1986. A robust layered control system for a mobile robot. *IEEE J Robot Autom* **2**:14–23. doi:10.1109/JRA.1986.1087032.

Brooks RA. 1990. Elephants don't play chess. *Robot Auton Syst* **6**:3–15. doi:10.1016/S0921-8890(05)80025-9.

Brooks RA. 1991. Intelligence without representation. *Artif Intell* **47**:139–159. doi:10.1016/0004-3702(91)90053-M.

Brunel N. 2000. Dynamics of sparsely connected networks of excitatory and inhibitory spiking neurons. *J Comput Neurosci* **8**:183–208.

Brunet T, Arendt D. 2016. From damage response to action potentials: early evolution of neural and contractile modules in stem eukaryotes. *Philos Trans R Soc B Biol Sci* **371**:20150043. doi:10.1098/rstb.2015.0043.

Bullock T, Perkel D. 1968. Neural coding: a report based on an NRP Work Session. *Neurosci Res Program Bull* **6**.

Burkhardt P, Colgren J, Medhus A, et al. 2023. Syncytial nerve net in a ctenophore adds insights on the evolution of nervous systems. *Science* **380**:293–297. doi:10.1126/science.ade5645.

Burnet FM. 1957. A modification of Jerne's theory of antibody production using the concept of clonal selection. *Aust J Sci* **20**:67–69.

Buzsáki G. 2011. Rhythms of the Brain, 1st ed. New York: Oxford University Press.

Buzsáki G. 2019. The Brain from Inside Out. New York: Oxford University Press. doi:10.1093/oso/9780190905385.001.0001.

Campbell DT. 1974. "Downward causation" in hierarchically organised biological systems. In Ayala FJ, Dobzhansky T, editors. Studies in the Philosophy of Biology: Reduction and Related Problems. London: Macmillan Education UK, pp. 179–186. doi:10.1007/978-1-349-01892-5_11,

Canguilhem G. 1966. Le Normal et le Pathologique. Paris: Presses Universitaires de France.

Chalmers DJ. 1996. Does a rock implement every finite-state automaton? *Synthese* **108**:309–333. doi:10.1007/BF00413692.

Chalmers DJ. 1997. The Conscious Mind: In Search of a Fundamental Theory, rev. ed. New York: Oxford University Press.

Chalmers DJ. 2011. A computational foundation for the study of cognition. *J Cogn Sci* **12**:323–357.

Chanauria N, Bharmauria V, Bachatene L, Cattan S, Rouat J, Molotchnikoff S. 2018. Sound induces change in orientation preference of V1 neurons: audio-visual cross-influence. *bioRxiv* 269589. doi:10.1101/269589.

Changeux J-P, Heidmann T, Patte P. 1984. Learning by selection. In Marler P, Terrace HS, editors. The Biology of Learning, Dahlem Workshop Reports. Berlin, Heidelberg: Springer, pp. 115–133. doi:10.1007/978-3-642-70094-1_6.

Chaudhuri R, Gerçek B, Pandey B, Peyrache A, Fiete I. 2019. The intrinsic attractor manifold and population dynamics of a canonical cognitive circuit across waking and sleep. *Nat Neurosci* **22**:1512–1520. doi:10.1038/s41593-019-0460-x.

Chemero A. 2011. Radical Embodied Cognitive Science. Cambridge, MA: MIT Press.

Cheng X, Ferrell JE. 2019. Spontaneous emergence of cell-like organization in Xenopus egg extracts. *Science* **366**:631–637. doi:10.1126/science.aav7793.

Chiel HJ, Beer RD. 1997. The brain has a body: adaptive behavior emerges from interactions of nervous system, body and environment. *Trends Neurosci* **20**:553–557. doi:10.1016/S0166-2236(97)01149-1.

Chomsky N. 2002. Syntactic Structures. Berlin, New York: Walter de Gruyter.

Cisek P. 2019. Resynthesizing behavior through phylogenetic refinement. *Atten Percept Psychophys* **81**:2265–2287. doi:10.3758/s13414-019-01760-1.

Clark A. 2013. Whatever next? Predictive brains, situated agents, and the future of cognitive science. *Behav Brain Sci* **36**:181–204. doi:10.1017/S0140525X12000477.

Clark A, Chalmers D. 1998. The extended mind. *Analysis* **58**:7–19.

Clark A, Toribio J. 1994. Doing without representing? *Synthese* **101**:401–431. doi:10.1007/BF01063896.

Cobb M. 2020. The Idea of the Brain: The Past and Future of Neuroscience. New York: Basic Books.

Cobb M. 2021. A brief history of wires in the brain. *Front Ecol Evol* **9**. doi:10.3389/fevo.2021.760269.

Colgin LL, Moser EI, Moser M-B. 2008. Understanding memory through hippocampal remapping. *Trends Neurosci* **31**:469–477. doi:10.1016/j.tins.2008.06.008.

Conant RC, Ross Ashby W. 1970. Every good regulator of a system must be a model of that system. *Int J Syst Sci* **1**:89–97. doi:10.1080/00207727008920220.

Constantinidis C, Klingberg T. 2016. The neuroscience of working memory capacity and training. *Nat Rev Neurosci* **17**:438–449. doi:10.1038/nrn.2016.43.

Copeland BJ. 2020. The Church-Turing thesis. In Zalta EN, editor. The Stanford Encyclopedia of Philosophy. Stanford, CA: Metaphysics Research Lab, Stanford University.

Cormen TH, Leiserson CE, Rivest RL, Stein C. 2009. Introduction to Algorithms, 3rd ed. Cambridge, MA: MIT Press.

Cortes C, Vapnik V. 1995. Support-vector networks. *Mach Learn* **20**:273–297. doi:10.1007/BF00994018.

Crick F. 1979. Thinking about the brain. *Sci Am* **241**:219–232. doi:10.1038/scientificamerican0979-219.

Crick F. 1989. Neural Edelmanism. *Trends Neurosci* **12**:240–248. doi:10.1016/0166-2236(89)90019-2.

Csikszentmihalyi M. 2008. Flow: The Psychology of Optimal Experience. New York: Harper Perennial Modern Classics.

Cummins R. 1991. Meaning and Mental Representation, reprint ed. Cambridge, MA: MIT Press.

Dacke M, Srinivasan MV. 2008. Evidence for counting in insects. *Anim Cogn* **11**:683–689. doi:10.1007/s10071-008-0159-y.

Danziger K. 1997. Naming the Mind: How Psychology Found Its Language, 1st ed. London: SAGE Publications.

Darwin C. 1859. On the Origin of Species by Means of Natural Selection, or Preservation of Favoured Races in the Struggle for Life. London: John Murray.

d'Avella A, Bizzi E. 2005. Shared and specific muscle synergies in natural motor behaviors. *Proc Natl Acad Sci* **102**:3076–3081. doi:10.1073/pnas.0500199102.

Deacon TW. 2012. Incomplete Nature: How Mind Emerged from Matter, 1st ed. New York: W. W. Norton.

Deco G, Jirsa VK, McIntosh AR. 2011. Emerging concepts for the dynamical organization of resting-state activity in the brain. *Nat Rev Neurosci* **12**:43–56. doi:10.1038/nrn2961.

Dehaene S. 2014. Consciousness and the Brain: Deciphering How the Brain Codes Our Thoughts. New York: Penguin Books.

Delgutte B, Joris PX, Litovsky RY, Yin TCT. 1999. Receptive fields and binaural interactions for virtual-space stimuli in the cat inferior colliculus. *J Neurophysiol* **81**:2833–2851. doi:10.1152/jn.1999.81.6.2833.

Deneve S. 2008. Bayesian spiking neurons I: inference. *Neural Comput* **20**:91–117. doi:10.1162/neco.2008.20.1.91.

Descartes R. 1637. Discours de la méthode pour bien conduire sa raison et chercher la vérité dans les sciences, plus La dioptrique, Les météores et La géométrie qui sont des essais de cette méthode. Leiden: Jan Maire.

Dewey J. 1896. The reflex arc concept in psychology. *Psychol Rev* **3**:357–370. doi:10.1037/h0070405.

Dörrbaum AR, Kochen L, Langer JD, Schuman EM. 2018. Local and global influences on protein turnover in neurons and glia. *eLife* **7**:e34202. doi:10.7554/eLife.34202.

Drew PJ, Abbott LF. 2006. Models and properties of power-law adaptation in neural systems. *J Neurophysiol* **96**:826–833. doi:10.1152/jn.00134.2006.

Dreyfus HL. 1978. What Computers Can't Do: The Limits of Artificial Intelligence, revised ed. New York: HarperCollins.

Dreyfus HL. 2012. A history of first step fallacies. *Minds Mach* **22**:87–99. doi:10.1007/s11023-012-9276-0.

Driscoll LN, Duncker L, Harvey CD. 2022. Representational drift: emerging theories for continual learning and experimental future directions. *Curr Opin Neurobiol* **76**:102609. doi:10.1016/j.conb.2022.102609.

Drossel B. 2021. Strong emergence in condensed matter physics. In Voosholz J, Gabriel M, editors. Top-Down Causation and Emergence, Synthese Library. Cham: Springer International Publishing, pp. 79–99. doi:10.1007/978-3-030-71899-2_4.

Dupuy J-P. 2013. Aux origines des sciences cognitives. Paris: La Découverte.

Dziri N, Lu X, Sclar M, et al. 2023. Faith and fate: limits of transformers on compositionality. doi:10.48550/arXiv.2305.18654.

Ebitz RB, Hayden BY. 2021. The population doctrine in cognitive neuroscience. *Neuron* **109**:3055–3068. doi:10.1016/j.neuron.2021.07.011.

Eccles JC. 1965. Conscious experience and memory. In Brain and Conscious Experience. Berlin: Springer, pp. 314–344. doi:10.1007/978-3-642-49168-9_14.

Eckert R. 1972. Bioelectric control of ciliary activity. *Science* **176**:473–481. doi:10.1126/science.176.4034.473.

Edelman GM. 1993. Neural Darwinism: selection and reentrant signaling in higher brain function. *Neuron* **10**:115–125. doi:10.1016/0896-6273(93)90304-A.

Egbert MD, Barandiaran XE, Paolo EAD. 2010. A minimal model of metabolism-based chemotaxis. *PLOS Comput Biol* **6**:e1001004. doi:10.1371/journal.pcbi.1001004.

Eigen M, Schuster P. 1978. The hypercycle. *Naturwissenschaften* **65**:7–41. doi:10.1007/BF00420631.

Eldar A, Elowitz MB. 2010. Functional roles for noise in genetic circuits. *Nature* **467**:167–173. doi:10.1038/nature09326.

Elices I, Kulkarni A, Escoubet N, Pontani L-L, Prevost AM, Brette R. 2023. An electrophysiological and kinematic model of Paramecium, the "swimming neuron." *PLOS Comput Biol* **19**:e1010899. doi:10.1371/journal.pcbi.1010899.

Erdin HO, Bickhard MH. 2018. Representing is something that we do, not a structure that we "use": reply to Gładziejewski. *New Ideas Psychol* **49**:27–37. doi:10.1016/j.newideapsych.2018.02.001.

Febvre-Chevalier C, Bilbaut A, Bone Q, Febvre J. 1986. Sodium-calcium action potential associated with contraction in the heliozoan Actinocoryne Contractilis. *J Exp Biol* **122**:177–192.

Fenk LM, Avritzer SC, Weisman JL, Nair A, Randt LD, Mohren TL, Siwanowicz I, Maimon G. 2022. Muscles that move the retina augment compound eye vision in Drosophila. *Nature* **612**:116–122. doi:10.1038/s41586-022-05317-5.

Festinger L. 1957. A Theory of Cognitive Dissonance, anniversary ed. Stanford, CA: Stanford University Press.

Festinger L, Riecken HW, Schachter S. 1956. When Prophecy Fails: A Social and Psychological Study of a Modern Group That Predicted the Destruction of the World. New York: Harper-Torchbooks.

Feyerabend P. 2010. Against Method. London: Verso.

Fields C, Levin M. 2018. Multiscale memory and bioelectric error correction in the cytoplasm–cytoskeleton–membrane system. *WIREs Syst Biol Med* **10**:e1410. doi:10.1002/wsbm.1410.

Fields RD. 2015. A new mechanism of nervous system plasticity: activity-dependent myelination. *Nat Rev Neurosci* **16**:756–767. doi:10.1038/nrn4023.

Fodor JA. 1980. The Language of Thought. Cambridge, MA: Harvard University Press.

Fodor JA. 1981. The mind–body problem. *Sci Am* **244**:114–123.

Franken TP, Joris PX, Smith PH. 2018. Principal cells of the brainstem's interaural sound level detector are temporal differentiators rather than integrators. *eLife* **7**:e33854. doi:10.7554/eLife.33854.

Freeman WJ. 1975. Mass Action in the Nervous System: Examination of the Neurophysiological Basis of Adaptive Behavior through the EEG. New York: Academic Press.

Friedman MB. 1975. Visual control of head movements during avian locomotion. *Nature* **255**:67–69. doi:10.1038/255067a0.

Friston K. 2009. The free-energy principle: a rough guide to the brain? *Trends Cogn Sci* **13**:293–301. doi:10.1016/j.tics.2009.04.005.

Friston K. 2010. The free-energy principle: a unified brain theory? *Nat Rev Neurosci* **11**:127–138. doi:10.1038/nrn2787.

Friston K, Thornton C, Clark A. 2012. Free-energy minimization and the dark-room problem. *Front Psychol* **3**. doi:10.3389/fpsyg.2012.00130.

Gallagher S. 2006. How the Body Shapes the Mind. New York: Clarendon Press.

Gallego JA, Perich MG, Miller LE, Solla SA. 2017. Neural manifolds for the control of movement. *Neuron* **94**:978–984. doi:10.1016/j.neuron.2017.05.025.

Gallistel CR. 2017. The coding question. *Trends Cogn Sci* **21**:498–508. doi:10.1016/j.tics.2017.04.012.

Gànti T. 2003. Chemoton Theory: Theory of Living Systems. New York: Springer Science and Business Media.

Gerland E, Traumüller F. 1899. Geschichte der physikalischen Experimentierkunst. Leipzig: W. Engelmann.

Gibson JJ. 1962. Observations on active touch. *Psychol Rev* **69**:477–491.

Gibson JJ. 1979. The Ecological Approach to Visual Perception. Boston: Routledge.

Gibson JJ. 1983. The Senses Considered as Perceptual Systems, revised ed. Westport, CT: Praeger.

Gilbert CD, Li W. 2013. Top-down influences on visual processing. *Nat Rev Neurosci* **14**:350–363. doi:10.1038/nrn3476.

Ginosar G, Aljadeff J, Las L, Derdikman D, Ulanovsky N. 2023. Are grid cells used for navigation? On local metrics, subjective spaces, and black holes. *Neuron* **111**:1858–1875. doi:10.1016/j.neuron.2023.03.027.

Giurfa M, Zhang S, Jenett A, Menzel R, Srinivasan MV. 2001. The concepts of "sameness" and "difference" in an insect. *Nature* **410**:930–933. doi:10.1038/35073582.

Goethals S, Brette R. 2020. Theoretical relation between axon initial segment geometry and excitability. *eLife* **9**:e53432. doi:10.7554/eLife.53432.

Goethals S, Sierksma MC, Nicol X, Réaux-Le Goazigo A, Brette R. 2021. Electrical match between initial segment and somatodendritic compartment for action potential backpropagation in retinal ganglion cells. *J Neurophysiol* **126**:28–46. doi:10.1152/jn.00005.2021.

Gomez-Marin A. 2017. Causal circuit explanations of behavior: are necessity and sufficiency necessary and sufficient? In Çelik A, Wernet M, editors. Decoding Neural Circuit Structure and Function. Cham: Springer, pp. 283–306. doi:10.1007/978-3-319-57363-2_11.

Gomez-Marin A, Ghazanfar AA. 2019. The life of behavior. *Neuron* **104**:25–36. doi:10.1016/j.neuron.2019.09.017.

Goodman DF, Benichoux V, Brette R. 2013. Decoding neural responses to temporal cues for sound localization. *eLife* **2**:2:e01312. doi:10.7554/eLife.01312.

Goodman DFM, Brette R. 2010. Spike-timing-based computation in sound localization. *PLoS Comput Biol* **6**:e1000993. doi:10.1371/journal.pcbi.1000993.

Gourévitch B, Brette R. 2012. The impact of early reflections on binaural cues. *J Acoust Soc Am* **132**:9–27. doi:10.1121/1.4726052.

Grothe B, Pecka M, McAlpine D. 2010. Mechanisms of sound localization in mammals. *Physiol Rev* **90**:983–1012. doi:10.1152/physrev.00026.2009.

Grubb MS, Shu Y, Kuba H, Rasband MN, Wimmer VC, Bender KJ. 2011. Short- and long-term plasticity at the axon initial segment. *J Neurosci* **31**:16049–16055. doi:10.1523/JNEUROSCI.4064-11.2011.

Guo C, Lee MJ, Leclerc G, Dapello J, Rao Y, Madry A, DiCarlo JJ. 2022. Adversarially trained neural representations may already be as robust as corresponding biological neural representations. *Proc 38th Int Conf Mach Learn*. Baltimore, MD: PMLR 162.

Hafting T, Fyhn M, Molden S, Moser M-B, Moser EI. 2005. Microstructure of a spatial map in the entorhinal cortex. *Nature* **436**:801–806. doi:10.1038/nature03721.

Hamada MS, Goethals S, de Vries SI, Brette R, Kole MHP. 2016. Covariation of axon initial segment location and dendritic tree normalizes the somatic action potential. *Proc Natl Acad Sci* **113**:14841–14846. doi:10.1073/pnas.1607548113.

Harnad S. 1990. The symbol grounding problem. *Phys Nonlinear Phenom* **42**:335–346. doi:10.1016/0167-2789(90)90087-6.

Harris KD. 2008. Stability of the fittest: organizing learning through retroaxonal signals. *Trends Neurosci* **31**:130–136. doi:10.1016/j.tins.2007.12.002.

Hasegawa K, Tsukahara Y, Shimamoto M, Matsumoto K, Nakaoka Y, Sato T. 1997. The Paramecium circadian clock: synchrony of changes in motility, membrane potential, cyclic AMP and cyclic GMP. *J Comp Physiol A* **181**:41–46. doi:10.1007/s003590050091.

Hendriks-Jansen H. 1996. Catching Ourselves in the Act: Situated Activity, Interactive Emergence, Evolution, and Human Thought. Cambridge, MA: Bradford Books.

Hennessey TM, Rucker WB, McDiarmid CG. 1979. Classical conditioning in paramecia. *Anim Learn Behav* **7**:417–423. doi:10.3758/BF03209695.

Hille B. 2001. Ion Channels of Excitable Membranes. Sunderland, MA: Sinauer Associates.

Hodgkin AL. 1964. The Conduction of the Nervous Impulse. Springfield, IL: C. C. Thomas.

Hodgkin AL, Huxley A. 1952. A quantitative description of membrane current and its application to conduction and excitation in nerve. *J Physiol Lond* **117**:500.

Hopfield JJ. 1982. Neural networks and physical systems with emergent collective computational abilities. *Proc Natl Acad Sci* **79**:2554.

Hornik K, Stinchcombe M, White H. 1989. Multilayer feedforward networks are universal approximators. *Neural Netw* **2**:359–366. doi:10.1016/0893-6080(89)90020-8.

Hosoya T, Baccus SA, Meister M. 2005. Dynamic predictive coding by the retina. *Nature* **436**:71–77. doi:10.1038/nature03689.

Hubel DH, Wiesel TN. 1959. Receptive fields of single neurones in the cat's striate cortex. *J Physiol* **148**:574–591. doi:10.1113/jphysiol.1959.sp006308.

Hubel DH, Wiesel TN. 1968. Receptive fields and functional architecture of monkey striate cortex. *J Physiol* **195**:215–243.

Hulse BK, Jayaraman V. 2020. Mechanisms underlying the neural computation of head direction. *Annu Rev Neurosci* **43**:31–54. doi:10.1146/annurev-neuro-072116-031516.

Hurley S. 2001. Perception and action: alternative views. *Synthese* **129**:3–40. doi:10.1023/A:1012643006930.

Huth AG, de Heer WA, Griffiths TL, Theunissen FE, Gallant JL. 2016. Natural speech reveals the semantic maps that tile human cerebral cortex. *Nature* **532**:453–458. doi:10.1038/nature17637.

Hutto DD, Myin E, Peeters A, Zahnoun F. 2018. The cognitive basis of computation: putting computation in its place. In Sprevak M, Colombo M, editors. The Routledge Handbook of the Computational Mind. London: Routledge, pp. 272–282.

Huygens C. 1690. Traité de la lumière. Leiden: Pieter van der Aa.

Izquierdo EJ, Beer RD. 2016. The whole worm: brain–body–environment models of C. elegans. *Curr Opin Neurobiol* **40**:23–30. doi:10.1016/j.conb.2016.06.005.

Jacob F. 1977. Evolution and tinkering. *Science* **196**:1161–1166.

Jaeger J, Monk N. 2015. Everything flows. *EMBO Rep* **16**:1064–1067. doi:10.15252/embr.201541088.

Jazayeri M, Movshon JA. 2006. Optimal representation of sensory information by neural populations. *Nat Neurosci* **9**:690–696. doi:10.1038/nn1691.

Jazayeri M, Ostojic S. 2021. Interpreting neural computations by examining intrinsic and embedding dimensionality of neural activity. *Curr Opin Neurobiol* **70**:113–120. doi:10.1016/j.conb.2021.08.002.

Jeffress LA. 1948. A place theory of sound localisation. *J Comp Physiol Psychol* **41**:35.

Jenkins WM, Masterton RB. 1982. Sound localization: effects of unilateral lesions in central auditory system. *J Neurophysiol* **47**:987–1016.

Jennings HS. 1906. Behavior of the Lower Organisms. New York, Columbia University Press, The Macmillan Company.

Johnson CH, Egli M, Stewart PL. 2008. Structural insights into a circadian oscillator. *Science* **322**:697–701. doi:10.1126/science.1150451.

Jonas H. 1953. A critique of cybernetics. *Soc Res* **20**:172–192.

Jonas H. 1966. The phenomenon of life: toward a philosophical biology. New York: Harper and Row.

Joris PX, Smith PH, Yin TC. 1998. Coincidence detection in the auditory system: 50 years after Jeffress. *Neuron* **21**:1235.

Joris PX, Yin TC. 1995. Envelope coding in the lateral superior olive. I. Sensitivity to interaural time differences. *J Neurophysiol* **73**:1043–1062.

Juarrero A. 1998. Causality as constraint. In van de Vijver G, Salthe SN, Delpos M, editors. Evolutionary Systems: Biological and Epistemological Perspectives on Selection and Self-Organization. Dordrecht, Netherlands: Springer, pp. 233–242. doi:10.1007/978-94-017-1510-2_17.

Juarrero A. 2002. Dynamics in Action: Intentional Behavior as a Complex System. Cambridge, MA: Bradford Books.

Juarrero A. 2023. Context Changes Everything: How Constraints Create Coherence. Cambridge, MA: MIT Press. doi:10.7551/mitpress/14630.001.0001.

Kaltenegger L. 2017. How to characterize habitable worlds and signs of life. *Annu Rev Astron Astrophys* **55**:433–485. doi:10.1146/annurev-astro-082214-122238.

Katz PS. 2007. Evolution and development of neural circuits in invertebrates. *Curr Opin Neurobiol* **17**:59–64. doi:10.1016/j.conb.2006.12.003.

Kauffman S, Logan RK, Este R, Goebel R, Hobill D, Shmulevich I. 2008. Propagating organization: an enquiry. *Biol Philos* **23**:27–45. doi:10.1007/s10539-007-9066-x.

Kauffman SA. 1986. Autocatalytic sets of proteins. *J Theor Biol* **119**:1–24. doi:10.1016/S0022-5193(86)80047-9.

Kauffman SA. 2019. A World beyond Physics: The Emergence and Evolution of Life. New York: Oxford University Press.

Keller EF, Segel LA. 1971. Model for chemotaxis. *J Theor Biol* **30**:225–234. doi:10.1016/0022-5193(71)90050-6.

Kim J, Ricci M, Serre T. 2018. Not-So-CLEVR: learning same–different relations strains feedforward neural networks. *Interface Focus* **8**:20180011. doi:10.1098/rsfs.2018.0011.

Kippert F. 1996. An ultradian clock controls locomotor behaviour and cell division in isolated cells of Paramecium tetraurelia. *J Cell Sci* **109**:867–873.

Knill DC, Pouget A. 2004. The Bayesian brain: the role of uncertainty in neural coding and computation. *Trends Neurosci* **27**:712–719. doi:10.1016/j.tins.2004.10.007.

Knoll G, Haacke-Bell B, Plattner H. 1991. Local trichocyst exocytosis provides an efficient escape mechanism for Paramecium cells. *Eur J Protistol* **27**:381–385. doi:10.1016/S0932-4739(11)80256-7.

Koch C. 2012. Consciousness: Confessions of a Romantic Reductionist. Cambridge, MA: MIT Press. doi:10.7551/mitpress/9367.001.0001.

Koumandou VL, Wickstead B, Ginger ML, van der Giezen M, Dacks JB, Field MC. 2013. Molecular paleontology and complexity in the last eukaryotic common ancestor. *Crit Rev Biochem Mol Biol* **48**:373–396. doi:10.3109/10409238.2013.821444.

Kovács IA, Barabási DL, Barabási A-L. 2020. Uncovering the genetic blueprint of the C. elegans nervous system. *Proc Natl Acad Sci* **117**:33570–33577. doi:10.1073/pnas.2009093117.

Krakauer JW, Ghazanfar AA, Gomez-Marin A, MacIver MA, Poeppel D. 2017. Neuroscience needs behavior: correcting a reductionist bias. *Neuron* **93**:480–490. doi:10.1016/j.neuron.2016.12.041.

Kuba H, Yamada R, Fukui I, Ohmori H. 2005. Tonotopic specialization of auditory coincidence detection in nucleus laminaris of the chick. *J Neurosci* **25**:1924–1934. doi:10.1523/JNEUROSCI.4428-04.2005.

Kucyi A, Davis KD. 2017. The neural code for pain: from single-cell electrophysiology to the dynamic pain connectome. *Neuroscientist* **23**:397–414. doi:10.1177/1073858416667716.

Kuhn TS. 1962. The Structure of Scientific Revolutions. Chicago: University of Chicago Press.

Kulesza RJ. 2007. Cytoarchitecture of the human superior olivary complex: medial and lateral superior olive. *Hear Res* **225**:80–90. doi:10.1016/j.heares.2006.12.006.

Kull K. 2014. Adaptive evolution without natural selection. *Biol J Linn Soc* **112**:287–294. doi:10.1111/bij.12124.

Kunita I, Kuroda S, Ohki K, Nakagaki T. 2014. Attempts to retreat from a dead-ended long capillary by backward swimming in Paramecium. *Front Microbiol* **5**.doi:10.3389/fmicb.2014.00270.

Kupiec JJ. 1997. A Darwinian theory for the origin of cellular differentiation. *Mol Gen Genet MGG* **255**:201–208. doi:10.1007/s004380050490.

Lakatos I. 1976. Falsification and the methodology of scientific research programmes. In Harding SG, editor. Can Theories Be Refuted? Dordrecht, Netherlands: Springer, Synthese Library. pp. 205–259. doi:10.1007/978-94-010-1863-0_14.

Lakoff G, Johnson M. 1980. Metaphors We Live By. Chicago: University of Chicago Press.

Land MF. 1969. Movements of the retinae of jumping spiders (Salticidae: dendryphantinae) in response to visual stimuli. *J Exp Biol* **51**:471–493. doi:10.1242/jeb.51.2.471.

Langdon C, Genkin M, Engel TA. 2023. A unifying perspective on neural manifolds and circuits for cognition. *Nat Rev Neurosci* **24**:363–377. doi:10.1038/s41583-023-00693-x.

Laudanski J, Zheng Y, Brette R. 2014. A structural theory of pitch. *eneuro*:ENEURO.0033-14.2014. doi:10.1523/ENEURO.0033-14.2014.

Laughlin S. 1981. A simple coding procedure enhances a neuron's information capacity. *Z Naturforsch [C]* **36**:910–912.

Laurent G. 2002. Olfactory network dynamics and the coding of multidimensional signals. *Nat Rev Neurosci* **3**:884.

LeCun Y. 2022. A path towards autonomous machine intelligence, version 0.9.2, 2022-06-27. *Open Review* **62**(1):1–62.

LeCun Y, Bengio Y. 1998. Convolutional networks for images, speech, and time series. In Michael A. Arbib, editor. The Handbook of Brain Theory and Neural Networks. Cambridge, MA: MIT Press, pp. 255–258.

LeCun Y, Bengio Y, Hinton G. 2015. Deep learning. *Nature* **521**:436–444. doi:10.1038/nature14539m.

LeMasson G, Marder E, Abbott LF. 1993. Activity-dependent regulation of conductances in model neurons. *Science* **259**:1915.

Le Mouel C, Brette R. 2017. Mobility as the purpose of postural control. *Front Comput Neurosci* **11**:67. doi:10.3389/fncom.2017.00067.

Le Mouel C, Brette R. 2019. Anticipatory coadaptation of ankle stiffness and sensorimotor gain for standing balance. *PLoS Comput Biol* **15**:e1007463. doi:10.1371/journal.pcbi.1007463.

Le Mouel C, Tisserand R, Robert T, Brette R. 2019. Postural adjustments in anticipation of predictable perturbations allow elderly fallers to achieve a balance recovery performance equivalent to elderly non-fallers. *Gait Posture* **71**:131–137. doi:10.1016/j.gaitpost.2019.04.025.

Levitan IB. 1994. Modulation of ion channels by protein phosphorylation and dephosphorylation. *Annu Rev Physiol* **56**:193–212. doi:10.1146/annurev.ph.56.030194.001205.

Lewis M, Mitchell M. 2024. Evaluating the robustness of analogical reasoning in large language models. doi:10.48550/arXiv.2411.14215.

Liao X, Vasilakos AV, He Y. 2017. Small-world human brain networks: perspectives and challenges. *Neurosci Biobehav Rev* **77**:286–300. doi:10.1016/j.neubiorev.2017.03.018.

Lodh S, Yano J, Valentine MS, Van Houten JL. 2016. Voltage-gated calcium channels of Paramecium cilia. *J Exp Biol* **219**:3028–3038. doi:10.1242/jeb.141234.

Lovelock JE. 1965. A physical basis for life detection experiments. *Nature* **207**:568–570. doi:10.1038/207568a0.

Lyon P. 2015. The cognitive cell: bacterial behavior reconsidered. *Front Microbiol* **6**:264. doi:10.3389/fmicb.2015.00264.

Lyon P, Keijzer F, Arendt D, Levin M. 2021. Reframing cognition: getting down to biological basics. *Philos Trans R Soc B Biol Sci* **376**:20190750. doi:10.1098/rstb.2019.0750.

Lyons NA, Kolter R. 2015. On the evolution of bacterial multicellularity. *Curr Opin Microbiol* **24**:21–28. doi:10.1016/j.mib.2014.12.007.

Macmillan NA, Creelman CD. 2005. Detection Theory: A User's Guide, 2nd ed. Mahwah, NJ: Lawrence Erlbaum Associates.

Mainen Z, Sejnowski T. 1995. Reliability of spike timing in neocortical neurons. *Science* **268**:1503–1506.

March A. 1962. The New World of Physics. New York: Random House.

Marcus G. 2012. Guitar Zero: The Science of Becoming Musical at Any Age, illustrated ed. New York: Penguin Books.

Marder E, Goaillard J-M. 2006. Variability, compensation and homeostasis in neuron and network function. *Nat Rev Neurosci* **7**:563–574. doi:10.1038/nrn1949.

Markram H. 2012. The human brain project. *Sci Am* **306**:50–55. doi:10.1038/scientificamerican0612-50.

Marr D. 1982. Vision, 1st ed. San Francisco: W. H. Freeman.

Martinac B, Machemer H. 1984. Effects of varied culturing and experimental temperature on electrical membrane properties in Paramecium. *J Exp Biol* **108**:179–194.

Martinez-Conde S, Macknik SL, Hubel DH. 2004. The role of fixational eye movements in visual perception. *Nat Rev Neurosci* **5**:229.

Mashour GA, Roelfsema P, Changeux J-P, Dehaene S. 2020. Conscious processing and the global neuronal workspace hypothesis. *Neuron* **105**:776–798. doi:10.1016/j.neuron.2020.01.026.

Maturana HR, Varela FJ. 1973. Autopoiesis and Cognition: The Realization of the Living, 1st ed. Dordrecht, Netherlands: D. Reidel Publishing.

Mayr E. 1961. Cause and effect in biology. *Science* **134**:1501–1506.

Mazor O, Laurent G. 2005. Transient dynamics versus fixed points in odor representations by locust antennal lobe projection neurons. *Neuron* **48**:661–673. doi:10.1016/j.neuron.2005.09.032.

McAlpine D, Jiang D, Palmer AR. 2001. A neural code for low-frequency sound localization in mammals. *Nat Neurosci* **4**:396.

McAnelly ML, Zakon HH. 2000. Coregulation of voltage-dependent kinetics of Na(+) and K(+) currents in electric organ. *J Neurosci Off J Soc Neurosci* **20**:3408–3414.

McCulloch WS, Pitts W. 1943. A logical calculus of the ideas immanent in nervous activity. *Bull Math Biophys* **5**:115–133. doi:10.1007/BF02478259.

McMullin B, Varela FJ. 1997. Rediscovering computational autopoeisis. In Husbands P, Harvey I, editors. Fourth European Conference on Artificial Life. Cambridge, MA: MIT Press/Bradford Books, pp. 38–48.

Meister M, Berry M. 1999. The neural code of the retina. *Neuron* **22**:435.

Merleau-Ponty M. 1942. La structure du comportement. [The structure of behavior.] Oxford, UK: Presses Universitaires.

Minsky ML, Papert SA. 1969. Perceptrons. Cambridge, MA: MIT Press.

Mitchell M. 2021. Why AI is harder than we think. arXiv:2104.12871. doi:10.48550/arXiv.2104.12871.

Monod J. 1970. Le Hasard et la Nécessité. Essai sur la Philosophie Naturelle de la Biologie Moderne. Paris: Editions du Seuil.

Montévil M, Mossio M. 2015. Biological organisation as closure of constraints. *J Theor Biol* **372**:179–191. doi:10.1016/j.jtbi.2015.02.029.

Moolenaar WH, De Goede J, Verveen AA. 1976. Membrane noise in Paramecium. *Nature* **260**:344–346. doi:10.1038/260344a0.

Moreno A, Mossio M. 2015. Biological Autonomy: A Philosophical and Theoretical Enquiry. Dordrecht, Netherlands: Springer.

Morgan F. 1990. Soap bubbles and soap films. *Ann N Y Acad Sci* **607**:98–106. doi:10.1111/j.1749-6632.1990.tb22749.x.

Morgan JL, Lichtman JW. 2013. Why not connectomics? *Nat Methods* **10**:494–500. doi:10.1038/nmeth.2480.

Moser EI, Kropff E, Moser M-B. 2008. Place cells, grid cells, and the brain's spatial representation system. *Annu Rev Neurosci* **31**:69–89. doi:10.1146/annurev.neuro.31.061307.090723.

Moser M-B, Rowland DC, Moser EI. 2015. Place cells, grid cells, and memory. *Cold Spring Harb Perspect Biol* **7**:a021808. doi:10.1101/cshperspect.a021808.

Muckli L, Naumer MJ, Singer W. 2009. Bilateral visual field maps in a patient with only one hemisphere. *Proc Natl Acad Sci* **106**:13034–13039. doi:10.1073/pnas.0809688106.

Nakaoka Y, Tokui H, Gion Y, Inoue S, Oosawa F. 1982. Behavioral adaptation of Paramecium caudatum to environmental temperature. *Proc Jpn Acad Ser B* **58**:213–217. doi:10.2183/pjab.58.213.

Naud R, Bathellier B, Gerstner W. 2014. Spike-timing prediction in cortical neurons with active dendrites. *Front Comput Neurosci* **8**:90.

Nave K. 2025. A Drive to Survive: The Free Energy Principle and the Meaning of Life. Cambridge, MA: MIT Press.

Newbury R, Collins J, He K, Pan J, Posner I, Howard D, Cosgun A. 2024. A review of differentiable simulators. *IEEE Access* **12**:97581–97604. doi:10.1109/ACCESS.2024.3425448.

Nicholson DJ. 2019. Is the cell really a machine? *J Theor Biol* **477**:108–126. doi:10.1016/j.jtbi.2019.06.002.

Nicholson DJ, Dupre J, editors. 2018. Everything Flows: Towards a Processual Philosophy of Biology, illustrated ed. Oxford: Oxford University Press.

Nissen SB, Perera M, Gonzalez JM, et al. 2017. Four simple rules that are sufficient to generate the mammalian blastocyst. *PLOS Biol* **15**:e2000737. doi:10.1371/journal.pbio.2000737.

Noble D. 2008. The Music of Life: Biology beyond Genes. Oxford: Oxford University Press.

Noble D. 2012. A theory of biological relativity: no privileged level of causation. *Interface Focus* **2**:55–64. doi:10.1098/rsfs.2011.0067.

Noë A. 2004. Action in Perception. Cambridge, MA: MIT Press.

Ohnishi Y, Huber W, Tsumura A, et al. 2014. Cell-to-cell expression variability followed by signal reinforcement progressively segregates early mouse lineages. *Nat Cell Biol* **16**:27–37. doi:10.1038/ncb2881.

Oizumi M, Albantakis L, Tononi G. 2014. From the phenomenology to the mechanisms of consciousness: Integrated Information Theory 3.0. *PLoS Comput Biol* **10**:e1003588. doi:10.1371/journal.pcbi.1003588.

Oka T, Nakaoka Y, Oosawa F. 1986. Changes in membrane potential during adaptation to external potassium ions in Paramecium caudatum. *J Exp Biol* **126**:111–117.

Ólafsdóttir HF, Bush D, Barry C. 2018. The role of hippocampal replay in memory and planning. *Curr Biol CB* **28**:R37–R50. doi:10.1016/j.cub.2017.10.073.

O'Leary T, Williams AH, Franci A, Marder E. 2014. Cell types, network homeostasis, and pathological compensation from a biologically plausible ion channel expression model. *Neuron* **82**:809–821. doi:10.1016/j.neuron.2014.04.002.

Olshausen BA, Field DJ. 2004. Sparse coding of sensory inputs. *Curr Opin Neurobiol* **14**:481–487. doi:10.1016/j.conb.2004.07.007.

O'Regan JK. 2011. Why Red Doesn't Sound Like a Bell: Understanding the Feel of Consciousness. Oxford: Oxford University Press.

O'Regan JK, Noë A. 2001. A sensorimotor account of vision and visual consciousness. *Behav Brain Sci* **24**:939–973. doi:10.1017/S0140525X01000115.

Pakan JM, Francioni V, Rochefort NL. 2018. Action and learning shape the activity of neuronal circuits in the visual cortex. *Curr Opin Neurobiol* **52**:88–97. doi:10.1016/j.conb.2018.04.020.

Palmer SE, Marre O, Berry MJ, Bialek W. 2015. Predictive information in a sensory population. *Proc Natl Acad Sci* **112**:6908–6913. doi:10.1073/pnas.1506855112.

Panzeri S, Harvey CD, Piasini E, Latham PE, Fellin T. 2017. Cracking the neural code for sensory perception by combining statistics, intervention, and behavior. *Neuron* **93**:491–507. doi:10.1016/j.neuron.2016.12.036.

Peluffo AE. 2015. The "Genetic Program": behind the genesis of an influential metaphor. *Genetics* **200**:685. doi:10.1534/genetics.115.178418.

Penfield W. 1975. The Mystery of the Mind. Princeton, NJ: Princeton University Press.

Perin R, Berger TK, Markram H. 2011. A synaptic organizing principle for cortical neuronal groups. *Proc Natl Acad Sci* **108**:5419–5424. doi:10.1073/pnas.1016051108.

Pessoa L. 2022. The Entangled Brain: How Perception, Cognition, and Emotion Are Woven Together. Cambridge, MA: MIT Press.

Pezzulo G, Cisek P. 2016. Navigating the affordance landscape: feedback control as a process model of behavior and cognition. *Trends Cogn Sci* **20**:414–424. doi:10.1016/j.tics.2016.03.013.

Pfeifer R, Bongard J. 2006. How the Body Shapes the Way We Think: A New View of Intelligence. Cambridge, MA: MIT Press.

Pinker S. 2003. The Blank Slate: The Modern Denial of Human Nature, reprint ed. London: Penguin Publishing Group.

Pittendrigh CS. 1958. Adaptation, natural selection, and behavior. *Behav Evol* **390**:416.

Platkiewicz J, Brette R. 2011. Impact of fast sodium channel inactivation on spike threshold dynamics and synaptic integration. *PLoS Comput Biol* **7**:e1001129. doi:10.1371/journal.pcbi.1001129.

Plattner H. 2020. Secretory mechanisms in Paramecium. In Lemos JR, Dayanithi G, editors. Neurosecretion: Secretory Mechanisms. Cham: Springer International Publishing, pp. 271–290. doi:10.1007/978-3-030-22989-4_13.

Poincaré H. 1968. La Science et l'Hypothèse. Paris: Flammarion.

Populin LC, Yin TC. 1998. Behavioral studies of sound localization in the cat. *J Neurosci* **18**:2147–2160.

Powers WT. 1973. Behavior: The Control of Perception. Oxford: Aldine.

Pradeu T. 2010. What is an organism? An immunological answer. *Hist Philos Life Sci* **32**:247–267.

Prast H, Philippu A. 2001. Nitric oxide as modulator of neuronal function. *Prog Neurobiol* **64**:51–68. doi:10.1016/S0301-0082(00)00044-7.

Putnam H. 1975. Philosophy and our mental life. In Putnam H, editor. Mind, Language and Reality: Philosophical Papers. Cambridge: Cambridge University Press, pp. 291–303.

Putnam H. 1991. Representation and Reality, reprint ed. Cambridge, MA: MIT Press.

Quiroga RQ. 2012. Concept cells: the building blocks of declarative memory functions. *Nat Rev Neurosci* **13**:587–597. doi:10.1038/nrn3251.

Rall W. 2011. Core conductor theory and cable properties of neurons. *Comp Physiol* 39–97.

Ramoino P, Candiani S, Pittaluga AM, et al. 2014. Pharmacological characterization of NMDA-like receptors in the single-celled organism Paramecium primaurelia. *J Exp Biol* **217**:463–471. doi:10.1242/jeb.093914.

Ramón y Cajal S. 1899. Textura del sistema nervioso del hombre y de los vertebrados. Madrid: Imprenta de Hijos de Nicolás Moya.

Ramsey W. 2017. Must cognition be representational? *Synthese* **194**:4197–4214. doi:10.1007/s11229-014-0644-6.

Ramsey WM. 2007. Representation Reconsidered. Cambridge: Cambridge University Press.

Rao RP, Ballard DH. 1999. Predictive coding in the visual cortex: a functional interpretation of some extra-classical receptive-field effects. *Nat Neurosci* **2**:79–87.

Rao RPN, Gklezakos DC, Sathish V. 2023. Active predictive coding: a unifying neural model for active perception, compositional learning, and hierarchical planning. *Neural Comput* **36**:1–32. doi:10.1162/neco_a_01627.

Rashevsky N. 1936. Mathematical biophysics and psychology. *Psychometrika* **1**:1–26. doi:10.1007/BF02287920.

Richard A, Boullu L, Herbach U, et al. 2016. Single-cell-based analysis highlights a surge in cell-to-cell molecular variability preceding irreversible commitment in a differentiation process. *PLOS Biol* **14**:e1002585. doi:10.1371/journal.pbio.1002585.

Rosen R. 1967. Optimality Principles in Biology. London: Butterworths.

Rosen R. 1985. Anticipatory Systems: Philosophical, Mathematical and Methodological Foundations, 1st ed. Oxford: Pergamon.

Rosen R. 2005. Life Itself—A Comprehensive Inquiry into the Nature, Origin, and Fabrication of Life, new ed. New York: Columbia University Press.

Rosenblatt F. 1958. The perceptron: a probabilistic model for information storage and organization in the brain. *Psychol Rev* **65**:386–408. doi:10.1037/h0042519.

Rosenblatt F. 1962. Principle of Neurodynamics. New York: Spartan Books.

Ros-Rocher N, Brunet T. 2023. What is it like to be a choanoflagellate? Sensation, processing and behavior in the closest unicellular relatives of animals. *Anim Cogn* **26**:1767–1782. doi:10.1007/s10071-023-01776-z.

Rossant C, Goodman DFM, Platkiewicz J, Brette R. 2010. Automatic fitting of spiking neuron models to electrophysiological recordings. *Front Neuroinformatics* **4**:2. doi:10.3389/neuro.11.002.2010.

Rossant C, Leijon S, Magnusson AK, Brette R. 2011. Sensitivity of noisy neurons to coincident inputs. *J Neurosci* **31**:17193–17206. doi:10.1523/JNEUROSCI.2482-11.2011.

Rowland DC, Roudi Y, Moser M-B, Moser EI. 2016. Ten years of grid cells. *Annu Rev Neurosci* **39**:19–40. doi:10.1146/annurev-neuro-070815-013824.

Rucci M, Iovin R, Poletti M, Santini F. 2007. Miniature eye movements enhance fine spatial detail. *Nature* **447**:851–854. doi:10.1038/nature05866.

Ruiz-Mirazo K, Moreno A. 2004. Basic autonomy as a fundamental step in the synthesis of life. *Artif Life* **10**:235–259. doi:10.1162/1064546041255584.

Rumelhart DE, Hinton GE, Williams RJ. 1986. Learning internal representations by error propagation. In Rumelhart DE, McClelland JL, editors. Parallel Distributed Processing. Cambridge, MA: MIT Press, pp. 318–362.

Rumsfeld D. 2011. Known and Unknown: A Memoir. New York: Penguin.

Sagan L. 1967. On the origin of mitosing cells. *J Theor Biol* **14**:225–274. doi:10.1016/0022-5193(67)90079-3.

Sakurai A, Newcomb JM, Lillvis JL, Katz PS. 2011. Different roles for homologous interneurons in species exhibiting similar rhythmic behaviors. *Curr Biol* **21**:1036–1043. doi:10.1016/j.cub.2011.04.040.

Sanes JR, Lichtman JW. 1999. Development of the vertebrate neuromuscular junction. *Annu Rev Neurosci* **22**:389–442. doi:10.1146/annurev.neuro.22.1.389.

Saxena S, Cunningham JP. 2019. Towards the neural population doctrine. *Curr Opin Neurobiol* **55**:103–111. doi:10.1016/j.conb.2019.02.002.

Schnapf JL, Kraft TW, Baylor DA. 1987. Spectral sensitivity of human cone photoreceptors. *Nature* **325**:439–441. doi:10.1038/325439a0.

Schneuwly S, Klemenz R, Gehring WJ. 1987. Redesigning the body plan of Drosophila by ectopic expression of the homoeotic gene Antennapedia. *Nature* **325**:816–818. doi:10.1038/325816a0.

Schönborn W, Dörfelt H, Foissner W, Krienitz L, Schäfer U. 1999. A fossilized microcenosis in Triassic amber. *J Eukaryot Microbiol* **46**:571–584. doi:10.1111/j.1550-7408.1999.tb05133.x.

Schoonover CE, Ohashi SN, Axel R, Fink AJP. 2021. Representational drift in primary olfactory cortex. *Nature* **594**:541–546. doi:10.1038/s41586-021-03628-7.

Schrödinger E. 1944. What Is Life? The Physical Aspect of the Living Cell. Cambridge: Cambridge University Press.

Searle JR. 1990. Is the Brain a Digital Computer? *Proc Addresses Am Philos Assoc* **64**:21–37. doi:10.2307/3130074.

Seriès P, Latham PE, Pouget A. 2004. Tuning curve sharpening for orientation selectivity: coding efficiency and the impact of correlations. *Nat Neurosci* **7**:1129–1135. doi:10.1038/nn1321.

Seung HS. 2003. Learning in spiking neural networks by reinforcement of stochastic synaptic transmission. *Neuron* **40**:1063–1073. doi:10.1016/S0896-6273(03)00761-X.

Seung S. 2012. Connectome: How the Brain's Wiring Makes Us Who We Are. Boston: Houghton Mifflin Harcourt.

Shackleton TM, Skottun BC, Arnott RH, Palmer AR. 2003. Interaural time difference discrimination thresholds for single neurons in the inferior colliculus of guinea pigs. *J Neurosci* **23**:716–724.

Shadlen MN, Newsome WT. 1994. Noise, neural codes and cortical organization. *Curr Opin Neurobiol* **4**:569–579.

Shagrir O. 2006. Why we view the brain as a computer. *Synthese* **153**:393–416. doi:10.1007/s11229-006-9099-8.

Shannon CE. 1948. A mathematical theory of communication. *Bell Syst Tech J* **27**:379–423.

Shapiro JA. 2022. Evolution: A View from the 21st Century. Fortified, 2nd ed. Chicago: Cognition Press.

Simoncelli EP. 2003. Vision and the statistics of the visual environment. *Curr Opin Neurobiol* **13**:144–149.

Singer W. 1999. Neuronal synchrony: a versatile code for the definition of relations? *Neuron* **24**:49–65.

Skottun BC. 1998. Sound localization and neurons. *Nature* **393**:531. doi:10.1038/31134.

Smeets JSJ, Horstman AMH, Schijns OEMG, et al. 2018. Brain tissue plasticity: protein synthesis rates of the human brain. *Brain* **141**:1122–1129. doi:10.1093/brain/awy015.

Smets BF, Barkay T. 2005. Horizontal gene transfer: perspectives at a crossroads of scientific disciplines. *Nat Rev Microbiol* **3**:675–678. doi:10.1038/nrmicro1253.

Somjen G. 1972. Sensory Coding in the Mammalian Nervous System. New York: Appleton-Century-Crofts.

Song S, Sjöström PJ, Reigl M, Nelson S, Chklovskii DB. 2005. Highly nonrandom features of synaptic connectivity in local cortical circuits. *PLoS Biol* **3**:e68. doi:10.1371/journal.pbio.0030068.

Sprevak M. 2018. Triviality arguments about computational implementation. In Sprevak M, Colombo M, editors. The Routledge Handbook of the Computational Mind. Abingdon, UK; New York: Routledge, pp. 175–191.

Stanley GB. 2013. Reading and writing the neural code. *Nat Neurosci* **16**:259–263. doi:10.1038/nn.3330.

Stensola H, Stensola T, Solstad T, Frøland K, Moser M-B, Moser EI. 2012. The entorhinal grid map is discretized. *Nature* **492**:72–78. doi:10.1038/nature11649.

Stepp N, Turvey MT. 2010. On strong anticipation. *Cogn Syst Res* **11**:148–164. doi:10.1016/j.cogsys.2009.03.003.

Sterling P, Laughlin S. 2017. Principles of Neural Design, reprint ed. Cambridge, MA: MIT Press.

Stokes MG. 2015. "Activity-silent" working memory in prefrontal cortex: a dynamic coding framework. *Trends Cogn Sci* **19**:394–405. doi:10.1016/j.tics.2015.05.004.

Sun Y-H, Chen S-P, Wang Y-P, Hu W, Zhu Z-Y. 2005. Cytoplasmic impact on cross-genus cloned fish derived from transgenic common carp (Cyprinus carpio) nuclei and goldfish (Carassius auratus) enucleated eggs. *Biol Reprod* **72**:510–515. doi:10.1095/biolreprod.104.031302.

Swadlow HA, Waxman SG. 2012. Axonal conduction delays. *Scholarpedia* **7**:1451. doi:10.4249/scholarpedia.1451.

Syka J, Straschill M. 1970. Activation of superior colliculus neurons and motor responses after electrical stimulation of the inferior colliculus. *Exp Neurol* **28**:384–392. doi:10.1016/0014-4886(70)90175-5.

Taube JS, Muller RU, Ranck JB. 1990. Head-direction cells recorded from the postsubiculum in freely moving rats. II. Effects of environmental manipulations. *J Neurosci* **10**:436–447. doi:10.1523/JNEUROSCI.10-02-00436.1990.

Telenczuk M, Fontaine B, Brette R. 2017. The basis of sharp spike onset in standard biophysical models. *PLOS ONE* **12**:e0175362. doi:10.1371/journal.pone.0175362.

Teller DY. 1984. Linking propositions. *Vision Res* **24**:1233–1246. doi:10.1016/0042-6989(84)90178-0.

Thompson E. 2004. Life and mind: from autopoiesis to neurophenomenology. A tribute to Francisco Varela. *Phenomenol Cogn Sci* **3**:381–398. doi:10.1023/B:PHEN.0000048936.73339.dd.

Thompson E. 2010. Mind in Life: Biology, Phenomenology, and the Sciences of Mind. Cambridge, MA: Belknap Press.

Thompson E, Palacios A, Varela FJ. 1992. Ways of coloring: comparative color vision as a case study for cognitive science. *Behav Brain Sci* **15**:1–26. doi:10.1017/S0140525X00067248.

Thomson E, Piccinini G. 2018. Neural representations observed. *Minds Mach* **28**:191–235. doi:10.1007/s11023-018-9459-4.

Tsai JJ, Koka K, Tollin DJ. 2010. Varying overall sound intensity to the two ears impacts interaural level difference discrimination thresholds by single neurons in the lateral superior olive. *J Neurophysiol* **103**:875–886. doi:10.1152/jn.00911.2009.

Turing, AM. 1948. Intelligent Machinery, National Physical Laboratory Report. Edinburgh, UK: Edinburgh University Press.

Turing AM. 1950. Computing machinery and intelligence. *Mind* **59**(236): 433–460.

Turnbaugh PJ, Ley RE, Hamady M, Fraser-Liggett CM, Knight R, Gordon JI. 2007. The Human Microbiome Project. *Nature* **449**:804–810. doi:10.1038/nature06244.

Turvey MT. 2018. Lectures on Perception. New York: Routledge.

Turvey MT, Fonseca ST. 2014. The medium of haptic perception: a tensegrity hypothesis. *J Mot Behav* **46**:143–187. doi:10.1080/00222895.2013.798252.

Tytell E, Holmes P, Cohen A. 2011. Spikes alone do not behavior make: why neuroscience needs biomechanics. *Curr Opin Neurobiol* **21**:816–822. doi:10.1016/j.conb.2011.05.017.

Uexküll J von. 1909. Umwelt und Innenwelt der Tiere. Berlin: J. Springer.

Uttal WR. 1973. The Psychobiology of Sensory Coding, 1st ed. New York: Harper & Row.

van Gelder T. 1995. What might cognition be, if not computation? *J Philos* **92**:345–381. doi:10.2307/2941061.

van Gelder T. 1998. The dynamical hypothesis in cognitive science. *Behav Brain Sci* **21**:615–628.

Varela FJ. 1979. Principles of Biological Autonomy. New York, Oxford: North-Holland.

Varela FJ, Frenk S. 1987. The organ of form: towards a theory of biological shape. *J Soc Biol Struct* **10**:73–83. doi:10.1016/0140-1750(87)90035-2.

Varela FJ, Maturana H. 1972. Mechanism and biological explanation. *Philos Sci* **39**:378–382.

Varela FJ, Maturana HR, Uribe R. 1974. Autopoiesis: the organization of living systems, its characterization and a model. *Biosystems* **5**:187–196. doi:10.1016/0303-2647(74)90031-8.

Villalobos M, Ward D. 2015. Living systems: autonomy, autopoiesis and enaction. *Philos Technol* **28**:225–239. doi:10.1007/s13347-014-0154-y.

von der Malsburg C. 1981. The Correlation Theory of Brain Function. Internal Report 81–2. Göttingen, Germany: Max-Planck-Institut für Biophysikalische Chemie.

von der Malsburg C. 1995. Binding in models of perception and brain function. *Curr Opin Neurobiol* **5**:520.

von der Malsburg C. 1999. The what and why of binding: the modeler's perspective. *Neuron* **24**:95–104.

von Neumann J. 1993. First draft of a report on the EDVAC. *IEEE Ann Hist Comput* **15**:27–75. doi:10.1109/85.238389.

Voss HU. 2000. Anticipating chaotic synchronization. *Phys Rev E* **61**:5115–5119. doi:10.1103/PhysRevE.61.5115.

Voss HU. 2016a. The leaky integrator with recurrent inhibition as a predictor. *Neural Comput* **28**:1498–1502. doi:10.1162/NECO_a_00859.

Voss HU. 2016b. Signal prediction by anticipatory relaxation dynamics. *Phys Rev E* **93**:030201. doi:10.1103/PhysRevE.93.030201.

Vyas S, Golub MD, Sussillo D, Shenoy KV. 2020. Computation through neural population dynamics. *Annu Rev Neurosci* **43**:249–275. doi:10.1146/annurev-neuro-092619-094115.

Waddington CH. 1942. Canalization of development and the inheritance of acquired characters. *Nature* **150**:563–565. doi:10.1038/150563a0.

Waddington CH. 1957. The Strategy of the Genes. London: George Allen & Unwin.

Wang C, Zhang K. 2020. Equilibrium states and their stability in the head-direction ring network. *Front Comput Neurosci* **13**:96. doi:10.3389/fncom.2019.00096.

Wang N, Tytell JD, Ingber DE. 2009. Mechanotransduction at a distance: mechanically coupling the extracellular matrix with the nucleus. *Nat Rev Mol Cell Biol* **10**:75–82. doi:10.1038/nrm2594.

Wark B, Lundstrom BN, Fairhall A. 2007. Sensory adaptation. *Curr Opin Neurobiol* **17**:423–429. doi:10.1016/j.conb.2007.07.001.

Weber SN, Sprekeler H. 2018. Learning place cells, grid cells and invariances with excitatory and inhibitory plasticity. *eLife* **7**:e34560. doi:10.7554/eLife.34560

Wehner R. 2020. Desert Navigator: The Journey of an Ant, Desert Navigator. Cambridge, MA: Harvard University Press. doi:10.4159/9780674247918.

Wehner R, Wehner S. 1990. Insect navigation: use of maps or Ariadne's thread? *Ethol Ecol Evol* **2**:27–48. doi:10.1080/08927014.1990.9525492.

Wen B, Wang GI, Dean I, Delgutte B. 2009. Dynamic range adaptation to sound level statistics in the auditory nerve. *J Neurosci* **29**:13797–13808. doi:10.1523/JNEUROSCI.5610-08.2009.

White JG, Southgate E, Thomson JN, Brenner S. 1986. The structure of the nervous system of the nematode Caenorhabditis elegans. *Philos Trans R Soc Lond B Biol Sci* **314**:1–340. doi:10.1098/rstb.1986.0056.

Wilson HR, Cowan JD. 1972. Excitatory and inhibitory interactions in localized populations of model neurons. *Biophys J* **12**:1–24. doi:10.1016/S0006-3495(72)86068-5.

Wood DC. 1973. Stimulus specific habituation in a protozoan. *Physiol Behav* **11**:349–354. doi:10.1016/0031-9384(73)90011-5.

Yong E. 2022. An Immense World: How Animal Senses Reveal the Hidden Realms around Us. New York: Random House.

Young JZ. 1936. The structure of nerve fibres in cephalopods and crustacea. *Proc R Soc Lond B Biol Sci* **121**:319–337. doi:10.1098/rspb.1936.0069.

Young JZ. 1939. Fused neurons and synaptic contacts in the giant nerve fibres of cephalopods. *Philos Trans R Soc B Biol Sci* **229**:465–503. doi:10.1098/rstb.1939.0003.

Zahnoun F. 2020. Explaining the reified notion of representation from a linguistic perspective. *Phenomenol Cogn Sci* **19**:79–96. doi:10.1007/s11097-018-9603-x.

Zahnoun F. 2021. On representation hungry cognition (and why we should stop feeding it). *Synthese* **198**:267–284. doi:10.1007/s11229-019-02277-8.

Zahnoun F. 2023. The Embodiment of Meaning: Why Matter Matters for Cognition and Experience. New York: Routledge.

Zbili M, Rama S, Yger P, et al. 2020. Axonal Na+ channels detect and transmit levels of input synchrony in local brain circuits. *Sci Adv* **6**:eaay4313. doi:10.1126/sciadv.aay4313.

Zutshi I, Apostolelli A, Yang W, et al. 2025. Hippocampal neuronal activity is aligned with action plans. *Nature* **639**:153–161. doi:10.1038/s41586-024-08397-7.

INDEX

A NOTE ON THE TYPE

This book has been composed in Arno, an Old-style serif typeface in the classic Venetian tradition, designed by Robert Slimbach at Adobe.

GPSR Authorized Representative: Easy Access System Europe - Mustamäe tee 50, 10621 Tallinn, Estonia, gpsr.requests@easproject.com

www.ingramcontent.com/pod-product-compliance
Lightning Source LLC
LaVergne TN
LVHW050955080826
845145LV00006B/1501

* 9 7 8 0 6 9 1 2 8 1 3 8 4 *